T0280144

Serial Communication Protocols and Standards

RS232/485, UART/USART, SPI, USB, INSTEON, Wi-Fi and WiMAX

RIVER PUBLISHERS SERIES IN COMMUNICATIONS

Series Editors:

ABBAS JAMALIPOUR
The University of Sydney
Australia

MARINA RUGGIERI
University of Rome Tor Vergata
Italy

JUNSHAN ZHANG
Arizona State University
USA

Indexing: All books published in this series are submitted to the Web of Science Book Citation Index (BkCI), to SCOPUS, to CrossRef and to Google Scholar for evaluation and indexing.

The "River Publishers Series in Communications" is a series of comprehensive academic and professional books which focus on communication and network systems. Topics range from the theory and use of systems involving all terminals, computers, and information processors to wired and wireless networks and network layouts, protocols, architectures, and implementations. Also covered are developments stemming from new market demands in systems, products, and technologies such as personal communications services, multimedia systems, enterprise networks, and optical communications.

The series includes research monographs, edited volumes, handbooks and textbooks, providing professionals, researchers, educators, and advanced students in the field with an invaluable insight into the latest research and developments.

For a list of other books in this series, visit www.riverpublishers.com

Serial Communication Protocols and Standards

RS232/485, UART/USART, SPI, USB, INSTEON, Wi-Fi and WiMAX

Dawoud Shenouda Dawoud

International University of East Africa
Uganda

Peter Dawoud

Microsoft
USA

Routledge
Taylor & Francis Group

LONDON AND NEW YORK

Published 2020 by River Publishers
River Publishers
Alsbjergvej 10, 9260 Gistrup, Denmark
www.riverpublishers.com

Distributed exclusively by Routledge
4 Park Square, Milton Park, Abingdon, Oxon OX14 4RN
605 Third Avenue, New York, NY 10017, USA

Serial Communication Protocols and Standards: RS232/485, UART/USART, SPI, USB, INSTEON, Wi-Fi and WiMAX / by Dawoud Shenouda Dawoud, Peter Dawoud.

Routledge is an imprint of the Taylor & Francis Group, an informa business

ISBN 978-87-7022-154-2 (print)

While every effort is made to provide dependable information, the publisher, authors, and editors cannot be held responsible for any errors or omissions.

To Nadia, Dalia, Dina, Bahya, and Adam

D.S.D.

To Dajana

P.D

Contents

Preface

Data communication standard comprises of two components: The "protocol" and "signal/data/port specifications for the devices or additional electronic circuitry involved." The protocol describes the format of the message and the meaning of each part of the message. This takes place by the hardware of the device. To connect any device to the bus, an external device must be used as an interface which will be responsible for putting the message in a form fulfilling all the electrical specifications of the port. For example, if the message is in the form of 0s and 1s, then it is the responsibility of the external interface to represent the binary 1 and binary 0 by certain well-defined voltage, for example, binary 1 by -12 V and binary 0 by $+12$ V, which is called "standard". The most famous serial communication standard is the RS-232.

Communication can be serial or parallel. Serial communication is used for transmitting data over long distances. It is much cheaper to run the single core cable needed for serial communication over a long distance than the multicore cables that would be needed for parallel communication. The same is in case of using wireless communication: Serial communication needs one channel, while parallel communication needs multichannel.

Serial communication can be classified in many ways, for example, synchronous and asynchronous. It can also be classified as simplex, duplex, and half duplex.

Because of the wide spread of serial communication in all fields of life from home automation to sensor and controller networks, a very large number of different standards and protocols for serial communication, ranging from the very simple to the seriously complicated have been developed during the last few decades. This large number of protocols was necessary to guarantee the optimum and the best performance in the targeted applications. It is important to match the right protocol with the right application. This reflects the importance for any communication engineer to have enough knowledge about the protocols available currently.

Some of the protocols that have been developed based on serial communication in the past few decades are:

1. SPI – Serial Peripheral Interface
2. eSPI Protocol
3. I2C – Inter-Integrated Circuit
4. I3C
5. CAN
6. FireWire
7. Ethernet
8. Universal Serial Bus (USB)
9. Microwire
10. 1-Wire
11. UART and USART

We can add to the above protocols the following:

12. Bluetooth
13. ZigBee
14. Z-Wave
15. Wi-Fi
16. WiMAX
17. INSTEON

The protocols can be divided widely into two groups: inter system protocols (e.g., UART Protocol, USART Protocol, and USB Protocol) and intra system protocols (e.g., I2C, I3C, CAN, and SPI) which can be used to build networks.

Serial communication, synchronous or asynchronous, is typically implemented with a recommended standard (RS). The standard usually defines signal levels, maximum bandwidth, connector pinout, supported handshaking signals, drive capabilities, and electrical characteristics of the serial lines.

Nowadays and in most cases, the standard is set by the Electronic Industries Association (EIA). The RS-232 and its variant are currently in use.

The main goal of the authors of this document is to give the reader enough knowledge about the above-mentioned protocols and standards. To achieve this target, the authors divided the document into two books: "Serial Communication Protocols and Standards" and "Microcontroller and Smart Home Networks."

The current book "Serial Communication Protocols and Standards" can be divided into the following sections:

Section 1: Serial communication (Chapter 1)

Section 2: UART and USART (Chapter 2)

Section 3: RS-232 standard and all its variants (Chapter 3, Chapter 4, and Chapter 5)

Section 4: SPI (Chapter 6)

Section 5: USB (Chapter 7 and Chapter 8)

Section 6: Wireless systems: Wi-Fi and WiMAX (Chapter 9 and Chapter 10)

Section 7: INSTEON technology

The remaining protocols are the subject of our book "Microcontroller and Smart Home Networks."

List of Figures

List of Tables

List of Abbreviations

AAA	Authentication, Authorization, and Accounting
AAS	Advanced Antenna Systems
ADSL	Asymmetric Digital Subscriber Loop
AES	Advanced Encryption Standard
AK	Authorization key
AKA	Authentication and Key Agreement
AODV	Ad hoc On Demand Distance Victor
AODVjr	Ad hoc On Demand Distance Vector junior
APL	Application Layer
APS	Application Support Sub-layer
ARQ	Automatic Repeat Request
ASN	Access Services Network
ASP	Application Service Provider
BPSK	Binary Phase Shift Keying
BS	Base Station
BWA	Broadband Wireless Access
CBC	Cipher Block Chaining
CCA	Clear Channel Assessment
CCK	Complementary Coded Keying
CD	Coordinator Device
CLEC	Competitive Local Exchange Carrier
CIR	Committed Information Rate
CSMA/CA	Carrier Sense Multiple Access with Collision Avoidance
CPS	Common Part Sublayer
CSMA/CD	Carrier Sense Multiple Access with Collision Detection (Ethernet)
DCF	Distributed Control Function
DES	Data (Digital) Encryption Standard
DH	Diffie-Hellman
DLL and DLH	Divisor Latches
DoS	Denial of Service

DSL	Digital Subscriber Line
DSR	Dynamic Source Routing
DSSS	Direct Sequence Spread Spectrum
EAP	Extensive Authentication Protocol
EDCA	Enhanced Distributed Control Access
ETSI	European Telecommunications Standards Institute
ED	End Device
ED-VO	Enhanced Version-Data Only (Data Optimized)
FCC	Federal Communications Commission
FCR	FIFO Control Register
FDD	Frequency Division Duplex
FDX	Full Duplex
FEC	Forward Error Correction
FFD	Fully Functional Device
FHSS	Frequency Hopping Spread Spectrum
HMAC	Hashed Message Authentication Code
Hz	Hertz (Prefix Kilo = Thousands, Mega = Millions, Giga = Billions)
GAP	Generic Access Profile
GFSK	Gaussian Frequency Shift Keying
HARQ	Hybrid-ARQ
HCI	Host Controller Interface
HIPERMAN	High-Performance Metropolitan Area Network
HSP	Headset Profile
HUMAN	High-speed Unlicensed Metropolitan Area Network
IEEE	Institute of Electrical and Electronic Engineers
IER	Interrupt Enable Register
IETF	Internet Engineering Task Force
IFS	Inter frame Spacing
IIR	Interrupt Identification Register
ILEC	Incumbent Local Exchange Carrier
IoT	Internet of Things
ISDN	Integrated Services Digital Network
ISM	Industrial, Scientific, and Medical
ITU	International Telecommunications Union
KEK	Key Encryption Key
L2CAP	Logical Link Control and Adaptation Protocol
LAN	Local Area Network
LCR	Line Control Register

LMP	Link Manager Protocol
LOS	Line of Sight
LQI	Link Quality Indication
LR	Location Register
LR-WPAN	Low-Rate Wireless Personal Area Network
LS	Least Squares
LSR	Line Status Register
LTE	Long Term Evolution
MAC	Media Access Control also Message Authentication Code
MANET	Mobile ad hoc Network
MBS	Multicast Broadcast Service
MCR	Modem Control Register
MDR	Mode Definition Register
MIC	Message Integrity Code
MIMO	Multiple Input-Multiple Output
MITM	Man-In-The-Middle
MMDS	Multi-channel Multipoint Distribution Service
MME	Multimedia Extensions (WME)
MMS	Multimedia Messaging Service
MPDU	MAC Protocol Data Unit
MS	Mobile Station
MS	Mobile Station
MSR	Modem Status Register
NIB	Network Layer Information Base
NLDE	Network Layer Data Entity
NLME	Network Layer Management Entity
NLOS	Non-Line-of-Sight
NWG	Network Working Group
NWK	Network Layer
OBEX	Object Exchange Protocol
OFDM	Orthogonal Frequency Division Multiplexing
PAN	Personal Area Network
PCF	Point Control Function
PHY	Physical Layer
PoP	Point of Presence
PKI	Public Key Infrastructure
PKM	Privacy and Key Management
PKMv1	Privacy and Key Management Version 1
PKMv2	Privacy and Key Management Version 2

PMK	Pairwise Master Key
PPDU	PHY Protocol Data Unit
PPP	Point to Point Protocol
PSDU	PHY Service Data Unit
RBR	Receiver Buffer Register
RFD	Reduced Functional Device
x-QAM	x-level Quadrature Amplitude Modulation
QoS	Quality of Service
QPSK	Quadrature Phase Shift Keying
RC4	Ron.s Code-4
REVID1&2	Revision Identification Registers
RFCOMM	Radio Frequency Communication
RSA	Rivest-Shamir-Adleman (encryption protocol)
SA	Security Association
SCR	Scratch Pad Register
SDAP	Service Discovery Application Protocol
SDDB	Service Discovery Database
SDP	Service Discovery Protocol
SFD	Start-of-Frame Delimiter
SIM	Subscriber Identity Module
SONET	Synchronous Optical Network Interface
SPP	Serial Port Profile
SS	Subscriber Station
SSID	Service Set Identifier
TDD	Time Division Duplex
TEK	Traffic Encryption Key
THR	Transmitter Holding Register
TKIP	Temporal Key Integrity Protocol
TTC	Time Triggered Communication
TTCAN	Time Triggered CAN
U-NII	Unlicensed National Information Infrastructure
VoIP	Voice over IP
VPN	Virtual Private Network
WBA	Wireless Broadband Access
WEP	Wired Equivalent Privacy
Wi-Fi	Wireless Fidelity
WiMAX	Worldwide Interoperability for Microwave Access
WISP	Wireless Internet Service Provider
WLAN	Wireless LAN

WMAN	Wireless Metropolitan Area Network
WME	Wi-Fi Multimedia Extensions
WPA	Wi-Fi Protected Access
WPAN	Wireless Personal Area Network
WPS	Wi-Fi Protected Setup
WRAN	Wireless Regional Area Network
WSM	Wi-Fi Scheduled Multimedia
WWAN	Wireless Wide Area Network
ZDO	ZigBee Device Objects
ZF	Zero Forcing
ZiCL	ZigBee Cluster Label
ZTC	ZigBee Trust Centre

1

Serial Communication

1.1 Introduction

Serial Data Communication

Data communication is one of the most challenging fields today as far as technology development is concerned. Data, essentially meaning information coded in digital form, that is, 0s and 1s, need to be sent from one point to the other either directly or through a network.

When many such systems need to share the same information or different information through the same medium, there arises a need for proper organization (rather, "socialization") of the whole network of the systems, so that the whole system works in a cohesive fashion.

Therefore, in order for a proper interaction between the data transmitter (the device that commences data communication) and the data receiver (the system that receives the data sent by a transmitter), there has to be some set of rules or "protocols" which all the interested parties must obey.

The requirement above finally paves the way for some *DATA COMMU-NICATION STANDARDS.* In general, communication standards incorporate both the software and hardware aspects of the system, while *buses* mainly define the cable characteristics for the same communication type. The standards define the communication capabilities of the data communication systems so that the systems are not vendor specific but for each system the user has the advantage of selecting the device and interface according to his own choice of make and range.

Depending on the requirement of applications, one has to choose the type of communication strategy. There are basically two major classifications, namely SERIAL and PARALLEL, each with its variants. **Serial communication** is the process of sending data one bit at a time, sequentially, over a communication channel or computer bus. This contrasts with parallel communication, where several bits are simultaneously sent on a link with several

1

parallel channels. Serial communication is, accordingly, relatively simple and has low hardware overhead compared with parallel interfacing. Due to that, serial communication is used extensively within the electronics industry. It is used for communicating between two or more devices. Normally, one device is a computer or a microcontroller, while the other device can be a modem, a printer, another computer, another microcontroller, a display unit, or a scientific instrument such as an oscilloscope or a function generator.

Serial communication is one of the most important applications of micro-controllers. Many microcontrollers possess the capabilities of serially communicating using one or more of the following techniques: USART/UART, SPI, I2C, and CAN.

Today, the most popular serial communications standard in use is certainly the EIA/TIA-232-E specification. This standard, which has been developed by the Electronic Industry Association and the Telecommunications Industry Association (EIA/TIA), is more popularly referred to simply as "RS-232" where "RS" stands for "recommended standard." In recent years, this suffix has been replaced with "EIA/TIA" to help identify the source of the standard. This book will use the common notation of "RS-232" while discussing the topic.

1.2 Data Communication Standard

Any data communication standard comprises of the two following components:

- The protocol.
- Signal/data/port specifications for the devices or additional electronic circuitry involved.

1.2.1 Communication Protocol

A **communication protocol** is a system of rules that allow two or more entities of a communications system to transmit information via any kind of variation of a physical quantity. The protocol defines the rules, syntax, semantics, and synchronization of communication and the possible error recovery methods. Protocols may be implemented by hardware, software, or a combination of both.

Communicating systems use well-defined formats for exchanging various messages. Each message has an exact meaning intended to elicit a response from a range of possible responses pre-determined for that particular situation. The specified behavior is typically independent of how it is to be implemented. Communication protocols have to be agreed upon by the parties involved and the possible responses.

Protocols should therefore specify rules governing the transmission: the format of the various exchanged messages.

1.2.1.1 Types of electronic communication protocols

Communication protocols are broadly classified into two types (Figure 1.1):

1. Inter-system Protocol
2. Intra-system Protocol

1. Inter-system Protocol: The inter system protocol is used to communicate between two different devices, for example, communication between computer and microcontroller kit. The communication is achieved through a inter bus system (Figure 1.2).

Different categories of inter-system protocol:
The inter-system protocol includes the following:

- UART protocol
- USART protocol
- USB protocol

These three protocols are described in detail in the following chapters; however, here they are summarized as follows:

UART (Universal Asynchronous Receiver–Transmitter): It is a serial communication with a two-wire protocol. The data cable signal lines are

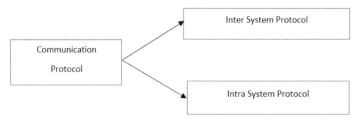

Figure 1.1   Classification of communication protocols.

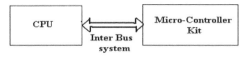

Figure 1.2   Inter system protocol.

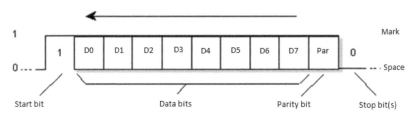

Figure 1.3 UART protocol data flow.

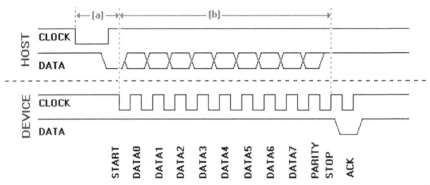

Figure 1.4 USART protocol data flow.

labeled as Rx and Tx. On the transmitting end, the UART takes bytes of data and send the individual bits in a sequential manner. The vice versa takes place on the receiving end. Figure 1.3 shows the UART protocol data flow.

The UART is a half-duplex protocol, transmitting and receiving the data but not at the same time, for example, Emails, SMS, and Walkie-talkie.

USART (Universal Synchronous/Asynchronous Receiver and Transmitter): It is a serial communication with a two-wire protocol. The data cable signal lines are labeled as Rx and Tx. This protocol is used for transmitting and receiving the data byte by byte along with the clock pulses. It is a full-duplex protocol, transmitting and receiving simultaneously with different baud rates. Different devices communicate with microcontroller using this protocol. Figure 1.4 shows USART protocol data flow.

USB (Universal Serial Bus): It is, as UART and USART, a serial communication with a two-wire protocol. The data cable signal lines are labeled as "D+" and "D−." USB protocol is used to communicate with the system peripherals. The protocol is used to send and receive the data serially to the host and peripheral devices. Communication using USB requires a driver

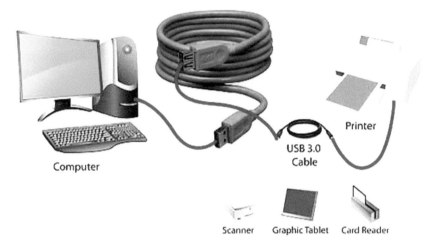

Figure 1.5 USB protocol communication.

Table 1.1 Summary of UART, USART, and USB

UART	USART	USB
1. UART stands for Universal Asynchronous Transmitter and Receiver.	USART stands for Universal Synchronous and Asynchronous Transmitter and Receiver.	USB stands for Universal Serial Bus.
2. It is a two-wire protocol Rx and Tx.	It is a two-wire protocol Rx and Tx.	It is a two-wire protocol D+ and D-.
3. It is transmitting and receiving packets of data byte by byte without classes' pulses.	It sends and receives a block of data along with classes' pulses.	It sends and receives data along with clock pulses.
4. It is a half-duplex communication.	It is a full-duplex communication.	It is a full-duplex communication.
5. It is slow compared with USART.	It is slow compared with USB.	It is fast compared with UART and USART.

software that is based on the functionality of the system. USB device can transfer data on the bus without any request on the host computer (Figure 1.5). Examples of USB are Mouse, Keyboard, Hubs, Switches, and pen drive.

Differences between the inter-system protocols:

Table 1.1 summarizes the differences between UART, USART, and USB protocols.

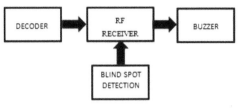

Figure 1.6 Inter-communication.

Figure 1.7 Intra-system protocol.

Example of inter-communication: Inter-vehicular communication:

Figure 1.6 is an example of inter-vehicular communication. Data from the RF transmitter of the first vehicle are decoded by the RF receiver in the second vehicle. The decoder converts the serial input into parallel output and then alerts the driver in the second vehicle. Using IR LED and photodiode, the blind spot of the vehicle is detected in order to inform the overtaking.

2. Intra-system Protocol: The intra system protocol is used to communicate between two devices within the circuit board (Figure 1.7). Using intra system protocol expands the peripherals of the microcontroller and increases the circuit complexity and power consumption. It is a very secure protocol to access the data.

The intra-system protocol includes the following:

- I2C protocol
- SPI protocol
- CAN protocol

These three protocols are summarized as follows:

I2C (also IIC or inter-integrated circuit): I2C requires only two wires to connect all peripherals to microcontroller. The two wires are labeled SDA (Serial Data Line) and SCL (Serial Clock Line) and they carry the information between devices. The protocol is a master to slave communication protocol. It is an address-based communication protocol: each slave has a unique address. Master device sends the address of the target slave device and read/write flag. Only the slave with its address matching with that included in the message will be enabled to receive the message and all the remaining slave devices will

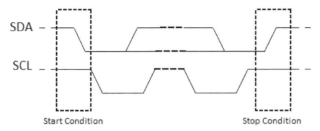

Figure 1.8 I2C protocol data flow.

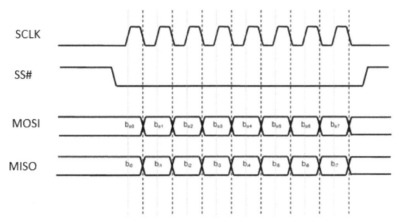

Figure 1.9 SPI protocol data flow.

be in disable mode. Once the address is matched, communication proceeds between the master and the slave devices, and transmitting and receiving of data begin. The transmitter (the master) sends 8-bit data, the receiver replies 1-bit of acknowledgement. When the communication is completed, the master issues the stop condition. Figure 1.8 shows the I2C protocol data flow.

SPI (serial peripheral interface): It is a serial communication protocol. This protocol is also called 4-wire protocol since it requires four wires for communication: MOSI, MISO, SS, and SCLK. It is a master and a slave protocol. The master first configures the clock using a frequency. The master then selects the particular slave device for communication by pulling the chip select button. That particular device is selected and starts the communication between the master and that particular slave. The master selects only one slave at a time. The protocol is a full-duplex communication protocol. Figure 1.9 shows the SPI protocol data flow.

Table 1.2   Summary of I2C, SPI, and CAN

I2C	SPI	CAN
I2C stands for Inter-Integrated Circuits	SPI stands for Serial Peripheral Interface	CAN stands for Controller Area Network
Developed by Phillips	Developed by Motorola	Developed by Robert Bosch
It is a half-duplex protocol	It is a full-duplex protocol	It is a full-duplex protocol
Synchronization	Synchronization	Synchronization
Two-wire protocol: SCL and SDL	Four-wire protocol: SCL, MISO/MOSI, and SS	Two-wire protocol: CAN H+ and CAN H-
Multi-master protocol	Single master protocol	Multi-muster protocol
Within the circuit board	Within the circuit board	Within two circuit boards

CAN (controller area network): It is a serial communication protocol that requires two wires: CAN High (H+) and CAN Low (H-). It was developed by Bosh Company in 1985 for vehicle networks. It is based on a message-oriented transmission protocol.

Differences between the intra-system protocols:

Table 1.2 summarizes the differences between I2C, SPI, and CAN protocols.

Example of intra-vehicular communication: Figure 1.10 shows a practical intra vehicular communication. In this figure, the ultrasonic sensor measures the distance of the vehicle from an object and the speed sensor senses the speed. This information is passed over to the receiver section through a CAN bus. Then, in the receiver section, the buzzer is activated indicating that the collision distance is reached, and the corresponding break light indicator is activated.

The distance will be displayed on LCD. If the speed is increasing, it will command for a gear shift, which in turn helps to increase the fuel efficiency of the vehicle. The LCD displays the speed of the vehicle. This information is given to the second vehicle through RF transmitter by encoding the data. Encoder is a parallel to serial converter. Both transmitter and receiver uses PIC 16F73.

1.2.2 Serial Data Communication Standards

At the beginning of this section, it has been mentioned that the data communication standard comprises of two components: The "protocol" and "signal/data/port specifications for the devices or additional electronic circuitry involved." The protocol, as explained above, describes the format of

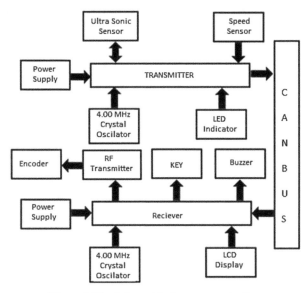

Figure 1.10 Intra-vehicular communication.

the message and the meaning of each part of the message. This takes place by the hardware of the device. To connect any device to the bus, an external device must be used as an interface which will be responsible for putting the message in a form fulfilling all the electrical specifications of the port. For example, if the message is in the form of 0s and 1s, then it is the responsibility of the external interface to represent the binary 1 and binary 0 by certain well-defined voltage, for example, binary 1 by -12 V and binary 0 by $+12$ V. The most famous serial communication standard is RS-232.

The above-mentioned communication protocols and also RS-232 and its deferent versions will be considered in detail in the chapters of this book. This book introduces to the reader enough knowledge that will let him/her to understand all the above-mentioned protocols and that will enable him/her to design applications that needed the use of any one of these protocols.

1.3 Serial and Parallel Communications

1.3.1 Serial Communication

In Telecommunication and Computer Science, serial communication is the process of sending/receiving data in one bit at a time in a sequential order over a computer bus or a communication channel (Figure 1.11).

Figure 1.11   Serial communication.

It is the simplest form of communication between a sender and a receiver. Because of the synchronization difficulties involved in parallel communication, along with cable cost, serial communication is considered best for long-distance communication.

The serial data communication is the most common *low-level* protocol for communicating between two or more devices. Normally, one device is a computer or a microcontroller, while the other device can be a modem, a printer, another computer, another microcontroller, a display unit, or a scientific instrument such as an oscilloscope or a function generator.

Serial data communication strategies and **standards** are used in situations having a limitation of the number of lines (channels) that can be secured for communication. This is the primary mode of transfer in long-distance communication. But it is also the situation in embedded systems where various subsystems share the communication channel and the speed is not a very critical issue.

All the data communication systems follow some specific set of *standards* defined for their communication capabilities so that the systems are not vendor specific but for each system the user has the advantage of selecting the device and interface according to his own choice of make and range.

Standards incorporate both the software and hardware aspects of the system, while *buses* mainly define the cable characteristics for the same communication type.

The most common serial communication system protocols can be studied under the following categories: *asynchronous, synchronous and bit-synchronous* communication standards.

In synchronous transmission, groups of bits are combined into frames, and frames are sent continuously with or without data to be transmitted. In asynchronous transmission, groups of bits are sent as independent units with start/stop flags and no data link synchronization, to allow for arbitrary size gaps between frames. However, start/stop bits maintain physical bit level synchronization once detected.

Examples of serial mode transmission include connections between a computer and a modem using the RS-232 interface. Although an RS-232 cable can theoretically accommodate 25 wires, all but two of these wires

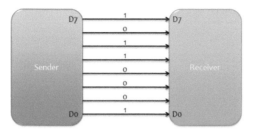

Figure 1.12   Parallel transfer.

are for overhead control signaling and not data transmission; the two data wires perform simple serial transmission in either direction. In this case, a computer may not be close to a modem, making the cost of parallel transmission prohibitive—thus, the speed of transmission may be considered less important than the economical advantage of serial transmission.

1.3.2 Parallel Communication

Parallel communication (Figure 1.12) is the process of sending/receiving multiple data bits (usually 8 bits or a byte/character) at a time through parallel channels and synchronized to a clock.

Parallel devices have a wider data bus than serial devices and can therefore transfer data in words of one or more bytes at a time. As a result, there is a speedup in parallel transmission bit rate over serial transmission bit rate. However, this speedup is a tradeoff versus cost since multiple wires cost more than a single wire, and as a parallel cable gets longer, the synchronization timing between multiple channels becomes more sensitive to distance. The timing for parallel transmission is provided by a constant clocking signal sent over a separate wire within the parallel cable; thus, parallel transmission is considered **synchronous**.

Examples of parallel mode transmission include connections between a computer and a printer (parallel printer port and cable). Most printers are within 6 m or 20 ft of the transmitting computer, and the slight cost for extra wires is offset by the added speed gained through parallel transmission of data.

1.3.2.1 Serial and parallel communication tradeoffs

Serial transmission via RS-232 is officially limited to 20 Kbps for a distance of 15 m or 50 ft. (Note: RS-232 is the subject of Chapter 3.) Depending on

Table 1.3 Serial versus parallel communication

Serial Communication	Parallel Communication
Sends data bit by bit at one clock pulse	Multiple data bits re-transceived at a time
Requires one wire to transmit the data	Requires "n" number of lines for transmitting "n" bits
Slow communication speed	Fast communication speed
Low installation cost	High installation cost
Preferred for long distance communication	Used for short distance communication
Example: Computer to computer	Example: Computer to multi-function printer

the type of media used and the amount of external interference present, RS-232 can be transmitted at higher speeds, or over greater distances, or both. Parallel transmission has similar distance-versus-speed tradeoffs, as well as a clocking threshold distance. Techniques to increase the performance of serial and parallel transmission (longer distance for same speed or higher speed for same distance) include using better transmission media, such as **fiber optics** or conditioned cables; implementing repeaters; or using shielded/multiple wires for noise immunity.

1.3.2.2 Serial versus parallel communication

Table 1.3 shows some of the basic differences between the two types of communications.

The basic differences mentioned in the table may lead to a conclusion that parallel communication is better than serious communication, especially if we consider the speed of transmission. Theoretically, the speed of a parallel link is equal to *bit rate x number of channels*. In practice, there are some factors that result on putting limitation on the speed of parallel link. These factors are "Clock Skew" and "Crosstalk". The limitation on the parallel communication speed may, in some applications, let the serial communication better. Let us consider the following terminologies:

1. **Bit rate or baud rate:** Bit rate and baud rate are the same when transmitting binary sequence, that is, the signal is either "0" or "1." It is the number of bits that are transmitted (sent/received) per unit time. The baud rate is given by (Figure 1.13):

   Baud rate $= 1/$bit time $= 1 / T_B$ (bits/ second)

For example, in a teletypewriter (TTY) system, T_B is 9.09 msec. This gives a baud rate of 110 bits/sec. Commonly used baud rates are 110,

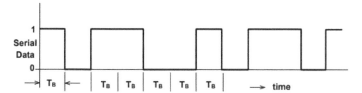

Figure 1.13 Serial data line during serial communication.

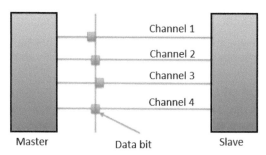

Figure 1.14 Clock skew.

150, 300, 600, 900, 1200, 2400, 4800, 9600,13.8 k,14.4 k,19.2 k, 38.4 k, and 57.8 k.

2. **Clock skew:** In a parallel circuit, clock skew is the *time difference* in the arrival of two sequentially adjacent registers. Figure 1.14 shows the meaning of "clock skew." The 8 bits are leaving the transmitter exactly at the same instant; during propagation on the communication channel, there is bound to be a time difference in the arrival of the eight bits to the receiver: there is difference between the arrival of the first bit and that of the second bit, and so on. This time difference is what is called "clock skew."

   Figure 1.14 illustrates this: There is a time lag in the data bits through different channels of the same bus. Clock skew is inevitable due to differences in physical conditions of the channels, like temperature, resistance, path length, etc.

3. **Crosstalk:** It is a phenomenon by which a signal transmitted on one channel of a transmission bus (or circuit) creates an undesired effect in another channel (or circuit). Crosstalk is usually caused by undesired capacitive, inductive, or conductive coupling from one circuit, part of a circuit, or channel, to another. It can be seen from the following diagram that clock skew and crosstalk are inevitable.

1.3.2.3 Major factors limiting parallel communication

Before the development of high-speed serial technologies, the choice of parallel links over serial links was driven by these factors:

1. **Speed**: Superficially, the speed of a parallel link is equal to *bit rate* × *number of channels*. In practice, clock skew reduces the speed of every link to the slowest of all of the links.
2. **Cable length**: Crosstalk creates interference between the parallel lines, and the effect only magnifies with the length of the communication link. This limits the length of the communication cable that can be used.

These two are the major factors, which limit the use of parallel communication.

1.3.2.4 Advantages of serial over parallel

Although a serial link may seem inferior to a parallel one, since it can transmit less data per clock cycle, it is often the case that serial links can be clocked considerably faster than parallel links in order to achieve a higher data rate. A number of factors allow serial to be clocked at a higher rate:

- **Clock skew** between different channels **is not an issue** (for unclocked asynchronous serial communication links).
- A **serial connection** requires **fewer interconnecting cables** (e.g., wires/fibers) and hence occupies less space. The extra space allows for better isolation of the channel from its surroundings.
- **Crosstalk** is not a much significant issue, because there are fewer conductors in proximity.

There are many different reasons to let serial interface better option. One of the most common reasons is the need to interface with a PC, during development and/or in the field. Most, if not all PCs, as opposed parallel one, have some sort of serial bus interface available to connect peripherals. For embedded systems that must interface with a general-purpose computer, a serial interface is often easier to use than the ISA or PCI expansion bus.

Another benefit of serial communication is the low pin counts. Serial communication can be performed with just one I/O pin, compared to eight or more for parallel communication. Many common embedded system peripherals, such as analog-to-digital and digital-to-analog converters, LCDs, and temperature sensors, support serial interfaces.

Serial buses can also provide for inter-processor communication-a network, if you will. This allows large tasks that would normally require larger processors to be tackled with several inexpensive smaller processors. Serial

Figure 1.15   Data transfer in serial communication.

interfaces allow processors to communicate without the need for shared memory and semaphores, and the problems they can create.

This is not to say that parallel buses have no use. For operational fetches, address and data buses, and other microprogram control, parallel buses have always been the clear winner. "Memory-mapping peripherals" is a technique commonly used for systems with address and data buses. This tendency allows parallel access to off-chip peripherals. However, with many 8-bit microcontrollers (let alone 8-pin) with no external address/data bus available for designs, memory-mapping is not an option.

1.4 How Are Data Sent Serially?

Data are normally stored in parallel form in microcontrollers, and any bit can be accessed irrespective of its bit number. For transmitting the data, it is transferred at first into an output buffer for transmission, and the data are still in parallel form. The output buffer is a Parallel-In-Serial-Out (PISO) shift register that receives the data in parallel form and output it in serial form, MSB (Most Significant Bit) first or LSB (Least Significant Bit) first as according to the protocol. Now these data are *transmitted in serial mode* (Figure 1.15).

When these data are received by another microcontroller, it goes at first to receiver buffer, the receiver buffer which is a **Serial-In-Parallel-Out** (**SIPO**) shift register. The SIPO converts the received data back into parallel data for further processing. The following diagram will make it clear (Figure 1.15).

1.5 Modes of Serial Transmission

1.5.1 Need for Synchronization

Transmitter and receiver synchronization: One of the main problems when two devices linked by a transmission medium wish to exchange data is that

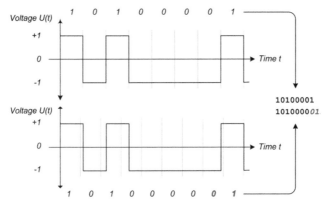

Figure 1.16   Synchronization problems.

of synchronizing the receiving device with the transmitting device. Typically, data are transmitted one bit at a time, and the data rate must be the same for both the transmitter and the receiver. The receiver must be able to recognize the beginning and end of a block of bits and know the time taken to transmit each bit, so that it can sample the line at the correct time to read each bit. When the sending device is transmitting a stream of bits, it uses an internal clock to control timing. If data are transmitted at 10 Kbps, a bit is transmitted every 0.1 milliseconds. The receiver attempts to sample the line at the center of each bit time, that is, at intervals of 0.1 milliseconds. If the receiver uses its own internal clock for timing, a problem will arise if the clocks in the transmitter and receiver are not synchronized. A drift of 1% will cause the first sample to be 0.01 of a bit time away from the center of the bit, so that after 50 or more samples, the receiver may be sampling at the wrong bit time. The smaller the timing difference, the later the error will occur, but if the transmitter sends a sufficiently long stream of bits, the transmitter and receiver will eventually be out of step. This fact shows the importance of synchronization between the transmitter and the receiver.

Figure 1.16 is an example of the problems that arise when the receiver's clock is running 12.5% ahead of time than the sender's one.

Example: Clocks running out of synchronization

Imagine a communication system with two data terminal equipments (DTEs) that have a major difference in the speed of their system clocks. The receiver's clock is running 12.5% ahead of time than the sender's one. If the sender transmits an 8-bit word, the receiver will interpret it as a 9-bit word. This is shown in Figure 1.16.

From Figure 1.16, it is clear that this difference in the speed of the clocks will not only increase the number of the received bits but also sample wrong bits. The conclusion should be that there can be no unambiguous interpretation of a common signal, if there is not a certain degree of synchronization of clocks.

Another well-known problem of time dispersion is called intersymbol interference.

Signals belonging to different symbols can be observed on the medium at the same time, leading to interpretation errors at the receiver's end.

Synchronization techniques will guide the receiving system in determining where data entities start and end and at which time interval the sampling result is least error prone.

It is possible to look to the bit and frame synchronization as a very basic mechanism of error control which will reduce the need for error control at higher levels.

Two approaches exist to solve the problem of synchronization: *asynchronous transmission* and *synchronous transmission*.

Each of the serial protocols that have been developed over the years as Universal Serial BUS (USB), Ethernet, SPI, I2C, and the serial standard, which will be considered in this book, can be sorted into one of these two modes: synchronous mode or asynchronous mode.

What is needed to decode the received signal? The receiver will have to determine where a signal cell (representing a bit) starts and ends in order to sample the signal as near at the middle of the signal as possible. It will have to know where a character or a byte starts or ends and, for packet-based transmission, where each message block starts or ends.

1.5.2 Modes of Transfer: Serial Transmission Modes

Over the years, dozens of serial protocols have been crafted to meet particular needs of embedded systems. USB (universal serial bus) and Ethernet are a couple of the more well-known computing serial interfaces. Other very common serial interfaces that include SPI, IIC, and the serial standard are what we are here to talk about today. Each of these serial interfaces can be sorted into one of the two groups or modes: synchronous or asynchronous.

- A **synchronous** serial interface always pairs its data line(s) with a clock signal, so all devices on a synchronous serial bus share a common clock.

This makes for a more straightforward, often faster serial transfer, but it also requires at least one extra wire between communicating devices. Examples of synchronous interfaces include SPI and I2C.

- **Bit-synchronization** is also a technique of synchronization in which during data transmission, sender and receiver should be synchronized at the bit level.

- **Asynchronous** means that data are transferred **without support from an external clock signal**. This transmission method is perfect for minimizing the required wires and I/O pins, but it does mean we need to put some extra effort into reliably transferring and receiving data. UART is the most common form of asynchronous transfer.

Asynchronous interface relies on four parameters, namely:

1. Baud rate control
2. Data flow control
3. Transmission and reception control
4. Error control.

This transmission method is perfect for minimizing the required wires and I/O pins, but it does mean we need to put some extra effort into reliably transferring and receiving data.

This chapter introduces the most common serial protocols of asynchronous transfer.

Asynchronous protocols are suitable for stable communication. These are used for long distance applications. UART is an example of asynchronous serial protocol, and examples of serial interface standards are RS-232, RS-422, and RS-485.

The next section introduces the topic of serial communication types, and the three sections after that discusses the three modes of operation: asynchronous data transfer, synchronous data transfer, and bit-synchronous operation.

1.5.3 Serial Communication Types

Serial communication can be a full-duplex or half-duplex transmission (Figure 1.17):

- Full-duplex means data can be sent and received simultaneously: data can be transmitted from the master to the slave, and from the slave to the master at the same time.

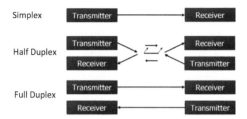

Figure 1.17 Types of transmission: Full-duplex and half-duplex transmissions.

- Half-duplex is when data can be sent or received, but not at the same time. This means that data transmission can occur in only one direction at a time, that is, either from the master to the slave, or from the slave to the master, but not both.

1.6 Asynchronous Data Transfer

1.6.1 Asynchronous Data Transfer

Data transfer is called asynchronous when data bits are not "synchronized" with a clock line, that is, there is no clock line at all.

Timing problems are avoided in case of asynchronous communication, by simply not sending long streams of bits. Data are transmitted one character (byte) at a time. Synchronization only needs to be maintained within each character, because the receiver can resynchronize at the beginning of each new character. When no characters are being transmitted, the line is idle (usually represented by a constant negative voltage). The beginning of a character is signaled by a start bit (usually a positive voltage), allowing the receiver to synchronized its clock with that of the transmitter. The rest of the bits that make up the character follow the start bit, and the last element transmitted is a stop bit. The transmitter then transmits the idle signal (which is usually the same voltage as the stop bit) until it is ready to send the next character, see Figures 1.18 and 1.3 (Figure 1.3 is part (a) of Figure 1.18).

Some of the features of asynchronous protocol are as follows:

- The protocol allows bits of information to be transmitted between two devices at an arbitrary point of time.
- The protocol defines that the data, more appropriately a "character," are sent as "frames" which in turn is a collection of bits.
  Definition: "Frame" is the minimum unit that can be transmitted in serial communication.

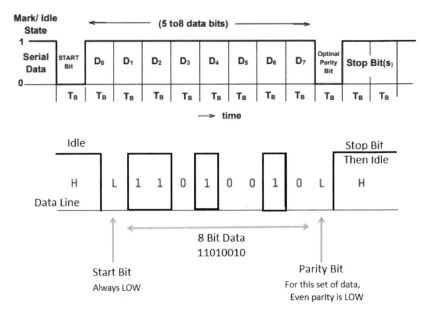

Figure 1.18   Asynchronous data transfer timing diagram.

- The protocol assumes that both the transmitter and the receiver are configured in the same way, that is, follow the same definitions for the start, stop, and the actual data bits.
- Both devices, namely the transmitter and the receiver, need to communicate at an agreed upon data rate *(baud rate)* such as **19.200 Kb/s or 115.200 Kb/s.**

The asynchronous protocol is *usually* as follows:

- The first bit is always the *START* bit (which signifies the start of communication on the serial line), followed by DATA bits (usually 8-bits), followed by a *STOP* bit (which signals the end of data packet). There may be a Parity bit just before the STOP bit. The Parity bit was earlier used for error checking, but is seldom used these days.
- The START bit is always low (0), while the STOP bit is always high (1). Figure 1.9 explains it.
- Data bits are a measurement of the actual data bits in a transmission. When the computer sends a frame of information, the amount of actual data may not be a full 8 bits. Asynchronous systems allow a number of variations including the number of *bits in a character. It allows* standard

values for frames as 5, 7, and 8 bits. Which setting you choose depends on what information you are transferring. For example, standard ASCII has values from 0 to 127 (7 bits). Extended ASCII uses 0 to 255 (8 bits). If the data you are transferring is standard ASCII, sending 7 bits of data per frame is sufficient for communication. A frame refers to a single byte transfer, including start/stop bits, data bits, and parity. Because the number of actual bits depends on the protocol selected, you can use the term "frame" to cover all instances.

- Stop bits are used to signal the end of communication for a single frame. Typical values are 1, 1.5, and 2 bits. Because the data are clocked across the lines and each device has its own clock, it is possible for the two devices to become slightly out of sync. Therefore, the stop bits not only indicate the end of transmission but also give the computers some room for error in the clock speeds. The more bits used for stop bits, the greater the lenience in synchronizing the different clocks, but the slower the data transmission rate.

- Parity is a simple form of error checking used in serial communication. There are four types of parity: even, odd, marked, and spaced. You can also use no parity. For even parity and odd parity, the serial port sets the parity bit (the last bit after the data bits) to a value to ensure that the transmission has an even or odd number of logic-high bits. For example, if the data value is 011, for even parity, the parity bit is 0 to keep the number of logic-high bits even. If the parity is odd, the parity bit is 1, resulting in 3 logic-high bits. Marked parity and spaced parity do not actually check the data bits but simply set the parity bit high for marked parity or low for spaced parity. This allows the receiving device to know the state of a bit so the device can determine if noise is corrupting the data or if the transmitting and receiving device clocks are out of sync.

Today, the most common standard has 8-bit characters, with 1 stop bit and no parity and this is frequently abbreviated as "*8-1-n*." A single *8-bit* character, therefore, consists of *10 bits on the line*, that is, One *Start* bit, Eight *Data* bits and One *Stop* bit (as shown in Figure 1.19).

For NRZ encoding, see latter, where line idle is encoded with 1, the start bit has the value 0 and the stop bit thus has the value 1. This variant, "8-1-n", of encoding adds at least 25% overhead to every byte transmitted, but that does not matter because asynchronous transmission is used mostly for peripheral devices such as keyboards or devices connected to the serial port of the computer.

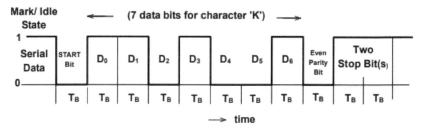

Figure 1.19   Serial data line to transmit character "K" in asynchronous mode.

Most important observation here is that the individual characters are *framed* (unlike all the other standards of serial communication), and *NO CLOCK* data are communicated between the two ends.

This protocol has been in use for many years and is used to connect PC peripherals such as *modems* and the applications include the classic *Internet dial-up* modem systems.

The typical data format *(known as "FRAME")* for asynchronous communication is shown in Figure 1.19.

Figure 1.19 shows an example of a serial data word that uses 7 data bits, an even parity bit and 2 STOP bits. The 7 data bits are the ASCII code to alpha numeric character "K" being transmitted.

Frame synchronization and data sampling points: As mentioned before, in case of asynchronous transmission, timing problems are avoided by simply dividing the data into characters (in general frames) and then transmitting one character (byte) at a time. Synchronization only needs to be maintained within each character, because the receiver can resynchronize at the beginning of each new character.

In other words, the receiver and transmitter in an asynchronous data transfer are not synchronized with respect to the time at which a character is transmitted; instead, once the transmitter starts to send a character, the receiver synchronizes itself with the bit times of the character in order to sample them at the correct time. The baud rate of the transmitter and the receiver is set to the same value. A start bit synchronizes the transmitter and the receiver. The receiver synchronizes its operation with the transmitter on the 1 to 0 transition of the data line. It waits one half a bit time, checks the input to make sure it is still logic 0 – and therefore a valid start bit – and begins sampling the data line at intervals equal to one bit time. The data line is sampled at the center of each transmitted bit. This eliminates errors that might occur if sampling takes place at the beginning of each bit time, since the leading or trailing edges of the transmission on the data line are distorted in transmission. It then samples the parity bit and stop bits.

The actual information, which the transmitter is sending to the receiver, is contained in the data bits. The data bits of each transmitted word are formed by the START and STOP bits. The receiver uses these framing bits as a means for determining which bits are the data bits.

The subject of frame synchronization and data sampling points will be discussed again in Chapter 2.

1.6.2 Interface Specifications for Asynchronous Serial Data Communication

The *serial port interface* for connecting two devices is specified by the **TIA** (*Telecommunications Industry Association*)/**EIA-232C** (*Electronic Industries Alliance*) standard published by the Telecommunications Industry Association; both the physical and electrical characteristics of the interfaces have been detailed in these publications.

RS-232, RS-422, RS-423, and RS-485 are each a recommended standard (RS-XXX) of the Electronic Industry Association (EIA) for asynchronous serial communication and have more recently been rebranded as *EIA-232, EIA-422, EIA-423, and EIA-485.*

It must be mentioned here that, although, some of the more advanced standards for serial communication like the USB and **FIREWIRE** are being

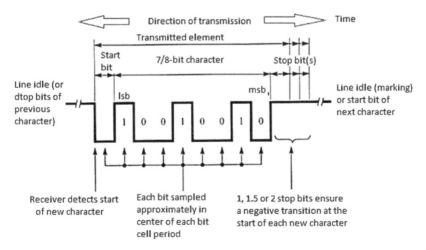

Figure 1.20 Asynchronous data transfer timing diagram.

popularized these days to fill the gap for high-speed, relatively short-run, heavy-data-handling applications, but still, the above four satisfy the needs of all those high-speed and longer run applications found most often in industrial settings for plant-wide security and equipment networking.

RS-232, RS-423, RS-422, and RS-485 specify the communication system characteristics of the hardware such as voltage levels, terminating resistances, and cable lengths. The standards, however, say nothing about the software protocol or how data are framed, addressed, checked for errors, or interpreted.

1.6.3 Application of Asynchronous Data Transmission: UART

The most common application of asynchronous data transmission is the Universal Asynchronous Receiver/Transmitter (UART). UART, as given in Chapter 2, is a hardware component of computer and microcontroller that is used in the serial communication subsystem, for example, in serial ports or internal modems.

The Universal Asynchronous Transmitter will transmit bytes in a serialized way by sending bit by bit to the receiver. At the receiver side, a Universal Asynchronous Receiver will receive the single bits and return full bytes.

1.6.3.1 Return-to-zero signaling

The UART, as given in Chapter-2, forms the data frames: Start bit, data bits, parity bit (if any), and the stop bit(s). An external unit must be used for signaling, for example, use of serial data standard RS-232.

The most popular variant of bit signaling is called return-to-zero signaling or pulse signaling. It is known that some media on the physical layer do not only support high and low values but also support positive and negative values. A signal may be represented by a positive high, negative high, or low value of voltage, amplitude, or phase shift.

Return-to-zero signaling essentially means that logic "1" is represented by a positive high value, and logic "0" is represented by a negative value. After the transmission, a low or no signal is sent; in other words, the signal returns to zero, that is, signal drops (returns) to zero between each pulse (Figure 1.21). This takes place even if a number of consecutive 0s or 1s occur in the signal. The signal is self-clocking. This means that a separate clock does not need to be sent alongside the signal, but suffers from using twice the bandwidth to achieve the same data rate as compared to non-return-to-zero format.

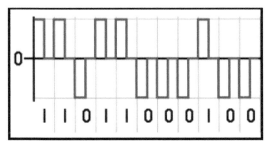

Figure 1.21 RTZ coding: The binary signal is encoded using rectangular pulse amplitude modulation with polar return-to-zero code.

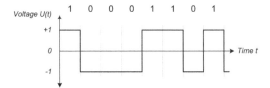

Figure 1.22 Return-to-zero encoding (voltage).

The "zero" between each bit is a neutral or rest condition, such as zero amplitude in pulse amplitude modulation (PAM), zero phase shift in phase-shift keying (PSK), or mid-frequency in frequency-shift keying (FSK). That "zero" condition is typically halfway between the significant condition representing a 1 bit and the other significant condition representing a 0 bit.

Although return-to-zero (RZ) contains a provision for synchronization, it still has a DC component resulting in "baseline wander" during long strings of 0 or 1 bits, just like the line code non-return-to-zero.

1.6.3.2 Return-to-zero signaling with different modulations

Let there be a byte with the bit sequence of 10110001. For frequency modulation, this means that there will be a frequency of 500 Hz for a high value (logical true) and a frequency of 250 Hz for a low value (logical false). Figure 1.22 is an illustration for "return-to-zero encoding (voltage)" as an example.

For amplitude shift keying, there will be a high voltage of 3 V for true and no voltage for false. Figure 1.23 illustrates this example.

As return-to-zero signaling relies on the availability of three signal states, it may not be available for every medium and keying scheme.

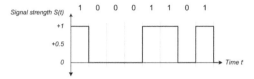

Figure 1.23   Return-to-zero encoding (abstract).

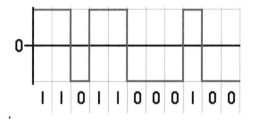

Figure 1.24   NRZ coding: The binary signal is encoded using rectangular pulse–amplitude modulation with polar NRZ(L) or polar non-return-to-zero-level code.

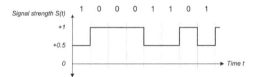

Figure 1.25   Encoding with NRZ: Non-return-to-zero encoding for amplitude shift modulation.

1.6.4 Non-return-to-zero Signaling

Non-return-to-zero (NRZ) code means that the signal will not be zero when data transmission is completed. In NRZ signaling, ones are represented by one significant condition, usually a positive voltage, while zeros are represented by some other significant condition, usually a negative voltage, with no other neutral or rest condition (Figure 1.24).

NRZ level itself is not a synchronous system but rather an encoding that can be used in either a synchronous or asynchronous transmission environment, that is, with or without an explicit clock signal involved.

As mentioned before, in case of "Non-return-to-zero Signaling", the signal will not return to line idle, even when only 0-bits are transmitted. A long series of logical ones will result in a long time span without any transition of amplitude or frequency, see Figure 1.25 as an example. The example shows the main problem that return-to-zero and non-return-to-zero encodings share.

For certain protocols, where long sequences of zeros or ones are transmitted, there will be long sequences without any change in signal. As the receiver's clock has no way to synchronize, it will run out of sync, and sampling of the incoming signal will lead to wrong results because the receiver fails to identify the bit boundaries.

Non-return-to-zero encoding and its derivatives should be used only when it can be guaranteed that the higher level protocols will not result in long sequences of identical bits. This behavior can be guaranteed for asynchronous transmission with start and stop bits, but not for synchronous transmission.

This limitation can be overcome by using the Manchester encoding, which we will cover in the next section.

1.6.5 Asynchronous transmission: Conclusion

Asynchronous data transmission with start and stop bits is a very simple way to transmit data. "Because asynchronous data are 'self-synchronizing', if there is no data to transmit, the transmission line can be idle."

The limitation of asynchronous communication is that errors cannot be detected, for example, when a stop bit is missed. Because of this fact, asynchronous commutation schemata can be found mostly in a single computer or between a computer and its peripheral modem or in direct terminal connections.

1.7 Synchronous Data Transfer

1.7.1 Synchronous Data Transfer

Synchronous data transfer is when the data bits are "synchronized" with a clock pulse. In other words, the receiver's clock is synchronized with the transmitter's clock.

Synchronous communication is used for applications that require higher data rates and greater error checking procedures. Character synchronization and bit duration are handled differently than asynchronous communications.

1.7.1.1 Bit duration

In case of asynchronous transmission, bit duration is fixed, and it is predefined by the bit rate. This matter simplifies the sampling instant of each bit. Similarly, the data are sent in the form of characters, and each character needs to start by 1 "START" bit and ends by 1 or 2 "STOP" bit, the matter that

Sync. Character	Sync. Character	Sync. Character	DATA	Sync. Character	DATA	DATA	Sync. Character	DATA

Figure 1.26   Typical sequence of data transmitted in synchronous mode.

helps in the character synchronization: once the receiver receives the start bit, it starts character assembly.

Bit duration in synchronous communications is not necessarily predefined at both the transmitting and receiving ends. Typically, a clock signal is provided in addition to the data signal. This clock signal will mark the beginning of a bit cell on a predefined transmission. The source of the clock is predetermined, and sometimes multiple clock signals are available. For example, if two nodes want to establish synchronous communications, point A could supply a clock to point B that would define all bit boundaries that A transmitted to B. Point B could also supply a clock to point A that would correspond to the data that A received from B. This example demonstrates how communications could take place between two nodes at completely different data rates.

1.7.1.2 Character synchronization
Character synchronization with synchronous communications is also very different from the asynchronous method of using start and stop bits to define the beginning and end of a character: Data words are transmitted continuously one after the other with no indication of character boundaries. When using synchronous communications, a predefined character or sequence of characters is used to let the receiving end know when to start character assembly.

The synchronizing character is an 8-bit character that would not normally appear in a stream of data words. A commonly used sync character is $00010110_2 = 16_{16}$ which is the ASCII code for the synchronizing character SYN. A typical synchronizing transmission sequence is shown in Figure 1.26.

Once the sync flag is received, the communication device will start character assembly. Sync characters are typically transmitted, while the communication line is idle or immediately before a block of information is transmitted.

To illustrate with an example, let's assume that we are communicating using 8 bits per character. Point A is receiving a clock from point B and sampling the receive data pin on every upward clock transition. Once point A receives the pre-defined bit pattern (sync flag), the next 8 bits are assembled into a valid character. The following 8 bits are also assembled into a character. This assembly will repeat until another predefined sequence of bits is received (either another sync flag or a bit combination that signals the end of the text, i.e., EOT). The actual sync flag and protocol varies depending on the sync format (SDLC, BISYNC, etc.).

In some more detail, in case of synchronous transmission, at the beginning of data transmission, only synchronizing characters are transmitted and only after that the actual data transmission begins. Therefore, to synchronize the transmitter with the starting of a character, the receiver operates in a hunt mode. Initially, all the bits of the data received are set "1." Then, the data line is sampled and the received bit is put on MSB shifting rest of the bits toward right. The data so formed are compared with the synchronizing character. If the data are different than the synchronizing character, the data line is again sampled and new data are formed and compared with the synchronizing character. In this way, the receiver operated in hunt mode—making bit by bit comparison of the input stream with the value of the synchronizing character until it detects the synchronizing character. Once the desired character is detected, the receiver treats each subsequent group of n-bits as a character. The transmitter continues to send synchronizing characters to maintain the synchronization, even if the sources of data character do not have data ready for transmission. In this case, the transmitter sends the synchronizing character continuously, and thus, the time interval between two characters is fixed. The clocks in the transmitter and receiver operate at exactly the same frequency and must be very stable to maintain synchronization for a long period of time. Typically, thousands of blocks if character can be sent without re-synchronizing the receiver.

1.7.1.3 Bit sampling

In case of asynchronous communication, as will be discussed in Chapter 2, the data bits are sampled in the middle of the data bit to ensure that the data received are correct, and this is implemented by giving a delay of $T_B/2$ after the START pulse is detected (T_B is the bit duration time). In case of synchronous communication, it is implemented with the help of common clock signal. The transmitter transmits the data bits on the TxD line at the falling edge of clock signal and the receiver samples on the RxD line at the rising edge of clock signal (Figure 1.27). Thus, the data bits are sampled in the middle of each bit. In this case, the baud rate is same as the frequency of clock signal. It is possible sometimes that the special SYN character may be present in the data itself, thus making it difficult to differentiate data with

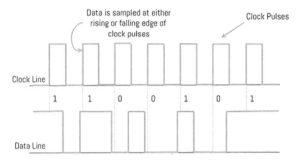

Figure 1.27   Synchronous data transfer timing diagram.

the synchronizing character. In such cases, instead of single synchronizing character, the two synchronizing characters SYN1 and SYN2 are transmitted together. The synchronization is also established when both the characters were detected together. Synchronous communication is typically used for high speed transfer of large block of data. This is because only 2% of the transmitted data is taken up by the synchronizing characters as compared to 20% for asynchronous communication.

In general, the concept for synchronous data transfer is simple and is as follows:

- Data bit sampling (or in other words, say, "recording") is done with respect to clock pluses, as can be seen in the timing diagrams (Figure 1.27).
- Since data are sampled depending upon clock pulses, and since the clock sources are very reliable, there is much less error in synchronous as compared to asynchronous.

For large blocks of data, synchronous transmission is much more efficient than asynchronous transmission, requiring much less overhead. The accuracy of the timing information allows much higher data rates. There is usually a minimum frame length, and each frame will contain the same amount of control information regardless of the amount of data in the frame.

Note: In the above discussion, we mentioned that in case of synchronous transmission, data are transmitted in a continuous stream, and the arrival time of each can be predicted by the receiver. In the above discussion, this is achieved by using separate timing circuit. It is also possible to achieve that by embedding the timing information in the signal itself. The latter can be achieved using bi-phase encoding (e.g., Manchester encoding). An embedded

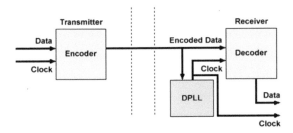

Figure 1.28 Use of embedded timing information.

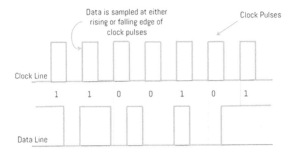

Figure 1.29 Synchronous data transfer timing diagram.

timing signal can be used by the receiver to synchronize with the transmitter using a *Digital Phase-Locked Loop* (DPLL) (Figure 1.28).

1.7.2 Bit-synchronous Operation

Bit-synchronous operation is a type of digital communication in which the data circuit-terminating equipment (DCE), data terminal equipment (DTE), and transmitting circuits are all operated in bit synchronism with a clock signal (Figure 1.29).

In bit-synchronous operation, clock timing is usually delivered at twice the modulation rate, and one bit is transmitted or received during each clock cycle.

Bit-synchronous operation is sometimes erroneously referred to as digital synchronization.

This requirement is fulfilled by using special encoding schemes that are a bit more complex than the NRZ-encoding variants that are normally used with asynchronous data transfer.

There are also ways to keep the clocks of two systems in synchronism, with or without directly sending the clock signal.

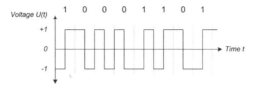

Figure 1.30   Signal values for Manchester encoding.

1.7.3 Manchester Encoding

The limiting factor at the physical layer is the fact that the clocks used for determining bit boundaries are never equally fast. This will lead for every encoding scheme which may create long sequences of equal signal values to an aberration of the two clocks and thus to bit errors when interpreting the signal.

Principle of operation: Manchester signaling means that every bit is represented by a change of value from high to low or low to high. The change of value happens exactly at the middle of the time span persevered for that bit. This makes determining the start and end of every bit very easy.

The following encoding convention has been accepted for the Manchester encoding:

Logic data: Physical signal
True: Transition from low to high
False: Transition from high to low

Example: Encoding with Manchester signaling: In this example, the bit sequence 00110001 should be encoded. Figure 1.30 will show the signal value curve for this example.

Comparing Manchester code with non-return-to-zero encoding or variants, it is possible to notice that for a given signaling rate, only half of the bit rate can be achieved, as one bit requires two signal elements, one high and one low.

The Manchester encoding is used in Ethernet networks.

Differential Manchester encoding: This variant of the Manchester signaling scheme also uses a transition in the middle of the bit for synchronization, but the representation of each bit depends on the signaling of the previous bit. If a 0 bit should be encoded, there will be a transition at the start of the bit and in the middle of the bit. If a 1 bit is encoded, there will be only a transition in the middle of the bit. Figure 1.31 is an example of differential Manchester scheme.

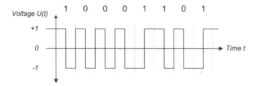

Figure 1.31 Differential Manchester encoding scheme.

Table 1.4 Differential Manchester encoding scheme

Previous Signal	Logical 0	Logical 1
High-low	High-low	Low-high
Low-high	Low-high	High-low

Figure 1.31 shows the encoding of the same bits as those in Figure 1.30. The differential Manchester encoding variant is used for token ring networks. Table 1.4 summarizes the differential Manchester encoding scheme.

1.8 Bit Synchronization

For synchronous transmission, data are not transferred byte-wise so there are no start or stop bits indicating the beginning or end of a character. Instead, there is a continuous stream of bits which have to be split up into bytes. Therefore, the receiver has to sample the received data in the right instant, and the sender's and receiver's clocks have to be kept in a synchronized state.

As the main task lies in synchronizing sender's and receiver's clocks, bit synchronization is also called clock synchronization and sometimes erroneously referred to as digital synchronization.

In general, **bit-synchronous operation** is a type of digital communication in which the data circuit-terminating equipment (DCE), data terminal equipment (DTE), and transmitting circuits are all operated in bit synchronism with a clock signal. In other words, during data transmission, sender and receiver should be synchronized at the bit level.

In bit-synchronous operation, clock timing is usually delivered at twice the modulation rate (see explanation next), and one bit is transmitted or received during each clock cycle.

1.8.1 Clock Encoding

The most self-evident way to accomplish clock synchronization is to send the clock signal to the receiver. This can be done by adding the signal of

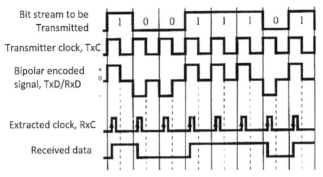

Figure 1.32   Bipolar clock encoding [From [halsall95], page 104, Figure 3.8].

the local clock to the encoded signal of the bit stream resulting in a bipolar encoded signal which the receiver will have to interpret. By using this bipolar encoding, it is not necessary to create an additional transmission line just for the clock signal.

Each bit span of the bipolar signal is dived in the middle by the signal shift of the clock. There are two possible values for each bit span: high-zero and low-zero, denoting logical one and logical zero. The received signal will contain enough information for the encoder as it can determine the length of a bit by the guaranteed signal change at the end of each bit, and it can determine the literal value by distinguishing between a positive or negative signal in the first half of the bit time.

A good example that describes bipolar clock encoding is given in Figure 1.32.

As can be seen in Figure 1.32, "bipolar clock encoding" signal of the transmitter clock is added to the bit stream that should be transmitted; the resulting signal is bipolar and contains a clock signal that can be extracted in order to encoding the bit stream.

This way of clock encoding is also a return-to-zero encoding, as it requires a medium capable of carrying bipolar encoded signals. This limitation can be overcome by using a digital phase-locked loop (DPLL).

1.8.2 Digital Phase-Locked Loop

The main idea of a digital phase-locked loop is that the receiver's clock is reasonably accurate, but should be resynchronized with the sender's clock whenever possible.

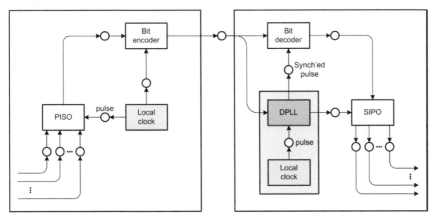

Figure 1.33   Digital phase-locked loop structure (FMC).

Unlike the direct clock encoding, there is no direct transmission of the clock signal, but it is possible to extract clock information from the received data signal.

It is important that there are enough bit transitions in the received data stream which indicate bit boundaries and make it possible to deduct the duration of a bit time as well as enabling the clock controller to reset the clock to a less diverged signal. This can be ensured by using a bit scrambler which removes long sequences of zeros or ones, but a more convenient way is to use an encoding scheme which ensures a sufficient number of bit transitions like the Manchester encoding.

Assume a system structure as described in Figure 1.33, where you have a digital clock that will have a sampling frequency that is at least 32 times as high as the bit rate of the incoming signal. This clock feeds its signal to the DPLL which also gets the received bit stream. The bit encoder encodes the bit stream and feeds it to the receiver's shift register where the serial to parallel transformation happens. The clock signal for this shift register comes from the DPLL.

The digital phase-locked loop has two inputs. The first input is the system clock which provides a signal to the DPLL in a well-defined frequency. The other input is the received bit stream, where every bit has at least one bit transition from 0 to 1 or from 1 to 0. This bit transition is either after around 16 clock signals, at the end of a bit time, in which case it will be ignored or after around 32 clock signals, in which it marks the middle of a bit time (for Manchester encoding) and will always occur.

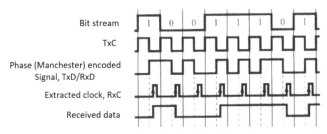

Figure 1.34   Manchester encoding for encoding a clock signal [From [halsall95], page 104, Figure 3.8].

If the clocks are in synchronism, there will be 32 clock signals between 2-bit middles.

The DPLL will feed its next signal after exactly 32 clock signals to the shift register.

Figure 1.34 uses Manchester encoding to encode the transmitter's clock signal in the bit stream without adding additional polarity to the transferred signal. The clock signal can be extracted nonetheless by using a digital phase-locked loop.

But if there are more than 32 clock signals, say 34, it means that the clock is two clock cycles ahead. In this case, the digital phase-locked loop will shorten the span until sending the next signal to the shift register by two clock signals and will feed the next signal after just 30 clock signals to the shift register.

The same procedure applies if there are less than 32 clock signals time span between two bits. The signaling span will be adjusted by waiting two clock cycles longer for giving the next impulse to the shift register and adding the delay to a total of 34 clock signals.

Example: Digital phase-locked loop with 8 clock cycles: In this example, there are only eight clock ticks for one bit time. The shaded area in Figure 1.35 marks the time span in which a signal transition from high to low or vice versa is expected. This is the time where the DPLL adjusts the clock signal.

Figure 1.36 focuses on two bits of the above bit stream. For the first bit, the clocks are in synchronism and no adjustment will take place, but for the second bit, the receiver's clock is one tick late, causing it to measure only seven ticks, while eight should have been there.

For the next bit, the clock will count only seven ticks (the missing tick will be omitted), which will make the clock going ahead. But going ahead for

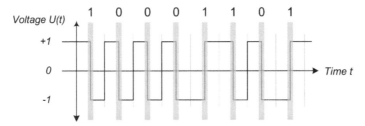

Figure 1.35   Manchester-encoded signal.

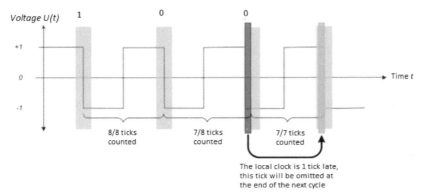

Figure 1.36   Manchester-encoded signal detailed view.

one time unit and being late for another will sum up to nothing. The clocks are in synchronism again.

A digital phase-locked loop will work with non-return-to-zero encodings as well as with non-return-to-zero inverted (NRZI) encodings like Manchester code or the differential Manchester code, but NRZI encodings require a higher signaling rate for the same bit rate and are therefore used especially in LANs.

1.9 Interface Specifications for Asynchronous Serial Data Communication

The *serial port interface* for connecting two devices is specified by the **TIA** (*Telecommunications Industry Association*)/**EIA-232C** (*Electronic Industries Alliance*) standard published by the Telecommunications Industry Association; both the physical and electrical characteristics of the interfaces have been detailed in these publications.

RS-232, RS-422, RS-423, and RS-485 are each a recommended standard (RS-XXX) of the Electronic Industry Association (EIA) for asynchronous serial communication and have more recently been rebranded as ***EIA-232, EIA-422, EIA-423, and EIA-485.***

It must be mentioned here that some of the more advanced standards for serial communication like the USB and FIREWIRE are being popularized these days to fill the gap for high-speed, relatively short-run, heavy-data-handling applications, but still, the above four satisfy the needs of all those high-speed and longer run applications found most often in industrial settings for plant-wide security and equipment networking.

RS-232, RS-423, RS-422, and RS-485 specify the communication system characteristics of the hardware such as voltage levels, terminating resistances, and cable lengths. The standards, however, say nothing about the software protocol or how data are framed, addressed, checked for errors, or interpreted.

The serial communication standards RS-xxx will be discussed in detail later.

1.10 Serial Communication Protocols

A variety of communication protocols have been developed based on serial communication in the past few decades. Some of them are as follows:

1. **SPI (Serial Peripheral Interface):** It is a synchronous serial communication interface specification used for short distance communication. It is a three-wire-based communication system. One wire each for master to slave and vice versa, and one for clock pulses. There is an additional, fourth wire, SS (slave select) line, which is mostly used when we want to send/receive data between multiple ICs. Theoretically, we can connect unlimited number of slaves, and practically, it depends on the load capacitance of the bus. SPI (serial peripheral interface) protocol sends and receives data in a continuous stream without any interruption. This protocol is recommended for high-speed data communication. The maximum speed it can provide is 10 Mbps. Unlike I2C, SPI has four wires. They are MOSI (Master Out Slave In), MISO (Master In Slave Out), Clock, and SS (Slave Select) signal.
2. **eSPI Protocol:** eSPI was developed by the Intel Corporation as the successor to the Low Pin Count (LPC) bus. eSPI stands for Enhanced Serial Peripheral Bus Interface—its primary function is to reduce the number of pins as compared to LPC.

Uses of the eSPI Protocol:

- Reduce the number of pins required on the motherboard
- Used in applications where real-time flash sharing is required

eSPI products offered by Total Phase:

3. **I2C (Inter-Integrated Circuit):** Pronounced eye-two-see or eye-square-see, this is an advanced form of USART. The transmission speeds can be as high as a whopping 400 KHz. The I2C bus has two wires – one for clock, and the other is the data line, which is bidirectional – this being the reason it is also sometimes (not always – there are a few conditions) called **Two-Wire Interface (TWI)**. It is a pretty new and revolutionary technology invented by Philips.

4. **CAN:** It is a serial communication protocol that requires two wires: CAN High (H+) and CAN Low (H-). It was developed by Bosch Company in 1985 for vehicle networks. It is based on a message-oriented transmission protocol.

5. **FireWire:** Developed by Apple, they are high-speed buses capable of audio/video transmission. The bus contains a number of wires depending upon the port, which can be either a 4-pin one, or a 6-pin one, or an 8-pin one.

6. **Ethernet:** Used mostly in LAN connections, the bus consists of 8 lines or 4 Tx/Rx pairs.

7. **Universal serial bus (USB):** This is the most popular of all. It is used for virtually all type of connections. The bus has 4 lines: V_{CC}, Ground, Data+, and Data−.

8. **Microwire:** Microwire is a three-wire synchronous interface developed by National Semiconductor and present on their COP8 processor family. Similar to SPI, microwire is a master/slave bus, with serial data out of the master (SO), serial data into the master (SI), and signal clock (SK). These correspond to SPI's MOSI, MISO, and SCK, respectively. There is also a chip select signal, which acts similar to SPI's SS. A full-duplex bus, microwire is capable of speeds of 625 Kbps and faster (capacitance permitting). Microwire devices from National Semiconductor come with different protocols, based on their data needs. Unlike SPI, which is based on an 8-bit byte, microwire permits variable length data and also specifies a "continuous" bitstream mode. Microwire has the same advantages and disadvantages as SPI with respect to multiple slaves, which require multiple chip select lines. In some instances, an SPI device will work on a microwire bus, as will a microwire device work on an SPI

bus, although this must be reviewed on a per-device basis. Both SPI and Microwire are generally limited to on-board communications and traces of no longer than 6 inches, although longer distances (up to 10 ft) can be achieved given proper capacitance and lower bit rates.

9. **1-Wire:** Dallas Semiconductor's 1-Wire bus is an asynchronous, master/slave bus with no protocol for multi-master. Like the I^2C bus, 1-Wire is half-duplex, using an open-drain topology on a single wire for bidirectional data transfer. However, the 1-Wire bus also allows the data wire to transfer power to the slave devices, although this is somewhat limited. Although limited to a maximum speed of 16 Kbps, bus length can be upwards of 1,000 ft, given the proper pull-up resistor.

10. **UART and USART:** UART stands for Universal Asynchronous Receiver Transmitter, whereas USART stands for Universal Synchronous Asynchronous Receiver Transmitter. They are basically just a piece of computer hardware that converts parallel data into serial data. The only difference between them is that UART supports only asynchronous mode, whereas USART supports both asynchronous and synchronous modes. Unlike Ethernet, Firewire, etc., there is no specific port for UART/USART. They are commonly used in conjugation with protocols like RS-232, RS-434, etc. (we have specific ports for these two!).

In *synchronous* transmission, the clock data are recovered separately from the data stream and no start/stop bits are used. This improves the efficiency of transmission on suitable channels since more of the bits sent are usable data and not character framing.

The USART has the following components:

- A clock generator, usually a multiple of the bit rate to allow sampling in the middle of a bit period
- Input and output shift registers
- Transmit/receive control
- Read/write control logic
- Transmit/receive buffers (optional)
- Parallel data bus buffer (optional)
- First-in, first-out (FIFO) buffer memory (optional)

11. **RS-232 (Recommended Standard 232):** The RS-232 is typically connected using a DB9 connector, which has 9 pins, out of which 5 are input, 3 are output, and one is ground. You can still find this so-called serial port in some old PCs. In our upcoming posts, we will discuss mainly about RS-232 and USART of AVR microcontrollers.

Table 1.5 Protocol comparison

Name	Sync/ Async	Type	Duplex	Max Devices	Max Speed (Kbps)	Max Distance (Kbps)	Pin Count (1)
RS-232	Async	Peer	Full	2	20(2)	30(3)	2(4)
RS-422	Async	Multi-drop	Half	10(5)	10,000	4,000	1(6)
RS-485	Async	Multi-point	Half	32(5)	10,000	4,000	2
I^2C	Sync	Multi-master	Half	−7	3,400	<>	2
SPI	Sync	Multi-master	Full	−7	>1,000	<>	3+1(8)
Microwire	Sync	Master/slave	Full	−7	>625	<>	3+1(8)
1-Wire	Async	Master/slave	Half	−7	16	1,000	1s

Notes

−1	Not including ground.
−2	Faster speeds available but not specified.
−3	Dependent on capacitance of the wiring.
−4	Software handshaking. Hardware handshaking requires additional pins.
−5	Device count given in unit loads (UL). More devices are possible if fractional-UL receivers/transmitters are used.
−6	Unidirectional communication only. Additional pins needed for each bidirectional communication.
−7	Limitation based on bus capacitance and bit rate.
−8	Additional pins needed for every slave if slave count is more than one.

Table 1.5 summarizes the characteristics of some of the above protocols.

UART, USART, and the serial communication protocols are the subject of the next chapter.

1.11 Electrical Interface Standards

Serial communication, synchronous or asynchronous, is typically implemented with a Recommended Standard (RS). Nowadays and in most cases, the standard is set by the Electronic Industries Association (EIA). The standard usually defines signal levels, maximum bandwidth, connector pin-out, supported handshaking signals, drive capabilities, and electrical characteristics of the serial lines. In this section, we briefly describe some of the more common communication standards. Chapter 3 will handle most of the electrical interface standards. The full specification for each standard is available from a number of engineering document dealers.

RS-232: RS-232 is the most widely used communication standard. This implementation has been defined and revised several times and is often referred to as RS-232C or EIA-232. The most common implementation of RS-232 is on a standard 25-pin D sub-connector, although the IBM PC computer defined the RS-232 port on a 9-pin D sub-connector. Both implementations are in wide spread use. RS-232 is capable of operating at data rates up to 20 Kbps / 50 ft. The absolute maximum data rate may vary due to line conditions and cable lengths. RS-232 often operates at 38.4 Kbps over very short distances. The voltage levels defined by RS-232 range from -12 to +12 V. RS-232 is a single-ended interface, meaning that a single electrical signal is compared to a common signal (ground) to determine binary logic states. A voltage of +12 V (usually +3 to +10 V) represents a binary 0 and -12 V (-3 to -10 V) denotes a binary 1.

RS-422: The RS-422 specification defines the electrical characteristics of balanced voltage digital interface circuits. RS-422 is a differential interface that defines voltage levels and driver/receiver electrical spec-ifications. On a differential interface, logic levels are defined by the difference in voltage between a pair of outputs or inputs. In contrast, a single-ended interface, for example, RS-232, defines the logic levels as the difference in voltage between a single signal and a common ground connection. Differential interfaces are typically more immune to noise or voltage spikes that may occur on the communication lines. Differential interfaces also have greater drive capabilities that allow for longer cable lengths. RS-422 is rated up to 10 Megabits/second and can have cabling 4000 ft long. RS-422 also defines driver and receiver electrical characteristics that will allow 1 driver and up to 32 receivers on the line at once. RS-422 signal levels range from 0 to +5 V. RS-422 does not define a physical connector.

RS-423: The RS-423 specification defines the electrical characteris-tics of unbalanced voltage digital interface circuits. The voltage levels defined by RS-423 range from −5 to +5 V. RS-423 is a single-ended interface, meaning that a single electrical signal is compared to a com-mon signal (ground) to determine binary logic states. A voltage of +5 V represents a binary 0 and −5 V denotes a binary 1. RS-423 is rated up to 100 K bits/second. RS-423 defines driver and receiver electrical characteristics. RS-423 does not define a physical connector.

RS-449: RS-449 (a.k.a. EIA-449) compatibility means that RS-422 signal levels are met, and the pin-out for the DB-37 connector is specified. The EIA (Electronic Industry Association) created the RS-449 specification to detail the pin-out and define a full set of modem control signals that can be used for regulating flow control and line status.

RS-485: RS-485 is backwardly compatible with RS-422; however, it is optimized for partyline or multi-drop applications. The output of the RS-422/485 driver is capable of being active (enabled) or tri-state (disabled). This capability allows multiple ports to be connected in a multi-drop bus and selectively polled. RS-485 allows cable lengths up to 4000 ft and data rates up to 10 Megabits/second. The signal levels for RS-485 are the same as those defined by RS-422. RS-485 has electrical characteristics that allow for 32 drivers and 32 receivers to be connected to one line. This interface is ideal for multi-drop or network environments. RS-485 tri-state driver (not dual-state) will allow the electrical presence of the driver to be removed from the line. The driver is in a tri-state or high impedance condition when this occurs. Only one driver may be active at a time, and the other driver(s) must be tri-stated. The output modem control signal "Request to Send (RTS)" controls the state of the driver. Some communication software packages refer to RS-485 as RTS enable or RTS block mode transfer. RS-485 can be cabled in two ways, two-wire mode and four-wire mode. Two-wire mode does not allow for full-duplex communication and requires that data be transferred in only one direction at a time. For half-duplex operation, the two transmit pins should be connected to the two receive pins (Tx+ to Rx+ and Tx- to Rx-). Four-wire mode allows full-duplex data transfers. RS-485 does not define a connector pin-out or a set of modem control signals. RS-485 does not define a physical connector.

RS-530: RS-530 (a.k.a. EIA-530) compatibility means that RS-422 signal levels are met, and the pin-out for the DB-25 connector is specified. The EIA (Electronic Industry Association) created the RS-530 specification to detail the pin-out and define a full set of modem control signals that can be used for regulating flow control and line status. The RS-530 specification defines two types of interface circuits: data terminal equipment (DTE) and data circuit-terminating equipment (DCE). The sealevel systems adapter is a DTE interface.

Current loop: This communication specification is based on the absence or presence of current, not voltage levels, over the communication lines. The logic of a current loop communication circuit is determined by the presence or absence of current (typically ± 20 mA). When referring to the specification, the current value is usually stated (i.e., 20 mA current loop). Current loop is used for point-to-point communication, and there are typically two current sources, one for transmit and one for receive. These two current sources may be located at either end of the communication line. To ensure a proper current path to ground, or loop, the cabling of two current loop communication ports will depend on the location of the current sources. Current loop is normally good for data rates up to 19.2 Kbps. This limitation is due to the fact that the drivers and receivers are usually optically isolated circuits that are inherently slower than non-isolated equivalent circuits.

MIL-188: This communication standard comes in two varieties, MIL-188/C and MIL-188/114. Both of these interfaces are military standards that are defined by the US Department of Defense. MIL-188/114 is a differential interface and MIL-188/C is an unbalanced or single-ended interface. Both MIL-188 interfaces are implemented on an RS-530 connector. MIL-188/C and MIL-188/114 have signal levels from +6 V to -6 V and are ideal for long distances at high speeds.

References

[1] "Interface Between Data Terminal Equipment and Data Circuit-Terminating Equipment Employing Serial Binary Data Interchange," TIA/EIA-232-F Standards, Electronics Industries Association Engineering Department.

[2] "Electrical Characteristics of Balanced Digital Interface Circuits," TIA/EIA-422-B Standards, Electronics Industries Association Engineering Department.

[3] "Standard for Electrical Characteristics of Generators and Receivers for Use in Balanced Digital Multipoint Systems," TIA/EIA-485-A Standards, Electronics Industries Association Engineering Department.

[4] "The I^2C Specification," Version 2.1, Philips Semiconductors.

[5] Aleaf, Abdul, "Microwire Serial Interface," Application Note AN-452, National Semiconductor.

[6] Goldie, John, "Summary of Well-known Interface Standards," Application Note AN-216, National Semiconductor.

[7] Nelson, Todd, "The Practical Limits of RS-485," Application Note AN-979, National Semiconductor.

[8] Wilson, Michael R., "TIA/EIA-422-B Overview," Application Note AN-1031, National Semiconductor.

[9] Goldie, John, "Ten Ways to Bulletproof RS-485 Interfaces," Application Note AN-1057, National Semiconductor.

[10] Martin H. Weik (2000). *Computer Science and Communication Dictionary, volume 2. Springer. ISBN 978-0-7923-8425-0.*

2

Universal Asynchronous Receiver/Transmitter (UART)

2.1 Introduction

The *Universal Asynchronous Receiver/Transmitter (UART)* controller is the key component of the serial communication subsystem of a microprocessor, microcontroller, and computer. It is a computer hardware device for asynchronous serial communication in which the data format and transmission speeds are configurable.

The UART device changes incoming parallel data to serial data which can be sent on a communication line. A second **UART** is used to receive the information which converts serial data to parallel. The **UART** performs all the tasks, timing, parity checking, etc. needed for the communication. UART does not generate the external signaling levels that are used between different equipment. An interface, external line drive, is used to convert the logic level signals of the UART to the external signaling levels. Examples of standards for voltage signaling are **RS-232, RS-422,** and **RS-485** from the **EIA**. For embedded system applications, UARTs are commonly used with RS-232. It is useful to communicate between microcontrollers and with PCs. **MAX 232** is one of the example ICs which provide RS-232 level signals. Besides the EIA RS standards, there is also a *CCITT* standard named *V.24* that resembles the specifications included in RS-232-C.

The UART structure contains several registers that are accessible to set or review the communication parameters that facilitate the use of **UART** in different environments. Using these registers, the communication speed (*baud rate*), the type of parity check, and the way incoming information is signaled to the running software are set according to the requirement of host processor.

Latest standard, UART employ First-In First-Out (*FIFO)* buffer for improved functional capability. UARTs are generally used to modem control functions. Using the different available functional control registers allows the designer to establish efficient interface with the host processor. Embedded processors employ built in hard core UART block while soft IP UART provides programmable and reconfigurable flexibilities which are very much advantageous in FPGA applications. Hardware descriptive languages (HDL) like Verilog can be used for the behavioral description of the UART.

As mentioned in Chapter 1, there are two types of serial transmission: synchronous and asynchronous. The UART, as it is clear from the name, uses asynchronous communication. On the other hand, when using synchronous transmission, it is called *Universal Synchronous-Asynchronous Receiver/Transmitter (USART)*.

2.2 Serial UART Types

PC compatible serial communication started with the **8250 UART** in the **IBM XT machines**. Then, UART is upgraded to **8250A, 8250B,** and then **16450** (manufactured by National Semiconductor) which is implemented in the **AT** machines. The higher bus speed could not be reached by the **8250** series but newer **16450** series were capable of handling a communication speed of 38.4 Kbps.

The main problem with the original series was the need to perform a software action for each single byte to transmit or receive. To overcome this problem, the 16550 was released which contained two on-board **FIFO buffers**, each capable of storing 16 bytes. One buffer for incoming bytes and one buffer for outgoing bytes (**16450** had 1 byte FIFO). This made it possible to increase the maximum reliable communication speed to 115.2 Kbps and use effectively in modems with on-board compression. The 16550 chip contained a firmware bug which made it impossible to use the buffers. The 16550A which appeared soon after was the first UART that was able to use its FIFO buffers. This made it possible to increase the maximum reliable communication speed to 115.2 Kbps. This speed was necessary to use effectively modems with on-board compression. A further enhancement introduced with the 16550 was the ability to use DMA (direct memory access) for the data transfer. Two pins were redefined for this purpose. DMA transfer is not used with most applications. Only special serial I/O boards with a high number of ports contain sometimes the necessary extra circuitry to make this feature work.

The most common **UART** used is **16550A**. Newer versions such as **16650** contain two 32-byte FIFO buffers, and on-board support for software flow control is the latest advancement in industry. Texas Instruments is developing the **16750 UART** which contains 64-byte FIFO buffers.

2.3 Serial Communication Terminologies

There are many terminologies, or "keywords" associated with serial communication. Some of these terminologies were considered and discussed in Chapter 1. Here some of the terminologies are briefly defined:

- **Asynchronous bus:** On an asynchronous bus, data are sent without a timing clock. A synchronous bus sends data with a timing clock.
- **Simplex communication:** It is a one-way communication technique. In this mode of serial communication, data can only be transferred from transmitter to receiver and not vice versa. Radio and television transmissions are the examples of simplex mode.
- **Full-duplex:** Full-duplex means data can be sent and received simultaneously: data can be transmitted from the master to the slave, and from the slave to the master at the same time. An example is the smartphone.
- **Half-duplex:** Half-duplex is when data can be sent or received, but not at the same time. This means that data transmission can occur in only one direction at a time, that is, either from the master to the slave, or the slave to the master, but not both. A good example is an internet. If a client (laptop) sends a request for a web page, the web server processes the application and sends back the information.
  For full-duplex and half-duplex, see Figure 1.17.
- **Master/slave:** Master/slave describes a bus where one device is the master and others are slaves. Master/slave buses are usually synchronous, as the master often supplies the timing clock for data being sent along in both directions.
- **Multi-master:** A multi-master bus is a master/slave bus that may have more than one master. These buses must have an arbitration scheme that can settle conflicts when more than one master wants to control the bus at the same time.
- **Point-to-point or peer interface:** Point-to-point or peer interfaces are where two devices have a peer relation to each other; there are no masters or slaves. Peer interfaces are most often asynchronous.
- **Multi-drop:** The term multi-drop describes an interface in which there are several receivers and one transmitter.

- **Multi-point**: Multi-point describes a bus in which there are more than two peer transceivers. This is different from a multi-drop interface as it allows bidirectional communication over the same set of wires.
- **MSB/LSB:** This stands for Most Significant Bit (or Least Significant Bit). Since data are transferred bit by bit in serial communication, one needs to know which bit is sent out first: MSB or LSB.
- **Baud rate:** Baud is synonymous to symbols per second or pulses per second. It is the unit of symbol rate, also known as *baud* or *modulation rate*. However, although technically incorrect, in the case of modem manufacturers baud commonly refers to bits per second.
- **Address:** If the application needs to send multiple data together over the same channel and/or the device the application needed is sharing the same channel space with other devices sending their own data, then the designer has to take care to properly address the data. The need for device address and the way of allocating it will be discussed in the following chapters.

Other Common Terminologies

The following terms are often used in serial communications (see Chapter 3):

- **DTE:** Data terminal equipment (DTE) is one of the two endpoints in a serial communication. An example would be a computer.
- **DCE:** Data communication equipment (DCE) is the other endpoint in a serial communication. Typically, it is a modem or serial terminal.
- **RS-232:** The original standard that defines hardware serial communications. It has since been renamed to TIA-232.
- **Serial port:** The serial port is a type of connection on PCs that is used for peripherals such as mice, gaming controllers, modems, and older printers. It is sometimes called a COM port or an RS-232 port, which is its technical name. There are two types of serial ports: DB9 and DB25. DB9 is a 9-pin connection, and DB25 is a 25-pin connection. A serial port can only transmit one bit of data at a time, whereas a parallel port can transmit many bits at once. The serial port is typically the slowest port on a PC. Most newer computers have replaced serial ports with much faster and more compatible USB ports
- **Serial cables:** There are several different kinds of serial cables. As given in Chapter 3, the two most common types are null-modem cables and standard RS-232 cables. The documentation for the hardware should describe the type of cable required.

These two types of cables, as shown in Chapter 3, differ in how the wires are connected to the connector. Each wire represents a signal, and each signal has a function. A standard serial cable passes all of the RS-232C signals straight through. For example, the "Transmitted Data" pin on the one end of the cable goes to the "Transmitted Data" pin on the other end. This is the type of cable used to connect a modem to a computer and is also appropriate for some terminals.

A null-modem cable switches the "Transmitted Data" pin of the connector on the one end with the "Received Data" pin on the other end. The connector can be either a DB-25 or a DB-9.

Straight cables and null-modem cables are given in detail in Chapter 3.

2.4 UART Timing: Serial Data Format and Asynchronous Serial Transmission

As mentioned in Chapter-1, asynchronous transmission does not need sending clock signal to send the data to the receiver. The sender and the receiver must agree on timing parameters in advance, and special bits such as start and stop bits are added to each word which is used to synchronize the sending and receiving units. The UART serial data format is the standard serial format discussed in Section 2.4.1. Figure 2.1, which in principle the same as Figure 1.9, shows the signaling as seen at a UART's transmit data (TXD) or receive data (RXD) pins. As given in Chapter 3, RS-232 is an interface between the device and the communication line. RS-232 bus drivers invert as well as level shift the waveform of the data format so that logic 1 is a negative voltage on the bus and logic 0 is a positive voltage.

When the line is idle, no data, its state is high voltage or powered. The data frame starts by the "**Start Bit**" at the beginning of each word (character

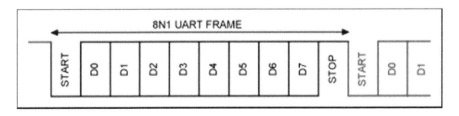

Typical UART data frame

Figure 2.1 Serial data format.

or byte) that is to be transmitted. The **Start Bit** indicates the start of the data transmission and it alerts the receiver that a word of data is about to be sent. Upon reception of start bit, the clock in the receiver goes into synchronization with the clock in the transmitter. The accuracy of these two clocks should not deviate more than 10% during the transmission of the remaining bits in the word.

The individual bits of the word of data are sent after the start bit. Least significant bit (LSB) is sent first. The transmitter does not know when the receiver has read at the value of the bit. The transmitter begins transmitting the next bit of the word on next clock edge. This continued till the transmitter completes transmitting the data bits. In Figure 2.1, the data consist of 8 bits (D0–D7), but it could be 5, 6, or 7 bits long.

Parity bit, which is an optional bit, is to be added when the entire data word has been sent. This bit can be used to detect errors at the receiver side. Then, one **Stop Bit** is sent by the transmitter to indicate the end of the valid data bits. Some protocols allow 1.5 or 2 STOP bits.

The UART transmitter and receiver must agree, before starting communication, upon the number of data bits per frame, if there is one or no parity bit and the number of stop bits. Line control register (LCR) is used for this purpose. Based on the settings chosen in the LCR, the UART transmitter sends the start bit, 5, 6, 7, or 8 data bits, one or no parity bit and 1, 1.5, or 2 stop bits.

The UART transmitter section, as given in Section 2.5.1, includes a transmitter hold register (THR) and a transmitter shift register (TSR). Modern UART devices include transmit and receive FIFO buffers. When the UART is in the FIFO mode, THR is a 16-byte FIFO. Line control register (LCR) is controlling the transmission section.

The UART receiver section includes a receiver shift register (RSR) and a receiver buffer register (RBR). When the UART is in the FIFO mode, RBR is a 16-byte FIFO. Receiver section control is a function of the UART line control register (LCR). Based on the settings chosen in LCR, the UART receiver accepts the following from the transmitting device: 1 START bit; 5, 6, 7, or 8 data bits; 1 PARITY bit, if parity is selected; and 1 STOP bit

Once the receiver section receives all of the bits in the data word, it can, based on the contents of LCR, check for the *parity bits*. (The contents of LCR reflect the agreement between the transmitter and the receiver about the parity bit, if it is used or not). Then, *stop bit* is encountered by receiver. A missing *stop bit* may result entire data to be garbage. This will cause a *framing error* and will be reported to the host processor when the data word is read. **Framing error** can be caused due to mismatch of transmitter and receiver clocks.

The UART automatically discards the Start, Parity and Stop bits irrespective of whether data is received correctly or not. If the sender and receiver are configured identically, these bits are not passed to the host. To transmit new word, the Start Bit for the new word is sent as soon as the Stop Bit for the previous word has been sent.

The transmission speed in asynchronous communication is measured by **baud rate**. A baud rate represents the number of bits that are actually being sent over the media. The baud rate includes the start, stop, and parity bits. The **bit rate** (**bits per second, bps**) represents the amount of data that are actually sent from the transmitting device to the other device. Speeds for UARTs are in bits per second (bit/s or bps), although often incorrectly called the baud rate. Standard baud rates are as follows: 110, 300, 1200, 2400, 4800, 9600, 14400, 19200, 28800, 38400, 57600, 76800, 115200, 230400, 460800, 921600, 1382400, 1843200, and 2764800 bit/s.

Importance of Baud Rate

- For two microcontrollers to communicate serially, they should have the *same* baud rate, else serial communication won't work. This is because when the designer sets a baud rate, he directs the microcontroller to transmit/receive the data at that particular rate. So, if the designer sets different baud rates, then the receiver might miss out the bits the transmitter is sending (because it is configured to receive data and process it with a different speed).
- Different baud rates are available for use. The most common ones are 2400, 4800, 9600, 19200, 38400, etc. It is impossible to choose any arbitrary baud rate, and there are some fixed values which the designer must use like 2400, 4800, etc.

2.4.1 Frame Synchronization and Data Sampling Points

When two UARTs communicate, it is a given that both transmitter and receiver know the signaling speed. The receiver does not know when a packet will be sent (no receiver clock since it is an asynchronous system). The receiver circuitry is correspondingly more complex than that of the transmitter. The transmitter simply has to output a frame of data at a defined bit rate. Contrastingly, the receiver has to recognize the start of the frame to

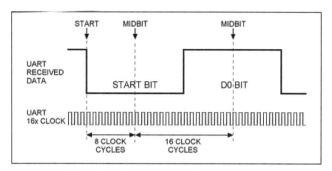

Figure 2.2 UART receive frame synchronization and data sampling points.

synchronize itself and therefore determine the best data-sampling point for the bit stream.

Figure 2.2 shows a common method used by a UART receiver to synchronize itself to a received frame. The receive UART uses a clock that is 16 times the data rate. A new frame is recognized by the falling edge at the beginning of the active-low START bit. This occurs when the signal changes from the active-high STOP bit or bus idle condition. The receive UART resets its counters on this falling edge, expects the mid-START bit to occur after 8 clock cycles, and anticipates the midpoint of each subsequent bit to appear every 16 clock cycles thereafter. The START bit is typically sampled at the middle of bit time to check that the level is still low and ensure that the detected falling edge was a START bit, not a noise spike. Another improvement is to sample the START bit three times (clock counts 7, 8, and 9, out of 16) instead of sampling it only at the midbit position (clock count 8 out of 16). The UART of most of the microcontrollers is using this technique.

Sometimes, the UART uses sampling rate which is 13 times (13x) the baud rate. In such cases, the sampling takes place on the 6th cycle.

2.4.2 Timing Accuracy

As mentioned before, the transmitter and the receiver have to agree upon the clock rate before starting communication. The absolute clock rate is unimportant for the purposes of accurate reception. The important matter is when there is transmit-receive clock mismatch. In this case, the following question may arise: how different the transmit and the receive UART clocks can be. The first point to understand toward answering this is that because the UART receiver synchronizes itself to the start of each and every frame, we only care about accurate data sampling during one frame. There is no buildup of error beyond a frame's STOP bit, which simplifies analysis because it is necessary only to consider one frame for the worst-case scenario.

2.4.3 Timing Error Due to Transmit-receive Clock Mismatch

In case of sampling each bit at the midpoint, no error will occur (Figure 2.3). If sampling takes place one-half a bit-period too early or too late, we will be sampling at the bit transition and have problems (Figure 2.3).

In reality, sampling close to the bit-transition point is unreliable. The primary reason for this is the finite (and typically slow) transmission rise and fall times. These times become even slower if overly capacitive cabling is used. A long bus incurs high attenuation, which reduces noise margin and makes it more important to sample when the bit level has settled.

It is difficult to quantitatively assess a worst-case acceptable sampling range across a bit's period. EIA/TIA-232-F does specify a 4% of bit-period maximum slew time for a transmission, but this is difficult to achieve for long runs at 192 Kbps. We consider two cases here: large slew time and normal slew time.

The large slew time case is shown in Figure 2.4. In the figure, this case is called "Nasty Scenario," and it can happen with a long capacitive RS-232 channel. In such cases, the capacitive load increases the slew rate. Because of the large slew time, the sampling is reliably possible within the middle 50% of the bit time.

The second case is shown by Figure 2.5 for a normal channel (link). With this normal channel, sampling can take place within the middle 75% of the bit time. This can be the case using a relatively short slew time (such as a meter-length bus) with buffered CMOS logic levels or an RS-485 differential pair within an equipment chassis. (RS-232 is the subject of Chapter 3.)

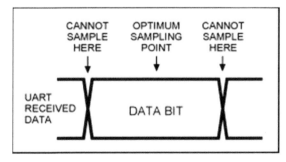

Figure 2.3   UART receive sampling range.

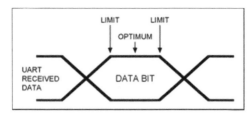

Figure 2.4   UART "nasty link" is sampled reliably within 50% of bit time.

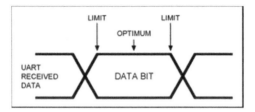

Figure 2.5   UART "normal link" is sampled reliably within 75% of bit time.

From Figures 2.4 and 2.5, it is possible to determine that the error budget is ±25% and ±37.5% from the optimal bit-center sampling point for the nasty and normal scenarios, respectively. This error is equivalent to ±4 or ±6 periods of the 16x UART receive clock. Another error to include in this budget is the synchronization error when the falling edge of the START bit is detected. The UART will most likely start on the next rising edge of its 16x clock after detecting the START bit. Since the 16x clock and the received data stream are asynchronous, the falling edge of the START bit could occur just after a 16x clock rising edge. The falling edge could, alternatively, occur just before the clock rising edge, but not with enough setup time to use it. This means that the UART has a ±1 bit error built in at the synchronization point. So our error budget reduces from ±4 or ±6 clock periods to ±3 or ±5 periods.

The above analysis presumed that short-term clock errors (essentially jitter) are very small, and therefore, we are only considering midterm and long-term errors. These errors point to a mismatch in the transmit UART and receive UART timing that is consistent during a frame. Since the timing is synchronized at the falling edge of the START bit, the worst-case timing error will be at the last data sampling point, which is the STOP bit. The optimum sampling point for the STOP bit is its bit center, which is calculated as:

(16 internal clock cycles per bit) × (1 start bit + 8 data bits + 1/2 a stop bit)

$$= (16) \times (9.5)$$
$$= 152 \text{ UART receive clocks after the original falling edge of the}$$
$$\text{START bit.}$$

It is possible now to calculate the allowable error as a percentage. For the normal scenario, the clock mismatch error can be $\pm 5/152 = \pm 3.3\%$. For the "nasty" scenario, it can be $\pm 3/152 = \pm 2\%$. As hinted earlier, although the problem will materialize at the receive end of the link, clock mismatch is actually a tolerance issue shared between the transmit and the receive UARTs. So presuming that both UARTs are attempting to communicate at exactly the same bit rate (baud), the allowable error can be shared, in any proportion, between the two UARTs.

Making use of the allowable error budget is helpful in systems where both ends of the link are being designed at the same time. This is partly because the tolerance of both ends will be known and partly because trade-offs and cost savings can be made. In general, a standard low-cost, ceramic resonator with $\pm 0.5\%$ accuracy and a further $\pm 0.5\%$ drift over temperature and life can be used for the clock source at both ends of the link. This meets the 2% "nasty" scenario discussed earlier. If the system uses a master controller (typically a microcontroller or a PC) with a standard 100 ppm crystal oscillator for the UART clock source, the link error can be cut approximately in half. Be careful with microcontrollers that synthesize baud frequencies for their internal UARTs. Depending on the choice of microcontroller clock, the baud rates may not be exact. If the error can be determined, it can be easily included in the link error budget.

2.5 UART Functional Block Diagram

The functional block diagram of any UART/USART device depends on the capabilities of the device. It is recommended, accordingly, for the user to consult the specification manual of the UART in use. Irrespective of the capabilities of the device, the structure of any UART consists of three main sections (Figure 2.6):

- Clock generator
- Transmitter section
- Receiver section

In its simplest form, the transmitter section and the receiver section consist of a number of registers each with its own function and control

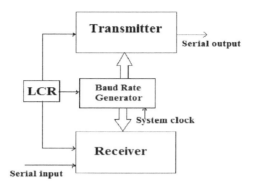

Figure 2.6   UART block diagram: three main sections.

logic circuits. Modern UART contains FIFO memories and DMA controller as part of the two sections. The two sections are commonly provided by a programmable baud-rate generator, clock generator. This generator can divide the UART input clock by divisor from 1 to 65535 producing a 16× reference clock or a 13× reference clock for the internal transmitter and receiver logic. The baud-rate generator is used for generating the speed when the transmitter section and receiver section have to transmit or receive the data. The speed ranges from 110 bps to 230400 bps. Typically, the baud rates of microcontrollers are 9600 to 115200.

To let the discussion useful, this section introduces the structure of a modern UART that has FIFO buffer memories and DMA controller. The registers and other control circuits in such modern UART cover those included in the structure of any UART device with less capabilities. Modern UART usually contains the following components:

- A clock generator, usually a multiple of the bit rate to allow sampling in the middle of a bit period.
- Input and output shift registers
- Transmit/receive control
- Read/write control logic
- Transmit/receive buffers (optional)
- System data bus buffer (optional)
- First-in, first-out (FIFO) buffer memory (optional)
- Signals needed by a third party DMA controller (optional)
- Integrated bus mastering DMA controller (optional)

Figure 2.7 gives the structure of 16550A, which is the most common UART and has all the capabilities except DMA.

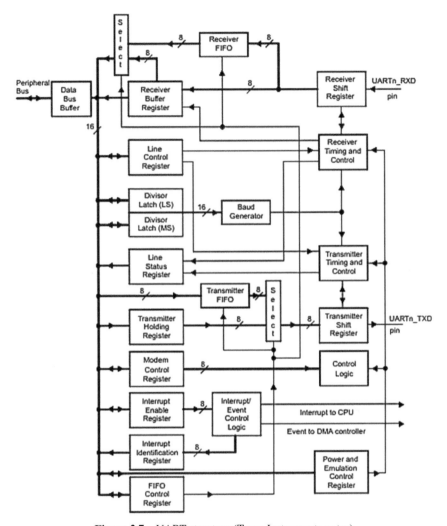

Figure 2.7 UART structure (Texas Instruments notes).

Next, the details of the three sections of the UART—clock generator, transmitter, and receiver – are discussed.

2.5.1 Baudrate Clock Generator

To generate the timing information mentioned in Section 2.3, each **UART** uses a programmable baud generator (clock generator). Figure 2.8 is a

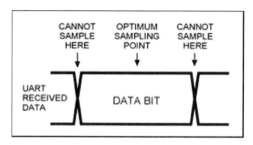

Figure 2.8   UART clock generation diagram.

conceptual clock generation diagram for the UART. The processor clock generator receives a signal from an external clock source and produces a UART input clock with a programmed frequency. The programmable baud generator takes the input clock and divides it by a divisor in the range between 1 and $(2^{16} - 1)$ to produce a baud clock (BCLK). The frequency of BCLK is 16 times ($16\times$) the baud rate (each received or transmitted bit lasts 16 BCLK cycles). Sometimes, BLCK is selected to be 13 times ($13\times$) the baud rate (each received or transmitted bit lasts 13 BCLK cycles). When the UART is receiving, the bit is sampled, as mentioned before, in the 8th BCLK cycle for $16\times$ over sampling mode and on the 6th BCLK cycle for $13\times$ oversampling mode. The $16\times$ or $13\times$ reference clock is selected by configuring the OSM_SEL bit in the mode definition register (MDR). The formula to calculate the divisor is:

Divisor = (UART Input Clock Frequency)/(Desired Baud Rate x16)
[MDR.OSM_SEL = 0]
Divisor = (UART Input Clock Frequency)/(Desired Baud Rate x13)
[MDR.OSM_SEL = 1]

Two 8-bit register fields (DLH and DLL), called divisor latches, hold this 16-bit divisor. DLH holds the most significant bits of the divisor, and DLL holds the least significant bits of the divisor. These register fields will be discussed in detail while discussing UART registers (Section 2.4.2). These divisor latches must be loaded during the initialization of the UART to ensure desired operation of the baud generator. Writing to the divisor latches results in two wait states being inserted during the write access while the baud generator is loaded with the new value.

Figure 2.9, which is an extended form of Figure 2.3, shows the **relationships between data bit, BCLK and UART input clock.**

n UART input clock cycles, where *n* = divisor in DLH:DLL

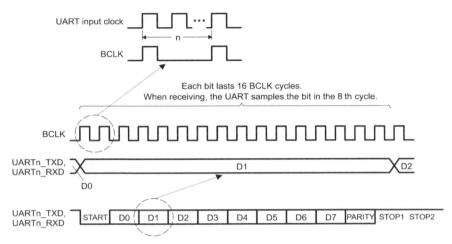

Figure 2.9 Relationships between data bit, BCLK and UART input clock.

Table 2.1 Baud rate examples for 150-MHz UART input clock and 16× oversampling mode

Baud Rate	Divisor Value	Actual Baud Rate	Error (%)
2400	3906	2400.154	0.01
4800	1953	4800.372	0.01
9600	977	9595.701	−0.04
19200	488	19211.066	0.06
38400	244	38422.131	0.06
56000	167	56137.725	0.25
128000	73	129807.7	0.33
3000000	3	3125000	4.00

Example baud rates and divisor values relative to a 150-MHz UART input clock and 16× oversampling mode are shown in Table 2.1.

2.5.2 UART Registers

Modern UART has 12 registers that are used to control the communication between the processor and the **UART.** Behavior of the communication can be changed by reading or writing registers. Each register is 8-bits wide. On PC compatible devices, the registers are accessible in the I/O address area. The function of each register is discussed here in brief. The 16550A is the most common UART at this moment. Newer versions are under development, including the 16650 which contains two 32-byte FIFO buffers and on-board support for software flow control. Texas Instruments is developing the 16750 which contains 64-byte FIFO buffers. It has the following registers:

1. RBR: "Receiver Buffer Register," or "Receiver Data Register (RDR)
2. THR: "Transmitter Holding Register" or "Transmit Data Register (TDR)
3. IER: "Interrupt Enable Register"
4. IIR: "Interrupt Identification Register"
5. FCR: "FIFO Control Register"
6. LCR: "Line Control Register"
7. MCR: "Modem Control Register"
8. LSR: "Line Status Register"
9. MSR: "Modem Status Register"
10. SCR: "Scratch Pad Register"
11. DLL and DLH: "Divisor Latches"
12. REVID1 and REVID2: "Revision Identification Registers"
13. PWREMU_MGMT: "Power and Emulation Management Register"
14. MDR: "Mode Definition Register"

Eight I/O bytes are used for each UART to access its registers. The following table shows where each register can be found. The base address used in the table is the lowest I/O port number assigned. The switch bit DLAB can be found in the line control register as bit 7 at I/O address base + 3.

Tables 2.2 and 2.3 give the location of each register and UART register to port conversion.

Before introducing the different registers, it is important to note that the register's names mentioned above are used by Texas Instrument and they are the most common names. Some microcontrollers and other UART devices may use other names. For example, the "Receiver Buffer Register" is called "Receiver Data Register" in some cases. The same comment is valid when we consider the names of the bits of the registers. The important matter here is the function of each register and each bit. The reader must consult the scientific manual of the UART device he is using to find the correspondences between the names.

2.5.2.1 RBR: receiver buffer register (also, receiver data register rdr) (read only)

The UART receiver section consists of a Receiver Shift Register (RSR) and a Receiver Buffer Register (RBR). The *Receiver Buffer Register (RBR)* contains the byte received if no FIFO is used, or the oldest unread byte with

Table 2.2 UART registers

Offset	Acronym	Register Description	Comment
0h	RBR	Receiver Buffer Register (also called Receiver Data Register RDR)	8 bit, Read Only
0h	THR	Transmitter Holding Register (also called Transmit Data Register TDR)	8 bit, Write Only
4h	IER	Interrupt Enable Register	8 bit, Read/Write
8h	IIR	Interrupt Identification Register (read only)	8 bit, Read Only
8h	FCR	FIFO Control Register (write only)	8 bit, Write Only
Ch	LCR	Line Control Register	8 bit, Read/Write
10h	MCR	Modem Control Register	8 bit, Read/Write
14h	LSR	Line Status Register	8 bit, Read Only
18h	MSR	Modem Status Register	8 bit, Read Only
1Ch	SCR	Scratch Pad Register	8 bit, Read/Write
20h	DLL	Divisor LSB Latch	8 bit, Read/Write
24h	DLM	Divisor MSB Latch	8 bit, Read/Write
28h	REVID1	Revision Identification Register 1	
2Ch	REVID2	Revision Identification Register 2	
30h	PWREMU_MGMT	Power and Emulation Management Register	
34h	MDR	Mode Definition Register	

FIFO buffers. If FIFO buffering is used, each new read action of the register will return the next byte, until no more bytes are present. Bit 0 in the *Line Status Register (LSR)* can be used to check if all received bytes have been read. This bit will change to zero if no more bytes are present.

The RSR receives serial data from the UARTn_RXD pin. Then, the RSR concatenates the data and moves them into the RBR (or the receiver FIFO). In the non-FIFO mode, when a character is placed in RBR and the receiver data-ready interrupt is enabled (DR = 1 in the IER), an interrupt is generated. This interrupt is cleared when the character is read from the RBR. In the FIFO mode, the interrupt is generated when the FIFO is filled to the trigger level selected in the FIFO control register (FCR), and it is cleared when the FIFO contents drop below the trigger level.

Table 2.3 UART register to port conversion

	DLAB = 0		DLAB = 1	
I/O Port	Read	Write	Read	Write
Base	**RBR** receiver buffer	**THR** transmitter holding	**DLL** divisor latch LSB	
base + 1	**IER** interrupt enable	**IER** interrupt enable	**DLM** divisor latch MSB	
base + 2	**IIR** interrupt identification	**FCR** FIFO control	**IIR** interrupt identification	**FCR** FIFO control
base + 3		**LCR** line control		
base + 4		**MCR** modem control		
base + 5	**LSR** line status	– factory test	**LSR** line status	– factory test
base + 6	**MSR** modem status	– notused	**MSR** modem status	– not used
base + 7		**SCR** scratch		

To access RBR, the following is considered:

- As shown in Table 2.3, RBR, THR, and DLL share one address. To read the RBR, write 0 to the DLAB bit in the LCR and read from the shared address. When DLAB = 0, writing to the shared address modifies thee THR. When DLAB = 1, all accesses at the shared address read or modify the DLL.
- The DLL also has a dedicated address. If the dedicated address is used, DLAB can = 0, so that RBR and THR are always selected at the shared address.

2.5.2.2 THR: transmitter holding register (also, transmitter data register, tdr) (write only)

The UART transmitter section consists of a transmitter hold register (THR) and a transmitter shift register (TSR). Transmitter control is a function of the line control register (LCR).

Transmitter Holding Register (THR) is used to buffer outgoing characters. The THR receives data from the internal data bus. When the TSR is idle, the UART then moves the data from the THR to the TSR. The UART serializes the data in the TSR and transmits the data on the TX pin. Without FIFO buffering, only one character can be stored. Otherwise, the amount of characters depends on the type of **UART**. To check if new information must be written to **THR,** bit 5 in the *Line Status Register (LSR)* can be used. Empty register is indicated by value 1. If FIFO buffering is used, the THR is a 16-byte FIFO and more than one character can be written to the transmitter holding register when the FIFO is empty. In this mode, the interrupt is generated when the transmitter FIFO is empty, and it is cleared when at least one byte is loaded into the FIFO.

THR access consideration:

- The RBR, THR, and DLL share one address. To load the THR, write 0 to the DLAB bit of the LCR and write to the shared address. When DLAB = 0, reading from the shared address gives the content of the RBR. When DLAB = 1, all accesses at the address read or modify the DLL.
- The DLL also has a dedicated address. If the dedicated address is used, DLAB can = 0, so that the RBR and the THR are always selected at the shared address.

2.5.2.3 IER: interrupt enable register (read/write)

In interrupt-driven configuration, the **UART** will signal each change by generating a processor interrupt. A software routine must be read interrupt signal to handle the interrupt and to check what state change was responsible for it. *Interrupt enable register (IER)* is used to enable the interrupt.

Interrupts are not generated, unless the UART is told to do so. This is done by setting bits in the IER, interrupt enable register. A bit value 1 indicates that an interrupt may take place.

IER access considerations:

- The IER and DLH share one address. To read or modify the IER, write 0 to the DLAB bit in the LCR. When DLAB = 1, all accesses at the shared address read or modify the DLH.
- The DLH also has a dedicated address. If the dedicated address is used, DLAB can = 0, so that the IER is always selected at the shared address.

The IER is shown in Figure 2.10 and described in Table 2.4.

2.5.2.4 IIR: interrupt identification register (read only)

The interrupt identification register (IIR) is a read-only register at the same address as the FIFO control register (FCR), which is a write-only register.

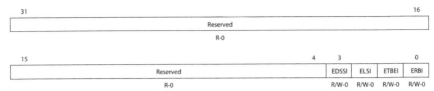

Legend: R = Read only; R/W = Read/Write; $-n$ = value after reset.

Figure 2.10   Interrupt enable register (IER).

Table 2.4   Interrupt enable register (IER) field descriptions

(a) Four description of the IER bits

IER : Interrupt enable register

Bit	Description
0	Received data available
1	Transmitter holding register empty interrupt enable
2	Receiver line status register change/Receiver line status interrupt enable
3	Modem status register change/Enable modem status interrupt
4	Sleep mode (16750 only)
5	Low power mode (16750 only)
6	Reserved
7	Reserved

(b) Interrupt enable register (IER) field descriptions

Bit	Field	Value	Description
6	Reserved	0	Reserved
5			Low power mode (16750 only)
4			Sleep mode (16750 only)
3	EDSSI	0	Enable modem status interrupt.
2	ELSI		Receiver line status interrupt enable. 0 = Receiver line status interrupt is disabled. 1 = Receiver line status interrupt is enabled.
1	ETBEI		Transmitter holding register empty interrupt enable. 0 = Transmitter holding register empty interrupt is disabled. 1 = Transmitter holding register empty interrupt is enabled.
0	ERBI		Receiver data available interrupt and character timeout indication interrupt enable. 0 = Receiver data available interrupt and character timeout indication interrupt is disabled. 1 = Receiver data available interrupt and character timeout indication interrupt is enabled.

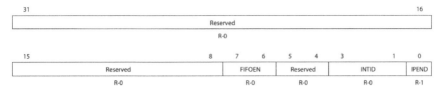

Legend: R = Read only; $-n$ = value after reset.

Figure 2.11 Interrupt identification register (IIR).

Table 2.5 Cause of the change: cause of the interrupt signal

IIR : Interrupt Identification Register					
Bit		Value		Description	Reset by
0		0		Interrupt pending	–
		1		No interrupt pending	–
1,2,3	Bit 3	Bit 2	Bit 1		
	0	0	0	Modem status change	MSR read
	0	0	1	Transmitter holding register empty	IIR read or THR write
	0	1	0	Received data available	RBR read
	0	1	1	Line status change	LSR read
	1	1	0	Character timeout (16550)	RBR read
4		0		Reserved	–
5		0		Reserved (8250, 16450, 16550)	–
		1		64-byte FIFO enabled (16750)	–
6,7	Bit 7	Bit 6			
	0	0		No FIFO	–
	1	0		Unusable FIFO (16550 only)	–
	1	1		FIFO enabled	–

A UART can generate a processor interrupt when a state change on the communication device occurs. One interrupt signal is used to call attention. This means that additional information is needed for the software before necessary actions can be performed. The IIR, interrupt identification register, is helpful in this situation. Its bits show the current state of the UART and which state change caused the interrupt to occur (see Table 2.5). IIR accordingly can be defined as follows:

The *Interrupt Identification Register (IIR)* bits show the current state of the **UART** and which state change caused the interrupt to occur. Based on bit values of the IIR, interrupt can be serviced.

Table 2.6   Interrupt identification register (IIR) field descriptions

Bit	Field	Value	Description
31–8	Reserved	0	Reserved
7–6	FIFOEN	0–3h	FIFOs enabled: 0= Non-FIFO mode. 1 = Reserved. 2 = Reserved. 3 = FIFO buffers are enabled. FIFOEN bit in the FIFO control register (FCR) is set to 1.
5–4	Reserved	0	Reserved
3–1	INTID	0–7h	Interrupt type: 0 = Reserved. 1 = Transmitter holding register empty (priority 3). 2 = Receiver data available (priority 2). 3 = Receiver line status (priority 1, highest). 4 = Reserved. 5 = Reserved. 6 = Character timeout indication (priority 2). 7 = Reserved.
0	IPEND		Interrupt pending. When any UART interrupt is generated and is enabled in IER, IPEND is forced to 0. IPEND remains 0 until all pending interrupts are cleared or until a hardware reset occurs. If no interrupts are enabled, IPEND is never forced to 0. 0 = Interrupts pending. 1 = No interrupts pending

When an interrupt is generated and enabled in the interrupt enable register (IER), the IIR indicates that an interrupt is pending in the IPEND bit and encodes the type of interrupt in the INTID bits. (See Table 2.5) – Interrupt Identification Register (IIR) Field Descriptions.

The IIR is shown in Figure 2.11 and described in Tables 2.5 and 2.6.

2.5.2.5 FCR: FIFO control register (write only)

The *FIFO control register (FCR)* is present starting with the **16550** series. The behavior of the FIFO buffers in the **UART** is controlled by this register.

The FIFO control register (FCR) is a write-only register at the same address as the interrupt identification register (IIR), which is a read-only register. Use the FCR to enable and clear the FIFO buffers and to select the receiver FIFO trigger level. The FIFOEN bit must be set to 1 before other FCR bits are written to or the FCR bits are not programmed.

15				8	7	6	5	4	3	2	1	0
		Reserved			RXFIFTL		Reserved		DMAMODE1[1]	TXCLR	RXCLR	FIFOEN
		R-0			R-0		R-0		W-0	WiC-0	W1C-0	W-0

Legend: R = Read only; W = Write only; W1C = Write 1 to clear (writing 0 has no effect); $-n$ = value after reset.

- Always write 1 to the DMAMODE1 bit. After a hardware reset, change the DMAMODE1 bit from 0 to 1. DMAMODE = 1 is required for proper communication between the UART and the DMA controller.

Figure 2.12   FCR register.

Table 2.7   FCR: FIFO control register

Bit	Value		Description
0	0		Disable FIFO buffers
	1		Enable FIFO buffers
1	0		–
	1		Clear receive FIFO
2	0		–
	1		Clear transmit FIFO
3	0		Select DMA mode 0
	1		Select DMA mode 1
4	0		Reserved
5	0		Reserved (8250, 16450, 16550)
	1		Enable 64 byte FIFO (16750)
6,7	Bit 7	Bit 6	Receive FIFO interrupt trigger level
	0	0	1 byte
	0	1	4 bytes
	1	0	8 bytes
	1	1	14 bytes

If a logical value 1 is written to bits 1 or 2, the function attached is triggered. The other bits are used to select a specific FIFO mode.

Access consideration: The IIR and FCR share one address. Regardless of the value of the DLAB bit, reading from the address gives the content of the IIR, and writing to the address modifies the FCR.

The IIR and FCR share one address. Regardless of the value of the DLAB bit, reading from the address gives the content of the IIR, and writing to the address modifies the FCR.

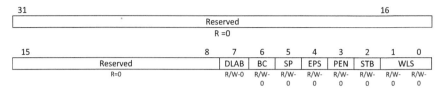

Legend: R = Read only; R/W = Read/Write; $-n$ = value after reset.

Figure 2.13   Line control register (LCR).

2.5.2.6 LCR: line control register (read/write)

The *Line Control Register (LCR)* is used at initialization to set the communication parameters such as parity, number of data bits, etc. In addition, the programmer can retrieve, inspect, and modify the content of the LCR. This eliminates the need for separate storage of the line characteristics in system memory.

The register also controls the accessibility of the **DLL** and **DLM** registers. These registers are mapped to the same I/O port as the RBR, THR, and IER registers. Because they are only accessed at initialization when no communication occurs, this register swapping has no influence on performance.

The LCR is shown in Figure 2.13, described in Table 2.8, and summarized in Table 2.9.

Some remarks about parity: The UART is capable of generating a trailing bit at the end of each data word which can be used to check some data distortion. Because only one bit is used, the parity system is capable of detecting only an odd number of false bits. If an even number of bits has been flipped, the error will not be seen.

When even parity is selected, the UART assures that the number of high bit values in the sent or received data is always even. Odd parity setting does the opposite. Using stick parity has very little use. It sets the parity bit to always 1 or always 0.

Common settings are as follows:

- 8 data bits, one stop bit, no parity
- 7 data bits, one stop bit, even parity

2.5.2.7 MCR: modem control register (read/write)

The **MCR, Modem Control Register,** is used to perform handshaking actions with the attached device. In the original UART series including the 16550, setting and resetting of the control signals must be done by software.

Table 2.8 Line control register (LCR) field descriptions

Bit	Field	Value	Description
31–8	Reserved	0	Reserved.
7	DLAB		Divisor latch access bit. The divisor latch registers (DLL and DLH) can be accessed at dedicated addresses or at addresses shared by RBR, THR, and IER. Using the shared addresses requires toggling DLAB to change which registers are selected. If the dedicated addresses are used, DLAB can = 0. 0 = Allows access to the receiver buffer register (RBR), the transmitter holding register (THR), and the interrupt enable register (IER) selected. At the address shared by RBR, THR, and DLL, the CPU can read from RBR and write to THR. At the address shared by IER and DLH, the CPU can read from and write to IER. 1 = Allows access to the divisor latches of the baud generator during a read or write operation (DLL and DLH). At the address shared by RBR, THR, and DLL, the CPU can read from and write to DLL. At the address shared by IER and DLH, the CPU can read from and write to DLH.
6	BC		Break control. 0 = Break condition is disabled. 1 = Break condition is transmitted to the receiving UART. A break condition is a condition in which the UARTn_TXD signal is forced to the spacing (cleared) state.
5	SP		Stick parity. The SP bit works in conjunction with the EPS and PEN bits. The relationship between the SP, EPS, and PEN bits is summarized in Table 2.9. 0 = Stick parity is disabled. 1 = Stick parity is enabled. When odd parity is selected (EPS = 0), the PARITY bit is transmitted and checked as set. When even parity is selected (EPS = 1), the PARITY bit is transmitted and checked as cleared.
4	EPS		Even parity select. Selects the parity when parity is enabled (PEN = 1). The EPS bit works in conjunction with the SP and PEN bits. The relationship between the SP, EPS, and PEN bits is summarized in Table 2.9. 0 = Odd parity is selected (an odd number of logic 1s is transmitted or checked in the data and PARITY bits). 1 = Even parity is selected (an even number of logic 1s is transmitted or checked in the data and PARITY bits).

(Continued)

Table 2.8   Continued

Bit	Field	Value	Description
3	PEN		Parity enable. The PEN bit works in conjunction with the SP and EPS bits. The relationship between the SP, EPS, and PEN bits is summarized in Table 2.9. 0 = No PARITY bit is transmitted or checked. 1 = Parity bit is generated in transmitted data and is checked in received data between the last data word bit and the first STOP bit.
2	STB		Number of STOP bits generated. STB specifies 1, 1.5, or 2 STOP bits in each transmitted character. When STB = 1, the WLS bit determines the number of STOP bits. The receiver clocks only the first STOP bit, regardless of the number of STOP bits selected. The number of STOP bits generated is summarized in Table 2.10. 0 = One STOP bit is generated. 1 = WLS bit determines the number of STOP bits: • When WLS = 0, 1.5 STOP bits are generated. • When WLS = 1 h, 2 h, or 3 h, 2 STOP bits are generated.
1–0	WLS	0–3h	Word length select. Number of bits in each transmitted or received serial character. When STB = 1, the WLS bit determines the number of STOP bits. 00 = 5 bits 01 = 6 bits 10 = 7 bits 11 = 8 bits

The new 16750 is capable of handling flow control automatically, thereby reducing the load on the processor.

In general, the modem control register (MCR) provides the ability to enable/disable the autoflow functions and enable/disable the loopback function for diagnostic purposes.

The MCR is shown in Figure 2.14 and described in Table 2.10.

The two auxiliary outputs are user definable. Output 2 is sometimes used in circuitry which controls the interrupt process on a PC. Output 1 is normally not used; however, on some I/O cards, it controls the selection of a second oscillator working at 4 MHz. This is mainly for MIDI purposes. Bit 0 is sometimes used to indicate that the data terminal is ready.

2.5.2.8 LSR: line status register (read only)

The *Line Status Register (LSR)* shows the current state of communication. Errors, the state of the receiver and transmit buffers, are available. Bits 1

Table 2.9   Relationship between different bits in LCR

			LCR: line control register
Bit	Value		Description
0,1	Bit 1	Bit 0	Data word length
	0	0	5 bits
	0	1	6 bits
	1	0	7 bits
	1	1	8 bits
2	0		1 stop bit
	1		1.5 stop bits (5 bits word)
			2 stop bits (6, 7, or 8 bits word)
3,4,5	Bit 5 Bit 4	Bit 3	
	X X	0	No parity
	0 0	1	Odd parity
	0 1	1	Even parity
	1 0	1	High parity (stick)
	1 1	1	Low parity (stick)
6	0		Break signal disabled
	1		Break signal enabled
7	0		DLAB: RBR, THR, and IER accessible
	1		DLAB: DLL and DLM accessible

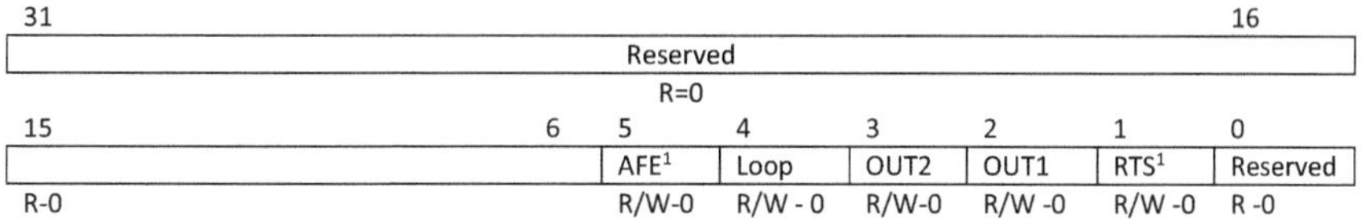

31								16
Reserved								
R=0								

15	6	5	4	3	2	1	0
		AFE[1]	Loop	OUT2	OUT1	RTS[1]	Reserved
R-0		R/W-0	R/W - 0	R/W-0	R/W -0	R/W -0	R -0

Legend: R = Read only; R/W = Read/Write; $-n$ = value after reset.

1. All UARTs do not support this feature. See the device-specific data manual for
   supported features. If this feature is not available, this bit is reserved and should be
   cleared to 0.

Figure 2.14   Modem control register (MCR).

through 4 record the error conditions that produce a receiver line status
interrupt.

The LSR is shown in Figure 2.15, described in Table 2.11, and summarized in Table 2.12.

Table 2.10 Modem control register (MCR) field descriptions

Bit	Field	Value	Description
31–6	Reserved	0	Reserved
5	AFE		Autoflow control enabled. Autoflow control allows the $\overline{UARTn\_RTS}$ and $\overline{UARTn\_CTS}$ signals to provide handshaking between UARTs during data transfer. When AFE = 1, the RTS bit determines the autoflow control enabled. Note that all UARTs do not support this feature. See the device-specific data manual for supported features. If this feature is not available, this bit is reserved and should be cleared to 0. 0 = Autoflow control is disabled. 1 = Autoflow control is enabled: • When RTS = 0, only UARTn_CTS is enabled. • When RTS = 1, UARTn_RTS and UARTn_CTS are enabled.
4	LOOP		Loopback mode enabled. LOOP is used for the diagnostic testing using the loopback feature. 0 = Loopback mode is disabled. 1 = Loopback mode is enabled. When LOOP is set, the following occurs: • The UARTn_TXD signal is set high. • The UARTn_RXD pin is disconnected • The output of the transmitter shift register (TSR) is looped back in to the receiver shift register (RSR) input.
3	OUT2	0	OUT2 Control Bit.
2	OUT1	0	OUT1 Control Bit.
1	RTS		RTS control. When AFE = 1, the RTS bit determines the autoflow control enabled. Note that all UARTs do not support this feature. See the device-specific data manual for supported features. If this feature is not available, this bit is reserved and should be cleared to 0. 0 = UARTn_RTS is disabled, and only UARTn_CTS is enabled. 1 = UARTn_RTS and UARTn_CTS are enabled.
0	Reserved	0	Reserved, sometimes used as "Data terminal ready"

Bit 5 and bit 6 both show the state of the transmitting cycle. The difference is that bit 5 turns high as soon as the transmitter holding register is empty, whereas bit 6 indicates that also the shift register which outputs the bits on the line is empty.

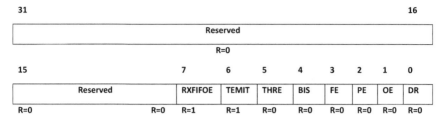

Legend: R = Read only; $-n$ = value after reset.

Figure 2.15 Line status register (LSR).

Table 2.11 MCR summary

MCR: Modem control register	
Bit	Description
0	Data terminal ready
1	Request to send
2	Auxiliary output 1
3	Auxiliary output 2
4	Loopback mode
5	Autoflow control (16750 only)
6	Reserved
7	Reserved

2.5.2.9 MSR: modem status register (read only)

The *Modem Status Register (MSR) provides information to the CPU* about the status of the four incoming modem control lines on the device. The information is split into two nibbles. The four most significant bits contain information about the current state of the inputs. The least four significant bits are used to indicate state changes. Each time the register is read, the four LSBs are reset.

The MSR is shown in Figure 2.16 and described in Tables 2.14 and 2.15.

2.5.2.10 SPR (or SCR) scratch pad register (read/write)

The scratch pad register (SCR) is intended for the programmer's use as a scratch pad. It temporarily holds the programmer's data without affecting UART operation.

The SCR was not present on the 8250 and 8250B UART. It can be used to store one byte of information. In practice, it has only limited use. The only

Table 2.12   Line status register (LSR) field descriptions

Bit	Field	Value	Description
31–8	Reserved	0	Reserved
7	RXFIFOE		Receiver FIFO error. **In non-FIFO mode:** 0 = There has been no error, or RXFIFOE was cleared because the CPU read the erroneous character from the receiver buffer register (RBR). 1 = There is a parity error, framing error, or break indicator in the receiver buffer register (RBR). **In FIFO mode:** 0 = There has been no error, or RXFIFOE was cleared because the CPU read the erroneous character from the receiver FIFO and there are no more errors in the receiver FIFO. 1 = At least one parity error, framing error, or break indicator in the receiver FIFO.
6	TEMT		Transmitter empty (TEMT) indicator. **In non-FIFO mode:** 0 = Either the transmitter holding register (THR) or the transmitter shift register (TSR) contains a data character. 1 = Both the transmitter holding register (THR) and the transmitter shift register (TSR) are empty. **In FIFO mode:** 0 = Either the transmitter FIFO or the transmitter shift register (TSR) contains a data character. 1 = Both the transmitter FIFO and the transmitter shift register (TSR) are empty.
5	THRE		Transmitter holding register empty (THRE) indicator. If the THRE bit is set and the corresponding interrupt enable bit is set (ETBEI = 1 in IER), an interrupt request is generated. **In non-FIFO mode:** 0 = Transmitter holding register (THR) is not empty. THR has been loaded by the CPU. 1 = Transmitter holding register (THR) is empty (ready to accept a new character). The content of THR has been transferred to the transmitter shift register (TSR). **In FIFO mode:** 0 = Transmitter FIFO is not empty. At least one character has been written to the transmitter FIFO. The transmitter FIFO may be written to if it is not full. 1 = Transmitter FIFO is empty. The last character in the FIFO has been transferred to the transmitter shift register (TSR).

(Continued)

Table 2.12 Continued

Bit	Field	Value	Description
4	BI		Break indicator. The BI bit is set whenever the receive data input (UARTn_RXD) was held low for longer than a full-word transmission time. A full-word transmission time is defined as the total time to transmit the START, data, PARITY, and STOP bits. If the BI bit is set and the corresponding interrupt enable bit is set (ELSI = 1 in IER), an interrupt request is generated. **In non-FIFO mode:** 0 = No break has been detected, or the BI bit was cleared because the CPU read the erroneous character from the receiver buffer register (RBR). 1 = A break has been detected with the character in the receiver buffer register (RBR). **In FIFO mode:** 0 = No break has been detected, or the BI bit was cleared because the CPU read the erroneous character from the receiver FIFO and the next character to be read from the FIFO has no break indicator. 1 = A break has been detected with the character at the top of the receiver FIFO.
3	FE		Framing error (FE) indicator. A framing error occurs when the received character does not have a valid STOP bit. In response to a framing error, the UART sets the FE bit and waits until the signal on the RX pin goes high. When the RX signal goes high, the receiver is ready to detect a new START bit and receive new data. If the FE bit is set and the corresponding interrupt enable bit is set (ELSI = 1 in IER), an interrupt request is generated. **In non-FIFO mode:** 0 = No framing error has been detected, or the FE bit was cleared because the CPU read the erroneous data from the receiver buffer register (RBR). 1 = A framing error has been detected with the character in the receiver buffer register (RBR). **In FIFO mode:** 0 = No framing error has been detected, or the FE bit was cleared because the CPU read the erroneous data from the receiver FIFO and the next character to be read from the FIFO has no framing error. 1 = A framing error has been detected with the character at the top of the receiver FIFO.

(Continued)

Table 2.12   Continued

Bit	Field	Value	Description
2	PE		Parity error (PE) indicator. A parity error occurs when the parity of the received character does not match the parity selected with the EPS bit in the line control register (LCR). If the PE bit is set and the corresponding interrupt enable bit is set (ELSI = 1 in IER), an interrupt request is generated. **In non-FIFO mode:** 0 = No parity error has been detected, or the PE bit was cleared because the CPU read the erroneous data from the receiver buffer register (RBR). 1 = A parity error has been detected with the character in the receiver buffer register (RBR). **In FIFO mode:** 0 = No parity error has been detected, or the PE bit was cleared because the CPU read the erroneous data from the receiver FIFO and the next character to be read from the FIFO has no parity error. 1 = A parity error has been detected with the character at the top of the receiver FIFO.
1	OE		Overrun error (OE) indicator. An overrun error in the non-FIFO mode is different from an overrun error in the FIFO mode. If the OE bit is set and the corresponding interrupt enable bit is set (ELSI = 1 in IER), an interrupt request is generated. **In non-FIFO mode:** 0 = No overrun error has been detected, or the OE bit was cleared because the CPU read the content of the line status register (LSR). 1= Overrun error has been detected. Before the character in the receiver buffer register (RBR) could be read, it was overwritten by the next character arriving in RBR. **In FIFO mode:** 0 = No overrun error has been detected, or the OE bit was cleared because the CPU read the content of the line status register (LSR). 1 = Overrun error has been detected. If data continues to fill the FIFO beyond the trigger level, an overrun error occurs only after the FIFO is full and the next character has been completely received in the shift register. An overrun error is indicated to the CPU as soon as it happens. The new character overwrites the character in the shift register, but it is not transferred to the FIFO.

(Continued)

Table 2.12 Continued

Bit	Field	Value	Description
0	DR		Data-ready (DR) indicator for the receiver. If the DR bit is set and the corresponding interrupt enable bit is set (ERBI = 1 in IER), an interrupt request is generated.
			In non-FIFO mode:
			0 = Data is not ready, or the DR bit was cleared because the character was read from the receiver buffer register (RBR).
			1 = Data is ready. A complete incoming character has been received and transferred into the receiver buffer register (RBR).
			In FIFO mode:
			0 = Data is not ready, or the DR bit was cleared because all of the characters in the receiver FIFO have been read.
			1 = Data is ready. There is at least one unread character in the receiver FIFO. If the FIFO is empty, the DR bit is set as soon as a complete incoming character has been received and transferred into the FIFO. The DR bit remains set until the FIFO is empty again.

Table 2.13 LSR register summary

LSR: Line status register	
Bit	Description
0	Data ready
1	Overrun error
2	Parity error
3	Framing error
4	Break signal received (Break Indicator)
5	THR is empty
6	THR is empty, and line is idle
7	Erroneous data in FIFO

real use is checking if the UART is an 8250/8250B or an 8250A/16450 series. Because the 8250 series are only found in XT, even this use of the register is not commonly seen anymore.

2.5.2.11 DLL and DLM: divisor latch registers (read/write)

The communication speed of the UART is changed using a programmable value stored in **Divisor Latch Registers *DLL*** and ***DLM*** which contain the least and most significant registers.

31	Reserved	16
	R-0	

15		8	7	6	5	4	3	2	1	0 6

Reserved	CD	RI	DSR	CTS	DCD	TERI	DDSR	DCTS
R-0	R-0	R-0	R-0	R-0	R-0	R-0	R-0	R-0

Legend: R = Read only; $-n$ = value after reset.

Figure 2.16   Modem status register (MSR).

Table 2.14   Modem status register (MSR) field descriptions

Bit	Field	Value	Description
31–8	Reserved	0	Reserved
7	CD	0	Complement of the Carrier Detect input. When the UART is in the diagnostic test mode (loopback mode MCR[4] = 1), this bit is equal to the MCR bit 3 (OUT2).
6	RI	0	Complement of the Ring Indicator input. When the UART is in the diagnostic test mode (loopback mode MCR[4] = 1), this bit is equal to the MCR bit 2 (OUT1).
5	DSR	0	Complement of the Data Set Ready input. When the UART is in the diagnostic test mode (loopback mode MCR[4] = 1), this bit is equal to the MCR bit 0 (DTR).
4	CTS	0	Complement of the Clear To Send input. When the UART is in the diagnostic test mode (loopback mode MCR[4] = 1), this bit is equal to the MCR bit 1 (RTS).
3	DCD	0	Change in DCD indicator bit. DCD indicates that the DCD input has changed state since the last time it was read by the CPU. When DCD is set and the modem status interrupt is enabled, a modem status interrupt is generated.
2	TERI	0	Trailing edge of RI (TERI) indicator bit. TERI indicates that the RI input has changed from a low to a high. When TERI is set and the modem status interrupt is enabled, a modem status interrupt is generated.
1	DDSR	0	Change in DSR indicator bit. DDSR indicates that the DSR input has changed state since the last time it was read by the CPU. When DDSR is set and the modem status interrupt is enabled, a modem status interrupt is generated.
0	DCTS	0	Change in CTS indicator bit. DCTS indicates that the CTS input has changed state since the last time it was read by the CPU. When DCTS is set (autoflow control is not enabled and the modem status interrupt is enabled), a modem status interrupt is generated. When autoflow control is enabled, no interrupt is generated.

Table 2.15 MSR field description – summary

Bit	Description
MSR: Modem status register	
0	Change in clear to send
1	Change in data set ready
2	Trailing edge ring indicator
3	Change in carrier detect
4	Clear to send
5	Data set ready
6	Ring indicator
7	Carrier detect

For generating DLR timing information, each UART uses an oscillator generating a frequency of about 1.8432 MHz. This frequency is divided by 16 to generate the time base for communication. Because of this division, the maximum allowed communication speed is 115200 bps. Modern UARTS like the 16550 are capable of handling higher input frequencies up to 24 MHz which makes it possible to communicate with a maximum speed of 1.5 Mbps. On PCs, frequencies higher than 1.8432 MHz are rarely seen because this would be software incompatible with the original XT configuration.

This 115200 bps communication speed is not suitable for all applications. To change the communication speed, the frequency can be further decreased by dividing it using a programmable value. For very slow communications, this value can go beyond 255. Therefore, the divisor is stored in two separate bytes, the divisor latch registers DLL and DLM which contain the least and the most significant bytes.

For error-free communication, it is necessary that both the transmitting and receiving UART use the same time base. Default values have been defined which are commonly used. The table shows the most common values with the appropriate settings of the divisor latch bytes. Note that these values only hold for a PC compatible system where a clock frequency of 1.8432 MHz is used.

These divisor latches must be loaded during initialization of the UART to ensure desired operation of the baud generator. Writing to the divisor latches results in two wait states being inserted during the write access while the baud generator is loaded with the new value.

Table 2.16   The divisor LSB latch

DLL and DLM: Divisor Latch Registers			
Speed (bps)	Divisor	DLL	DLM
50	2,304	0x00	0x09
300	384	0x80	0x01
1,200	96	0x60	0x00
2,400	48	0x30	0x00
4,800	24	0x18	0x00
9,600	12	0x0C	0x00
19,200	6	0x06	0x00
38,400	3	0x03	0x00
57,600	2	0x02	0x00
115,200	1	0x01	0x00

Access considerations:

- The RBR, THR, and DLL share one address. When DLAB = 1 in the LCR, all accesses at the shared address are accesses to the DLL. When DLAB = 0, reading from the shared address gives the content of the RBR, and writing to the shared address modifies the THR.
- The IER and DLH share one address. When DLAB = 1 in the LCR, accesses to the shared address read or modify the DLH. When DLAB = 0, all accesses at the shared address read or modify the IER.

The DLL and DLH also have dedicated addresses. If dedicated addresses are used, the DLAB bit can be kept cleared, so that the RBR, THR, and IER are always selected at the shared addresses.

The divisor LSB latch (DLL) is shown in Table 2.16.

2.5.2.12 Revision identification registers (REVID1 and REVID 2)

Revision identification registers (REVID1 and REVID2) contain peripheral identification data for the peripheral.

2.5.2.13 Power and emulation management register (PWREMU_MGMT)

The power and emulation management register (PWREMU_MGMT) is shown in Figure 2.17 and described in Table 2.17.

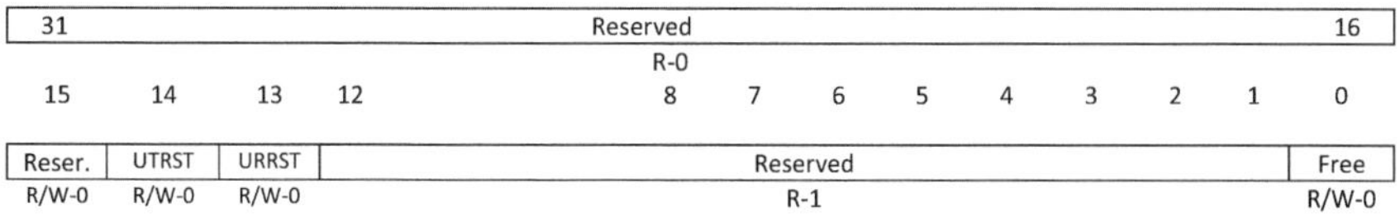

31	Reserved	16
	R-0	

15	14	13	12	8	7	6	5	4	3	2	1	0

Reser.	UTRST	URRST	Reserved	Free
R/W-0	R/W-0	R/W-0	R-1	R/W-0

Legend: R = Read only; R/W = Read/Write; $-n$ = value after reset.

Figure 2.17 Power and emulation management register (PWREMU_MGMT).

Table 2.17 Power and emulation management register (PWREMU_MGMT) field descriptions

Bit	Field	Value	Description
31–16	Reserved	0	Reserved
15	Reserved	0	Reserved. This bit must always be written as 0.
14	UTRST		UART transmitter reset. Resets and enables the transmitter. 0 = Transmitter is disabled and in reset state. 1 = Transmitter is enabled.
13	URRST		UART receiver reset. Resets and enables the receiver. 0 = Receiver is disabled and in reset state. 1 = Receiver is enabled.
12–1	Reserved	1	Reserved
0	FREE		Free-running enable mode bit. This bit determines the emulation mode functionality of the UART. When halted, the UART can handle register read/write requests, but does not generate any transmission/reception, interrupts or events. 0 = If a transmission is not in progress, the UART halts immediately. If a transmission is in progress, the UART halts after completion of the one-word transmission. 1 = Free-running mode is enabled. UART continues to run normally.

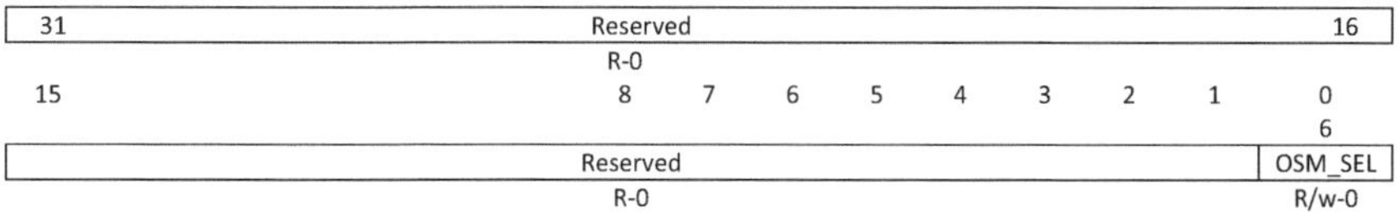

31	Reserved	16
	R-0	

15	8	7	6	5	4	3	2	1	0 6

Reserved	OSM_SEL
R-0	R/w-0

Legend: R = Read only; R/W = Read/Write; $-n$ = value after reset.

Figure 2.18 Mode definition register (MDR).

2.5.2.14 Mode definition register (MDR)

The mode definition register (MDR) determines the oversampling mode for the UART. The MDR is shown in Figure 2.18 and described in Table 2.18.

Table 2.18   Mode definition register (MDR) field descriptions

Bit	Field	Value	Description
31-1	Reserved	0	Reserved
0	OSM_SEL		Oversampling mode select. $0 = 16\times$ oversampling. $1 = 13\times$ oversampling.

2.6 UART in Operation

2.6.1 Transmission

The UART transmitter section includes a transmitter hold register (THR) and a transmitter shift register (TSR). When the UART is in the FIFO mode, THR is a 16-byte FIFO into which data are written by the host processor. Transmitter section control is a function of the UART line control register (LCR). Based on the settings chosen in LCR, the UART transmitter sends the following to the receiving device:

1. 1 START bit
2. 5, 6, 7, or 8 data bits
3. 1 PARITY bit (optional)
4. 1, 1.5, or 2 STOP bits

THR receives data from the internal data bus, and when TSR is ready, the UART moves the data from THR to TSR. The UART serializes the data in TSR and transmits the data on the UARTn_TXD pin. In the non-FIFO mode, if THR is empty and the THR empty interrupt is enabled in the interrupt enable register (IER), an interrupt is generated. This interrupt is cleared when a character is loaded into THR. In the FIFO mode, the interrupt is generated when the transmitter FIFO is empty, and it is cleared when at least one byte is loaded into the FIFO.

The block representation of serial data transmission is depicted in Figure 2.19.

Transmit FIFO: The FIFO is 8-bit by 32-word. It receives control signals from the serial transmit block. The data on signal data_bus is written into its buffer. At the same time, the write pointer is incremented. The data are read onto FIFO and the read pointer is reset when the read pointer has reached its maximum. The write pointer is cleared when the write pointer has reached its maximum. The tx_en is set low when the FIFO is full.

Serial Transmit Block: This component is responsible for serial transmission of data onto tx. It generates the requisite control signals for reading and

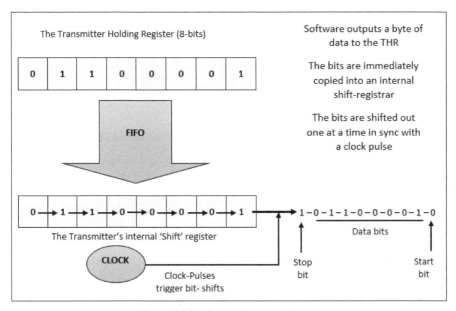

Figure 2.19   Serial data transmission.

writing the transmit FIFO. This signal is used as an enable by the transmit data counter and the transmit block. The transmit data counter keeps count of the number of data bits transmitted onto tx. These signals are provided by the transmit control block. The parity counter counts the number of bits that were high in the 8 bits of data being transmitted. The transmit control block controls the whole process of transmission. It is modeled in the form of a state machine.

2.6.2 UART Receiver

The block representation of serial data reception is depicted in Figure 2.20. The serial receive block can also has a FIFO buffer. The block checks for the parity and the validity of the data frame on the *rx* input and then writes correct data into its buffers. It also sets the signal ***byte_ready*** low if its FIFO is empty.

Serial Receive Block: Serial data are received by this component. It generates the requisite control signals for reading and writing the receive FIFO. It generates required sample clock to sample the incoming data and determine the baud rate of the incoming data.

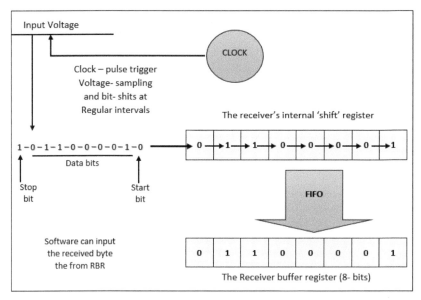

Figure 2.20   Serial data reception.

Receive FIFO: The FIFO is 8-bit wide and 32-byte deep. It receives control signals from the serial receive block. The data are received from the receive block written into its buffer. The write pointer is cleared when the write pointer reaches its maximum limit before further increment.

Reception: The UART receiver section includes a receiver shift register (RSR) and a receiver buffer register (RBR). When the UART is in the FIFO mode, RBR is a 16-byte FIFO (receiver FIFO). Timing is supplied by the receiver clock. Receiver section control is a function of the UART line control register (LCR). Based on the settings chosen in LCR, the UART receiver accepts the following from the transmitting device:

- 1 START bit
- 5, 6, 7, or 8 data bits
- 1 PARITY bit (optional)
- 1 STOP bit (any other STOP bits transferred with the above data are not detected)

RSR receives the data bits from the UARTn_RXD pin. Then, RSR concatenates the data bits and moves the resulting value into RBR (or the receiver FIFO). The UART also stores three bits of error status information next to each received character to record a parity error, framing error, or break.

In the non-FIFO mode, when a character is placed in RBR and the receiver data-ready interrupt is enabled in the interrupt enable register (IER), an interrupt is generated. This interrupt is cleared when the character is read from RBR. In the FIFO mode, the interrupt is generated when the FIFO is filled to the trigger level selected in the FIFO control register (FCR), and it is cleared when the FIFO contents drop below the trigger level.

2.7 UART Errors

Overrun error: An "overrun error" occurs when the UART cannot process the byte that just came in before the next one arrives. The host processor must service the UART in order to remove characters from the buffer. If the host processor (CPU or DMA, for example) does not service the UART quickly enough and the buffer becomes full, then Overrun Error will occur.

Underrun error: An "underrun error" occurs when the UART transmitter has completed sending a character and the transmit buffer is empty. In asynchronous modes, this is treated as an indication that no data remains to be transmitted, rather than an error, since additional stop bits can be appended. This error indication is commonly found in USARTs, since an underrun is more serious in synchronous systems.

Framing error: A "framing error" occurs when the designated "start" and "stop" bits are not valid. Start bit acts as a reference for the remaining bits. If the data line is not in the expected state (high) when the "stop" bit is expected (according to the number of data and parity bits for which the UART is set), the UART will signal a Framing Error. A "break" condition on the line is also signaled as a framing error.

Parity error: A "parity error" occurs when the number of "active" bits does not agree with the specified parity configuration of the UART. Parity errors occur when parity is implemented on the data link and there is a corruption that causes a parity mismatch in the received data. Use of a parity bit is optional, so this error will only occur if parity-checking has been enabled.

Break condition: A break condition occurs when the receiver input is at the "space" (logic low, i.e., '0') level for longer than some duration of time, typically, for more than a character time. This is not necessarily an error, but appears to the receiver as a character of all zero bits with a framing error. The term "break" derives from current loop signaling, which was the traditional signaling used for teletypewriters and which will be discussed in Chapter

3. The "spacing" condition of a current loop line is indicated by no current flowing, and a very long period of no current flowing is often caused by a break or other fault in the line.

Some equipment will deliberately transmit the "space" level for longer than a character as an attention signal. When signaling rates are mismatched, no meaningful characters can be sent, but a long "break" signal can be a useful way to get the attention of a mismatched receiver to do something (such as resetting itself). Computer systems can use the long "break" level as a request to change the signaling rate, to support dial-in access at multiple signaling rates. The DMX512 protocol uses the break condition to signal the start of a new packet.

2.8 Universal Synchronous and Asynchronous Receiver–Transmitter (USART)

A **universal synchronous and asynchronous receiver–transmitter** (**USART**) is a type of a serial interface device that can be programmed to communicate asynchronously or synchronously with serially connected devices. A USART provides serial data communication from the serial port and over RS-232 standardized protocol. A USART is also known as a serial communications interface (SCI).

A USART is similar to a universal asynchronous receiver/transmitter (UART), as each supports and provides serial communication. Both work by receiving parallel data from the central processing unit (CPU) and converting it to serial data for transmission to a serial port/connection. Similarly, it receives serial data from the serial connection/port, converts it to parallel data, and sends it to the CPU. However, UARTs only support asynchronous serial communication.

The USART is embedded on an integrated circuit (IC) or the motherboard and can be configured for synchronous and asynchronous transfer mode (ATM).

Synchronous serial data transmission will be discussed in detail in Chapter 5 while discussing SPI.

2.8.1 Purpose and History of USART

The USART's synchronous capabilities were primarily intended to support synchronous protocols like IBM's synchronous transmit–receive (STR), binary synchronous communications (BSC), synchronous data link control (SDLC), and the ISO-standard high-level data link control (HDLC) synchronous link-layer protocols, which were used with synchronous voice-frequency modem. These protocols were designed to make the best use of bandwidth when modems were analog devices. In those times, the fastest asynchronous voice-band modem could achieve at most speeds of 300 bit/s using frequency-shift keying (FSK) modulation, while synchronous modems could run at speeds up to 9600 bit/s using phase-shift keying (PSK) modulation. Synchronous transmission used only slightly over 80% of the bandwidth of the now more familiar asynchronous transmission, since start and stop bits were unnecessary. Those modems are obsolete, having been replaced by modems which convert asynchronous data to synchronous forms, but similar synchronous telecommunications protocols survive in numerous block-oriented technologies such as the widely used IEEE 802.2 (Ethernet) link-level protocol. USARTs are still sometimes integrated with MCUs. USARTs are still used in routers that connect to external CSU/DSU devices, and they often use either Cisco's proprietary HDLC implementation or the IETF standard point-to-point protocol (PPP) in HDLC-like framing as defined in RFC 1662.

2.8.2 Operation

The operation of a USART is closely related to the protocol in use. This section only provides a few general notes.

- USARTs in synchronous mode transmit data in frames. In synchronous operation, characters must be provided on time until a frame is complete; if the controlling processor does not do so; this is an *"underrun error,"* and transmission of the frame is aborted.
- USARTs operating as synchronous devices used either character-oriented or bit-oriented mode. In character (STR and BSC) modes, the device relied on particular characters to define frame boundaries; in bit (HDLC and SDLC) modes, earlier devices relied on physical-layer signals, while later devices took over the physical-layer recognition of bit patterns.
- A synchronous line is never silent; when the modem is transmitting, data are flowing. When the physical layer indicates that the modem is active, a USART will send a steady stream of padding, either characters or bits as appropriate to the device and protocol.

2.9 External Interface: Introduction to Serial Data Standards RS-232

The external signaling levels that are used between different equipment are not generated by UART. An interface is used to convert the logic level signals of the UART to the external signaling levels. Examples of standards for voltage signaling are **RS-232, RS-422,** and **RS-485** from the **EIA**. For embedded system applications, UARTs are commonly used with RS-232. It is useful to communicate between microcontrollers and also with PCs. **MAX 232** is one of the example ICs which provide RS-232 level signals.

2.10 Advantages and Disadvantages of UARTs

No communication protocol is perfect, but UARTs are pretty good at what they do. Here are some pros and cons to help you decide whether or not they fit the needs of your project:

Advantages

- Only two wires are used.
- No clock signal is necessary.
- It has a parity bit to allow for error checking.
- The structure of the data packet can be changed as long as both sides are set up for it being a well-documented and widely used method.

Disadvantages

- The size of the data frame is limited to a maximum of 9 bits.
- It doesn't support multiple slave or multiple master systems.
- The baud rates of each UART must be within 10% of each other.

2.11 Example of USART: Microprocessor 8251 USART

In the following, we consider the microprocessor 8251 USAER as an example. The universal synchronous asynchronous receiver transmitter (USART) acts as a mediator between microprocessor and peripheral to transmit serial data into parallel form and vice versa.

It takes data serially from peripheral (outside devices) and converts into parallel data. After converting the data into parallel form, it transmits it to the CPU.

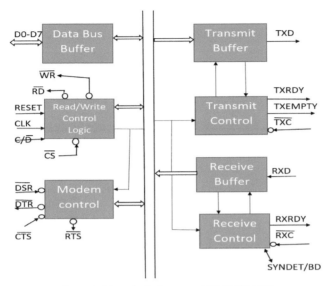

Figure 2.21   Block diagram of 8251 USART.

Similarly, it receives parallel data from microprocessor and converts it into serial form. After converting data into serial form, it transmits it to outside device (peripheral).

The block diagram of 8251 USART is shown in Figure 2.21.

It contains the following blocks:

Data bus buffer: This block helps in interfacing the internal data bus of 8251 to the system data bus. The data transmission is possible between 8251 and CPU by the data bus buffer block.

Read/Write control logic: It is a control block for overall device. It controls the overall working by selecting the operation to be done. The operation selection depends upon the input signals as given in the table below:

$\overline{CS}$	C/D	$\overline{RD}$	$\overline{WR}$	Operation
1	X	X	X	Invalid
0	0	0	1	DataCPU 8251
0	0	1	0	DataCPU 8251
0	1	0	1	Status wordCPU 8251
0	1	1	0	Control wordCPU 8251

In this way, this unit selects one of the three registers—data buffer register, control register, status register.

Modem control (modulator/demodulator): A device converts analog signals into digital signals and vice versa and helps the computers to communicate over telephone lines or cable wires. The following are active-low pins of Modem.

- DSR: Data set ready signal is an input signal.
- DTR: Data terminal ready is an output signal.
- CTS: It is an input signal that controls the data transmit circuit.
- RTS: It is an output signal that is used to set the status RTS.

Transmit buffer: This block is used for parallel to serial converter that receives a parallel byte for conversion into serial signal and further transmission onto the common channel.

- TXD: It is an output signal; if its value is one, the transmitter will transmit the data.

Transmit control: This block is used to control the data transmission with the help of following pins:

- TXRDY: It means transmitter is ready to transmit data character.
- TXEMPTY: An output signal which indicates that TXEMPTY pin has transmitted all the data characters and transmitter is empty now.
- TXC: An active-low input pin that controls the data transmission rate of transmitted data.

Receive buffer: This block acts as a buffer for the received data.

- RXD: An input signal that receives the data.

Receive control: This block controls the receiving data.

- RXRDY: An input signal indicates that it is ready to receive the data.
- RXC: An active-low output signal that controls the data transmission rate of received data.
- SYNDET/BD: An input or output terminal. External synchronous mode—input terminal, and asynchronous mode—output terminal.

3

Serial Data Standards RS-232

3.1 Introduction

In Chapter 1, we introduced the concept of serial communications, the placing of one bit after another on a single media channel. Serial communication, as mentioned before, is the most prevalent form of data communication.

In Chapter 2, we introduced UART and USART as the most widely used serial communication protocol in use to communicate between devices. The external signaling levels that are used between different equipment are not generated by UART: UART does not give any electrical characteristics. An interface is used to convert the logic level signals of the UART to the external signaling levels. The interface used is called *Serial Data Standard*. Examples of standards for voltage signaling are **RS-232, RS-422,** and **RS-485** from the **EIA**. For embedded system applications, UARTs are commonly used with RS-232. It is useful to communicate between microcontrollers and also with PCs. **MAX 232** is one of the example ICs which provide RS-232 level signals. Besides the EIA RS standards, there is also a *CCITT* standard named *V.24* that resembles the specifications included in RS-232-C. Figure 3.1 is a block diagram explaining the implementation of RS-232 and its function as an interface for UART.

Figure 3.1 shows that the RS-232 interface works in combination with UART (universal asynchronous receiver/transmitter). It is a piece of integrated circuit integrated inside the processor or controller. It takes bytes and transmits the individual bits in a sequential fashion in a frame. A frame is a defined structure, carrying meaningful sequence of bit or bytes of data. The frame of asynchronous system introduced in Section 3.3. The frame has a start bit followed by 5–8 data bits, a parity bit (optional), and a stop bit. Once data are changed into bits, separate line drivers are used to convert the logic level of UART to RS-232 logic. Finally, the signals are transferred along the interface cable at the specified voltage level of RS-232.

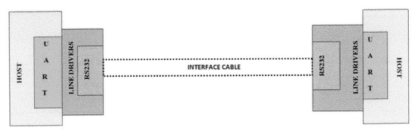

Figure 3.1   Block diagram explaining implementation of RS-232 in devices.

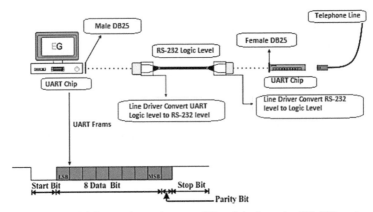

Figure 3.2   Serial data exchange between PC and device using RS-232 protocol.

At the other end, the receiver end, the line driver interface converts it into UART compatible logic levels. At the destination, a second UART reassembles the bits into bytes. This is how RS-232 made the data exchange compatible and reliable.

Figure 3.2 shows how this entire arrangement works.

As shown in Figure 3.3, the equipment at the far end of the connection is named the DTE (Data Terminal Equipment, usually a computer or terminal) device, has a male DB25 connector, and utilizes 22 of the 25 available pins for signals or ground (DB25 and DB9 connectors are discussed latter in the chapter). Equipment at the near end of the connection (the telephone line interface) is named the DCE (Data Circuit-terminating Equipment, usually a modem) device, has a female DB25 connector, and utilizes the same 22 available pins for signals and ground. DCE devices are sometimes called "Data Communications Equipment" instead of Data Circuit-terminating Equipment. The cable linking DTE and DCE devices is a parallel straight-through cable with no cross-overs or self-connects in the connector hoods. If all devices exactly

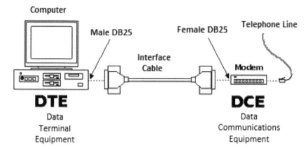

Figure 3.3 RS-232 (EIA232) communication function and connector types for a personal computer and modem.

followed this standard, all cables would be identical, and there would be no chance that an incorrectly wired cable could be used. This drawing shows the orientation and connector types for DTE and DCE devices:

Although serial communication is slower than parallel communication, which allows the transmission of an entire byte at once, it is simpler and can be used over longer distances because of lower power consumption. For example, the IEEE 488 standard for parallel communication requires that the cabling between equipment can be no more than 20 m total, with no more than 2 m between any two devices. RS-232/RS-485/RS-449 cabling, however, can extend 1200 m or greater.

Serial data transmission standards including RS-232, RS-422, RS-423, and RS-485 were widely used for many data links, proving effective connectivity for the day. Although not nearly as widely used today, they can still be found in some areas.

Another element needed for completing the communication between the two communicating ends is the cables. Serial data transmission links used cables with a variety of different functions to enable data to be sent reliably between two equipment. Although a number of different wires were often used within a cable, they did not require nearly as many as the systems that sent parallel data.

This chapter focuses on three EIA/TIA serial standards for 232, 422, and 485 as well as the ancillary 423, 449, and 530 standards. The chapter will cover also the 20 mA current loops. Since the United States now sits on the international standards committees, the EIA/TIA standards have their equivalency in ISO standards, and indeed, most have been changed to meet the ISO standards.

It is important to mention here that serial interface to the PC is being accomplished by newer and much faster serial technologies. Chapter 4 will focus on four PC-based standards, USB 2.0, Firewire (IEEE 1394), SATA, and PCIe, that will impact industrial use and applications.

3.2 Serial Data Standards Background

The first of the RS standards was RS232, or more correctly RS-232. This was developed in 1962 when the need for forms of transmitting data from modems attached telephone lines to remote communications equipment became apparent.

The "RS" stands for "Recommended Standard," although later these standards were formally adopted by the EIA/TIA in the USA.

The EIA is the abbreviation of "Electrical Industries Association" and the TIA is the abbreviation of "Telecommunications Industries Association." Once RS-232 was established, an equivalent standard was written for the ITU (International Telecommunications Union) to provide a more international standard. This would enable the same standards to be used worldwide and also give manufacturers access to a global market using just one product. This standard was known as V.24 and is totally compatible with RS-232.

With RS-232 well established and the need for faster communications over longer distances, further standards beyond RS-232 were introduced. Although a number of standards were introduced, the most widely used are RS-422 and RS-485.

3.2.1 RS-232 Standard Development and Timeline

In terms of the RS-232 timeline, the RS-232 standard for data communications was devised in 1962 when the need to be able to transmit data along a variety of types of line started to grow. The idea for a standard had grown out of the realization in the USA that a common approach was required to allow interoperability. As a result, the Electrical Industries Association (EIA) in the USA created a standard for serial data transfer or communication known as RS-232. It defined the electrical characteristics for transmission of data between a Data Terminal Equipment (DTE) and the Data Communications Equipment (DCE). Normally, the data communication equipment is the modem (modulator/demodulator) which encodes the data into a form that can be transferred along the telephone line. A DTE could be a computer.

Formally, RS-232 is specified as the interface between DTE equipment and DCE equipment using serial binary data exchange (Figure 3.4).

In Figure 3.4, the DTE (computer) transmits the information serially to the other end equipment DCE (modem). In this case, DTE sends binary data

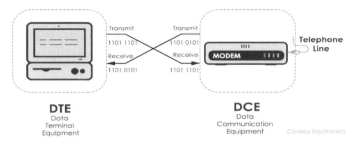

Figure 3.4 Communication between DTE and DCE.

"11011101" to the DCE device and DCE sends binary data "11010101" to the DTE device.

As will be discussed latter, RS-232 describes the common voltage levels, electrical standards, operation mode, and number of bits to be transferred from DTE to DCE. This standard is used for the transmission of information exchange over the telephone lines.

The RS-232 standard underwent several revisions. The C issue, known as RS-232C, was issued in 1969 to accommodate the electrical characteristics of the terminals and devices that were being used at the time.

The RS-232 standard underwent further revisions, and in 1986, Revision D was released (often referred to as RS-232D). This revision of the RS-232 standard was required to incorporate various timing elements and to ensure that the RS-232 standard harmonized with the CCITT standard V.24, while still ensuring interoperability with older versions of RS-232 standard.

Further updates and revisions have occurred since then and the current version is TIA-232-F issued in 1997 under the title: "*Interface between Data Terminal Equipment and Data Circuit-Terminating Equipment Employing Serial Binary Data Interchange.*"

The name of the RS-232 standard has changed during its history several times as a result of the sponsoring organization. As a result, it has variously been known as EIA RS-232, EIA-232, and most recently as TIA-232.

3.2.2 RS-232 Standard Variants

There are number of different specifications and standards that relate to RS-232. A description of some of the RS-232 standards and the various names and references used are given below:

- *RS-232* or RS232 is the most widely used serial standard that is in use. Many laptop computers incorporate a serial interface, and it was also used on many printers, although much less so now. Because RS-232 signals share a common ground, they are sensitive for interference, and therefore, the distance is limited to 15 m. RS-232 can only be used for point-to-point connections (one sender, one recipient).

- *EIA/TIA-232:* This reference to the RS-232 standard includes the names of the first and current sponsoring organizations, the Electronic Industries Alliance (EIA) and the Telecommunications Industry Alliance (TIA).

- *RS-232-C:* This was the designation given to the release of RS-232 standard updated in 1969 to incorporate many of the device characteristics.

- *RS-232-D:* This was the release of the RS-232 standard that occurred in 1986. It was revised to incorporate various timing elements and to ensure that the RS-232 standard harmonized with the CCITT standard V.24.

- *RS-232-F:* This version of the RS-232 standard was released in 1997 to accommodate further revisions to the standard. It is also known as TIA-232-F.

- *V10:* V.10 is an ITU standard or recommendation that was first released in 1976 for unbalanced data communication circuits for data rates up to 100 Kbps. It can inter-work with V.28 provided that the signals do not exceed 12 V. Using a 37-pin ISO 4902 connector, it is actually compatible with RS-423.

- *V24:* The International Telecommunications Union (ITU)/ Consultative Committee for International Telephone and Telegraph (CCITT) of the ITU developed a standard known as ITU v.24, often just written as V24. This standard is compatible with RS-232, and its aim was to enable manufacturers to conform to global standards and thereby allow products that would work in all countries around the world. It is entitled "List of definitions for interchange circuits between data terminal equipment (DTE) and data circuit-terminating equipment (DCE)."

- *V28:* V.28 is an ITU standard defining the electrical characteristics for unbalanced double current interchange circuits, that is, a list of definitions for interchange circuits between data terminal equipment (DTE) and data circuit-terminating equipment (DCE).

Other Standards

- *RS-422 or RS422:* This standard gives a much higher data rate than RS-232. RS-422 (also RS-449) uses signals with reference to ground. It uses differential transmission techniques. By twisting the two wires and

making a "twisted pair," any noise or interference picked up on one wire
will also be picked up on the other, because both wires pick up the same
interference; the RS-422 (also RS-449) differential interface just shifts in
voltage level with reference to ground, but does not change with respect
to each other. The receivers are only looking at the difference in voltage
level of each wire to the other not to ground.

- *RS-423:* RS-423 is not a commonly used enhanced version of the RS-
  232 standard. The only difference is that it can be used for greater distances. The interface looks like it uses differential signals, but in fact, all
  B pins are connected to ground. RS-423 can be used for point-to-point
  or multidrop (one sender, up to 10 recipients) communication.
- *RS-449:* RS-449 is not a commonly used enhanced version of the RS-
  232 standard. Another name for RS-449 is V.11. Like RS-485 and RS-
  422, it uses differential signals over twisted pair cables which reduce
  errors caused by noise or interference. RS-449 can be used for point-to-
  point or multidrop (one sender, up to 10 recipients) communication.
- *RS-485* or RS485 is a standard that allows high-speed data transmission
  along with multiple transmitters and receivers and this makes it able to
  be incorporated as a network solution. *RS-485* is the most enhanced RS-
  232-based serial interface available. It can handle high speeds and long
  cable distances up to 4000 ft. Like RS-449 and RS-422, it uses differential signals over twisted pair cables which reduce errors caused by
  noise or interference. RS-485 uses multipoint technology; it is possible
  to connect 32 devices on a single RS-485 bus. Connection speeds can be
  as high as 35 Mbps.

Each of these standards meets a different requirement. RS-232 is still
being very widely used despite the fact that it has been in use for over
40 years. Other serial communication standards have been introduced more
recently, for example, I2C, which provides higher levels of performance that
is very useful in many applications. In spite of that, RS-232 is still in use. The
reason is that RS-232 signals spread over longer distances when compared
to I2C and serial TTL signals. Moreover, it has better noise immunity. It
is proven to be compatible across different manufacturers for interfacing
computer and modems.

Many of the mentioned RS-232 variants are discussed in this chapter.
Table 3.1 shows a comparison of three of these standards.

Comparison of serial data standards: Table 3.1 is a comparison between
the three standards: RS-232, RS-422, and RS-485.

Table 3.1   Comparison of serial data standard

Common RS Series Serial Data Transmission Standards			
Parameter	RS-232	RS-422	RS-485
Cabling	Single ended	Differential	Differential
Number of devices	One transmit and one receive	Five transmitters and ten receivers	32 transmitters and 32 receivers
Communication mode	Full duplex	Full duplex/half duplex	Half duplex
Maximum distance	50 ft at 19.2 Kbps	4000 ft at 100 Kbps	4000 ft at 100 Kbps
Maximum data rate	19.2 Kbps at 50 ft	10 Mbps at 50 ft	10 Mbps at 50 ft
Signaling mode	Unbalanced	Balanced	Balanced
Mark (1)	-5 to -15 V	2 V to 6 V max. (B>A)	1.5 V to 5 V max. (B>A)
Space (0)	$+5$ to $+15$ V	2 V to 6 V max. (A>B)	1.5 V to 5 V max. (A>B)
Output current capability	500 mA	150 mA	250 mA

3.3 RS-232 Serial Data Transmission: Data Timing

The working of RS-232 can be understood by the protocol format. As RS-232 is a point-to-point asynchronous communication protocol, it sends data in a single direction. Here, no clock is required for synchronizing the transmitter and receiver. The data format is initiated with a start bit followed by 7-bit (or 8-bit, Figure 3.5) binary data, parity bit, and stop bit which are sent one after another.

Protocol format:

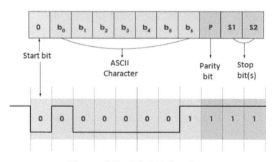

Figure 3.5   RS-232 framing.

Start bit: The data are sent serially on RS-232; each bit is sent one after the next because there is only one data line in each direction. This mode of data transmission also requires that the receiver knows when the actual data bits are arriving so that it can synchronize itself to the incoming data.

Taking into consideration that the data line has two states, on and off, and that an idle line is always on. To achieve synchronization, when the instrument or computer wants to send data, it sets the line to off, that is, logic 0: this is the **Start bit**. The bits immediately after the start bit are therefore the data bits.

Data: The start bit is followed by the **data** itself and there are normally seven or eight bits (LSB sent first and MSB sent last). The receiver obviously has to know how many data bits to expect, and there are often small dual in line switches either on the back of the equipment or inside it to set this information.

Data on RS-232 are normally sent using ASCII (American Standard Code for Information Interchange). However other codes including the Murray Code or EBCDIC (Extended Binary Coded Decimal Interchange Code) can be used equally well.

Parity bit: After the data itself, a **parity bit** is sent. Parity is the state of being either odd or even. In serial communication, parity may be used to check for errors in the transmission of data. When performing a parity check, the instrument or PC sending messages counts the number of 1's in a group of data bits. Depending on the result, the value of another bit, the Parity Bit, is set. The device receiving the data also counts the 1's and checks whether the Parity Bit is as it should be. Parity is a rudimentary **error checking** mechanism. It can detect a single bit in error in a transmitting message, but if 2 bits happened to be wrong it would not pick this up. It also provides no help as to which bit is wrong (parity is not **error correcting**). Other error checking mechanisms may be used.

To perform a parity check, the computer and the instrument must obviously agree on how they are calculating the Parity Bit. Are they setting it on for an even or odd number of 1's? When a device uses Even Parity, the data bits and the parity bit will always contain an even number of 1's. The reverse is true for Odd Parity. Since the parity bit is optional and can be Odd or Even, it requires setting defining the parity type.

Note: **Mark and Space Parity:** Two other parity options often available in driver software are Mark and Space. These aren't effective in error checking. Mark means the device always sets the Parity Bit to 1 and Space always to 0.

Stop bit: Finally, a stop bit is sent. The Stop Bit is present to allow the instrument and computer to re-synchronize should anything go wrong: noise

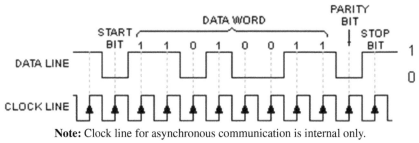

Note: Clock line for asynchronous communication is internal only.

Figure 3.6   Example of data frame.

on the line masking the start bit, for example. The period of time between the start and stop bit is constant, according to the baud rate and number of data and parity bits. The stop bit is always on. If the receiver detects an off value when the stop bit should be present, it knows there has been an error.

Setting of stop bit: The stop bit is not actually 1 bit but a minimum length of time the line must be on at the end of each data transmission. On PCs, this is normally equal to 1 or 2 bits, and the user must specify this in the driver software. Although 1 stop bit is most common, selecting 2 will at worst slow the message down slightly. (You might see an option to set the stop bit to 1.5. This is only used when the number of data bits is less than 7. If this is the case, then ASCII characters cannot be transmitted and so 1.5 is rarely used.)

Note: RS-232 data transmission is normally asynchronous. However, transmit and receive speeds must obviously be the same. A certain degree of tolerance is allowed. Once the start bit is sent, the receiver will sample the center of each bit to see the level. Within each data word, the synchronization must not differ by more than half a bit length; otherwise, the incorrect data will be seen. Fortunately, this is very easy to achieve with today's accurate bit or baud rate generators.

Example: Figure 3.6 is an example showing how the data frame is composed of and synchronized with the clock signal. This example uses an 8-bit word with even parity and 1 stop bit also referred to as an 8E1 setting.

3.4 RS-232 Serial Interface (also known as V24)

RS-232 (Recommended Standard-232), also known as V24, is a standard interface standardized by EIA which provides a useful means of serial data communication for many years. RS-232 (ANSI/EIA-232 Standard) is the serial connection historically found on IBM-compatible PCs. It is used for many purposes, such as connecting a mouse, printer, or modem, as well as industrial instrumentation. RS-232 describes the physical interface and protocol for relatively low-speed serial data communication between computers and related devices. In other words, RS-232 is the interface that the computer uses to communicate and exchange data with modem and other serial devices. It is limited to point-to-point connections between PC serial ports and devices, which means that RS-232 protocol can have only one device connected to each port.

The RS-232 interface is intended to operate over distances of up to 15 meters. This is because any modem is likely to be near the terminal. Data rates are also limited. The maximum for RS-232C is 19.2 k baud or bits per second although slower rates are often used.

Because of improvements in line drivers and cables, applications often increase the performance of RS-232 beyond the distance and speed listed in the standard. In theory, it is possible to use any baud rate, but there are a number of standard transmission speeds used.

RS-232/V24 was found in many areas from computers to remote terminals and many more. It was an effective way of providing serial data connectivity and as such it was widely used.

Typically, RS-232/RS-485 is used to transmit American Standard Code for Information Interchange (ASCII) data (see Figure 3.5). Although National Instruments serial hardware can transmit 7-bit as well as 8-bit data, many applications use 7-bit data. Seven-bit ASCII can represent the English alphabet, decimal numbers, and common punctuation marks. It is a standard protocol that virtually all hardware and software understand. Serial communication is completed using three transmission lines: (1) ground, (2) transmit, and (3) receive. Because RS-232/RS-485 communication is asynchronous, the serial port can transmit data on one line while receiving data on another line. Other lines are available for handshaking, but are not required. The important serial characteristics are baud rate, data bits, stop bits, and parity. These parameters must match to allow communication between a serial device and a serial port on your computer.

The RS-232/V24 serial interface communication standard is still used in some instances, especially in existing installations, although its use is now decreasing as Ethernet and other standards take its place.

3.4.1 RS–232 Specifications: RS-232 Interface Basics

RS-232 is a "complete" standard. This means that the standard sets out to ensure perfect compatibility between the host and peripheral systems by specifying:

1. common voltage and signal levels
2. common pin wiring configurations
3. a minimal amount of control information between the host and peripheral systems.

To meet the above three criteria, RS-232 standard covers the following areas:

1. Electrical characteristics This includes:

   a. voltage level (Logic Level)
   b. slew rate
   c. voltage withstand level

2. Functional characteristics

   a. This covers the functions of each circuit in the interface

3. Mechanical characteristics. This includes:

   a. pluggable connectors
   b. pin identification.

Figure 3.7 summarizes the standard RS-232 specifications with their values and examples.

This section deals with detailed understanding of each characteristic on the standard.

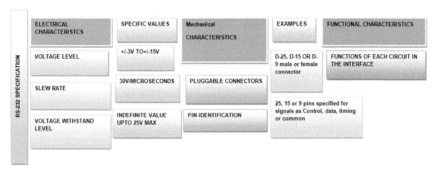

Figure 3.7   Block diagram summarizing standard RS-232 specifications.

3.4.2 Electrical Characteristics: RS-232 Signal Levels and Level Conversion

The electrical characteristics section of the RS-232 standard includes specifications on voltage levels, rate of change of signal levels, slew rate, and line impedance (Figure 3.7). The RS-232 standard besides defining the levels for the lines it includes also a mode of operation for the handshaking.

In the following, the main characteristics that define the electrical characteristics are introduced. This includes:

a. voltage level or logic level
b. slew rate
c. line impedance
d. short-circuit tolerance
e. fail-safe signals
f. mode of operation
g. baud rate
h. requirements needed while changing the logic state.

3.4.2.1 Voltage level or logic level

The voltage levels are one of the main items in the specification. By defining the voltage levels, any RS-232 system can be assured of its correct operation. If the voltages fall within the defined levels, then the receivers are able to correctly detect the data that are being transmitted or the state of the other lines. If the lines fall outside the required limits, then there can be uncertainty and data errors. Besides the line voltage, it is also necessary to define the voltage states for the control signals as these are widely used within RS-232. "**Voltage level**" or "**logic level**" means the range of voltage over which a high bit (1) and a low bit (0) is accepted in a particular IC, gate, etc. Various logic (voltage) levels have been standardized, for example, for TTL, LVTTL, and also RS-232.

The original RS-232 standard was defined in 1962. As this was before the days of TTL logic, it should not be surprising that the standard does not use 5 V and ground logic levels. Instead, a high level for the driver output is defined as being +5 to +15 V, and a low level for the driver output is defined as being between −5 and −15 V. The receiver logic levels were defined to provide a 2 V noise margin. As such, a high level for the receiver is defined as +3 to +15 V and a low level is −3 to −15 V. Figure 3.8 illustrates the logic levels defined by the RS-232 standard. It is necessary to note that for RS-232 communication, a low level (−3 to −15 V) is defined as logic 1 and

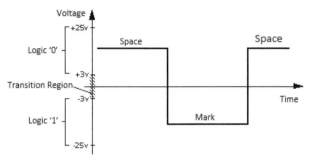

Figure 3.8   Logic states are assigned to the voltage ranges shown here. Note that this is a "negative logic" convention, which is the reverse of that used in most modern digital designs.

Table 3.2   Line voltage levels

RS-232 Signal Line Voltage Levels	
Signal Voltage Levels (volts)	Logical State
-3 to -25	1
+3 to +25	0

Table 3.3   Voltage states for the control signals as these are widely used within RS-232

RS-232 Control Line Voltage Levels	
Control Voltage Levels (volts)	Logical State
−3 to −25	Off
+3 to +25	On

is historically referred to as "marking." Likewise, a high level (+3 to +15 V) is defined as logic 0 and is referred to as "spacing." The signal level between +3 V and −3 V represents transition region. Table 3.2 gives the RS-232 line voltage levels and Table 3.3 gives the control signals.

Note: Most contemporary applications will show an open-circuit signal voltage of −8 to −14 V for logic "1" (mark), and +8 to +14 V for logic "0" (space). Voltage magnitudes will be slightly less when the generator and receiver are connected (when the DTE and DCE devices are connected with a cable).

The extreme voltages of an RS-232 signal help to make it less susceptible to noise, interference, and degradation. This means that an RS-232 signal can generally travel longer physical distances than their TTL counterparts, while still providing a reliable data transmission.

With such RS-232 line voltage levels, and keeping in mind that RS-232 is an interface between UART/USART and external devices, RS-232

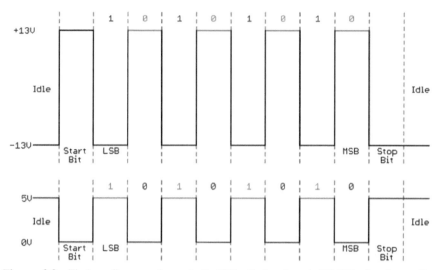

Figure 3.9 Timing diagram shows both TTL (bottom) and RS-232 signals sending 0b01010101.

must guarantee the compatibility between voltages on the two sides. In many cases, this matter needs to use "level conversion" circuits. "Level Conversion" circuits will be discussed in Section 3.7. To discuss such "level conversion" circuits, it is important to start here by considering the voltage (logic) level of TTL and LVTTL logic families.

TTL Logic Level: TTL (Transistor-Transistor Logic) is the most widely used logic. TTL is mostly used in ICs and gates, like 74xx logic gates. A major drawback of the TTL logic is that most of the devices working on the TTL logic consume a lot of current, even individual gates may draw up to 3–4 mA. In TTL logic, a HIGH (or 1) is +5 V, whereas a LOW (or 0) is 0 V. But since attaining exact +5 V and 0 V is practically not possible every time; various IC manufacturers define TTL logic level range differently, but the usual accepted range for a HIGH is within +3.5 ∼ +5.0 V, and the range for a LOW is 0 ∼ +0.8 V. Figure 3.9 is a timing diagram showing both TTL and RS-232 signals while sending the binary 0101010101.

LVTTL Logic Level: LVTTL (Low Voltage Transistor–Transistor Logic) is increasingly becoming popular these days, because of the nominal HIGH voltages, and hence lesser power consumption. By lowering the power supply from 5 V to 3.3 V, switching power reduces by almost 60%.There are several transistors and gates, which work on LVTTL logic. Atmel's Atmega

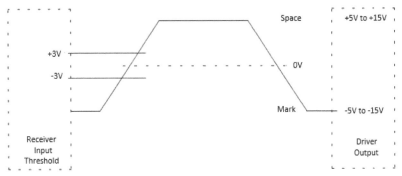

Figure 3.10   Logic level and slew rate.

microcontrollers are designed to work on both, LVTTL and TTL, depending upon the VCC supplied to the IC. In LVTTL logic, a LOW is defined for voltages 0 V $\sim$ 1.2 V, and a High for voltages 2.2 V $\sim$ 3.3 V, making 1.2 V $\sim$ 2.2 V undefined.

3.4.2.2 Slew rate

The RS-232 standard also limits the maximum slew rate at the driver output. This limitation was included to help reduce the likelihood of crosstalk between adjacent signals. The slower the rise and fall time, the smaller the chance of crosstalk. With this in mind, the maximum slew rate allowed is 30 V/μs. Additionally, a maximum data rate of 20 kbps has been defined by the standard, again with the purpose of reducing the chance of crosstalk. Figure 3.10 shows the logic state and the rise and fall time.

3.4.2.3 Line impedance

The impedance of the interface between the driver and receiver has also been defined. The load seen by the driver is specified to be 3 kΩ to 7 kΩ (Figure 3.11). For the original RS-232 standard, the cable between the driver and the receiver was also specified to be a maximum of 15 m in length. This part of the standard was changed in revision "D" (EIA/TIA–232–D). Instead of specifying the maximum length of cable, a maximum capacitive load of 2500 pF was specified which is clearly a more adequate specification. The maximum cable length is determined by the capacitance per unit length of the cable which is provided in the cable specifications.

Figure 3.11 is the equivalent circuit for an RS-232 (EIA232) signal line and applies to signals originating at either the DTE or DCE side of the connection. "C_o" is not specified in the standard, but is assumed to be small

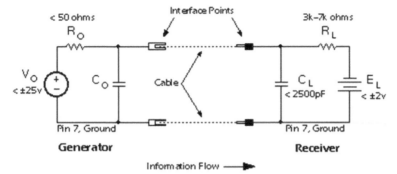

Figure 3.11   Equivalent circuit of RS-232 line.

and to consist of parasitic elements only. "R_o" and "V_o" are chosen so that the short-circuit current does not exceed 500 mA. The cable length is not specified in the standard; acceptable operation is experienced with cables that are less than 25 ft in length.

It is important to note here that the equivalent circuit of Figure 3.11 represents all signal lines, regardless of whether they provide data, timing, or control information.

3.4.2.4 Short-circuit tolerance
The generator (Figure 3.11) is designed to withstand an open-circuit (uncon-nected) condition or short-circuit condition between its signal conductor and any other signal conductor, including ground, without sustaining damage to itself or causing damage to any associated circuitry. The receiver is also designed to accept any signal voltage within the range of ±25 V without sustaining damage.

3.4.2.5 Fail-safe signals
Four signals are intended to be fail-safe in that during power-off or cable-disconnected conditions, they default to logic "1" (negative voltage). They are as follows:

- Request to Send: Default condition is deasserted.
- Sec. Request to Send: Default condition is deasserted.
- DTE Ready: Default condition is DTE not ready.
- DCE Ready: Default condition is DCE not ready.

Note specifically that if the cable is connected but the power is off in the generator side, or if the cable is disconnected, there should be adequate bias voltage in the receiver to keep the signal above +3 V (logic "0") to ensure that the fail-safe requirement is met.

3.4.2.6 Mode of operation
The RS-232 devices work on **single-ended signaling** (two wire). This means one wire transmits an altering voltage and another wire is connected to ground. Single-ended signals are affected by the noise induced by differences in ground voltages of the driver and receiver circuits. The advantage of the single-ended technique is it requires fewer wires to transmit information.

3.4.2.7 Baud rate
It is the number of binary bits transferred per second. RS-232 supports baud rates from 110 to 230400. Commonly, the baud rate with 1200, 4800, 9600, and 115200 are used. It determines the speed at which data are to be sent from the transmitter to the receiver.

Note: The baud rate has to be same at both the transmitter side and receiver side.

3.4.2.8 Requirements needed while changing the logic state
Changes in signal state from logic "1" to logic "0" or vice versa must abide by several requirements as follows:

1 – Signals that enter the transition region during a change of state must move through the transition region to the opposite signal state without reversing direction or reentering.
2 – For control signals, the transit time through the transition region should be less than 1 ms.
3 – For data and timing signals, the transit time through the transition region should be:

   o less than 1 ms for bit periods greater than 25 ms,
   o 4% of the bit period for bit periods between 25 ms and 125 μs,
   o less than 5 μs for bit periods less than 125 μs.

   The rise and fall times of data and timing signals ideally should be equal, but in any case vary by no more than a factor of three.
4 – The slope of the rising and falling edges of a transition should not exceed 30 V/μs. Rates higher than this may induce crosstalk in adjacent conductors of a cable.

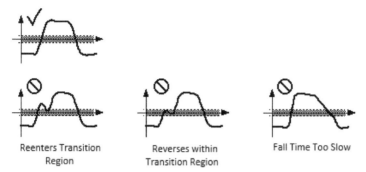

Reenters Transition Region	Reverses within Transition Region	Fall Time Too Slow

Figure 3.12 An acceptable pulse (top) moves through the transition region quickly and without hesitation or reversal. Defective pulses (bottom) could cause data errors.

3.4.3 Functional characteristics

The second aspect of operation that is covered by the standard concerns the functional characteristics of the interface. This essentially means that RS-232 has defined the function of the different signals that are used in the interface. These signals are divided into four different categories: common, data, control, and timing.

Table 3.4 illustrates the signals that are defined by the RS-232 standard. As can be seen from the table, there is an overwhelming number of signals defined by the standard. The standard provides an abundance of control signals and supports a primary and secondary communication channels. Fortunately, few applications, if any, require all of these defined signals. For example, only eight signals are used for a typical modem. Some simple applications may require only four signals (two for data and two for handshaking), while others may require only data signals with no handshaking. Examples of how the RS-232 standard is used in some "real-world" applications are discussed later in this chapter.

The complete list of defined signals is given next and it is included here as a reference, but it is beyond the scope of this chapter to review the functionality of all of these signals.

There are few terms used in the table like loop back, off or on hook, and secondary channels.

Loop back: It is a method to perform transmission test of the lines at the switching center. Loop back allows user to test their own network to ensure it is functioning properly.

Table 3.4   RS-232 defined signals

Circuit Mnemonic	Circuit Name*	Circuit Direction	Circuit Type
AB	Signal Common	–	Common
BA	Transmitted Data (TD)	To DCE	Data
BB	Received Data (RD)	From DCE	
CA	Request to Send (RTS)	To DCE	
CB	Clear to Send (CTS)	From DCE	
CC	DCE Ready (DSR)	From DCE	
CD	DTE Ready (DTR)	To DCE	
CE	Ring Indicator (RI)	From DCE	
CF	Received Line Signal Detector** (DCD)	From DCE	Control
CG	Signal Quality Detector	From DCE	
CH	Data Signal Rate Detector from DTE	To DCE	
CI	Data Signal Rate Detector from DCE	From DCE	
CJ	Ready for Receiving	To DCE	
RL	Remote Loopback	To DCE	
LL	Local Loopback	To DCE	
TM	Test Mode	From DCE	
DA	Transmitter Signal Element Timing from DTE	To DCE	
DB	Transmitter Signal Element Timing from DCE	From DCE	Timing
DD	Receiver Signal Element Timing from DCE	From DCE	
SBA	Secondary Transmitting Data	To DC	Data
SBB	Secondary Received Data	From DCE	
SCA	Secondary Request to Send	To DCE	Control
SCB	Secondary Clear to Send	From DCE	
SCF	Secondary Received Line Signal Detector	From DCE	

*Signals with abbreviations in parentheses are the eight most commonly used signals.
**This signal is more commonly referred to as Data Carrier Detect (DCD).

Off hook: A condition that occurs when a telephone or other user instrument is in use that is during dialing or communicating. It was originally referred for telephones that have separate earpiece (receiver) which hangs on the switch hook until user wants to use it.

Secondary channel: These are the data channels and have same capability as the first one. For example, Secondary Transmitted Data (STD), Secondary Received Data (SRD), Secondary Request To Send (SRTS), Secondary Clear To Send (SCTS), and Secondary Carrier Detect (SDCD).

3.4.3.1 RS-232 signal (lines) definitions and functions

Signal functions in the EIA232 standard can be subdivided into six categories. These categories are summarized below, after which each signal is described.

1 – Signal ground and shield.
2 – Primary communication channel. This is used for data interchange and includes flow control signals.
3 – Secondary communication channel. When implemented, this is used for control of the remote modem, requests for retransmission when errors occur, and governance over the setup of the primary channel.
4 – Modem status and control signals: These signals indicate modem status and provide intermediate checkpoints as the telephone voice channel is established.
5 – Transmitter and receiver timing signals: If a synchronous protocol is used, these signals provide timing information for the transmitter and receiver, which may operate at different baud rates.
6 – Channel test signals: Before data are exchanged, the channel may be tested for its integrity, and the baud rate automatically adjusted to the maximum rate that the channel can support.

Group 1: Signal Ground and Shield

Pin 7, Pin 1, and the **shell** are included in this category. Cables provide separate paths for each, but internal wiring often connects pin 1 and the cable shell/shield to signal ground on pin 7. This signal (called also protective ground) ensures that both equipment are at the same earth potential. This is very useful when there is a possibility that either equipment is not earthed. The signal ground (Pin 7) is used as the return for the digital signals travelling along the data link. It is important that large currents that are not part of the signaling do not flow along this line; otherwise, data errors may occur.

Pin 7 – Ground: All signals are referenced to a common ground, as defined by the voltage on pin 7. This conductor may or may not be connected to protective ground inside the DCE device. The existence of a defined ground potential within the cable makes the EIA232 standard different from a balanced differential voltage standard, such as EIA530, which provides far greater noise immunity.

Group 2: Primary Communications Channel

Pin 2 – Transmitted Data (TxD): This signal is active when data is transmitted from the DTE device to the DCE device. When no data is transmitted, the signal is held in the mark condition (logic "1", negative voltage).

Note: Pin 2 on the DCE device is commonly labeled "Received Data," although by the EIA232 standard it should still be called Transmitted Data because the data are thought to be destined for a remote DTE device.

Pin 3 – Received Data (RxD): This signal is active when the DTE device receives data from the DCE device. When no data is transmitted, the signal is held in the mark condition (logic "1", negative voltage).

Note: Pin 3 on the DCE device is commonly labeled "Transmitted Data", although by the EIA232 standard it should still be called Received Data because the data are thought to arrive from a remote DTE device.

Pin 4 – Request to Send (RTS): This signal is asserted (logic "0", positive voltage) to prepare the DCE device for accepting transmitted data from the DTE device. Such preparation might include enabling the receive circuits or setting up the channel direction in half-duplex applications. When the DCE is ready, it acknowledges by asserting Clear to Send.

Note: Pin 4 on the DCE device is commonly labeled "Clear to Send", although by the EIA232 standard it should still be called Request to Send because the request is thought to be destined for a remote DTE device.

Pin 5 – Clear to Send (CTS): This signal is asserted (logic "0", positive voltage) by the DCE device to inform the DTE device that transmission may begin. RTS and CTS are commonly used as handshaking signals to moderate the flow of data into the DCE device.

Note: Pin 5 on the DCE device is commonly labeled "Request to Send", although by the EIA232 standard it should still be called Clear to Send because the signal is thought to originate from a remote DTE device.

Group 3: Secondary Communications Channel

There are two types of lines that are specified in the RS-232 specification. There are the primary channels (group 2) that are normally used and operate at the normal or higher data rates. However, there is also provision for a secondary channel (group 3) for providing control information. If it is used, it will usually send data at a much slower rate than the primary channel.

As the secondary lines are rarely used or even implemented on equipment, manufacturers often use these connector pins for other purposes. In view of this, it is worth checking that the lines are not being used for other purposes before considering using them. When the secondary system is in use, the handshaking signals operate in the same way as for the primary circuit.

Pin 13 – Secondary Clear to Send (SCTS)

Pin 14 – Secondary Transmitted Data (STxD)

Pin 16 – Secondary Received Data (SRxD)

Pin 19 – Secondary Request to Send (SRTS)

These signals are equivalent to the corresponding signals in the primary communications channel. The baud rate, however, is typically much slower in the secondary channel for increased reliability.

Group 4: Modem Status and Control Signals

Pin 6 – DCE Ready (DSR): When originating from a modem, this signal is asserted (logic "0", positive voltage) when the following three conditions are all satisfied:

1 – The modem is connected to an active telephone line that is "off-hook".
2 – The modem is in data mode, not voice or dialing mode.
3 – The modem has completed dialing or call setup functions and is generating an answer tone.

If the line goes "off-hook," a fault condition is detected, or a voice connection is established, the DCE Ready signal is deasserted (logic "1", negative voltage).

IMPORTANT: If DCE Ready originates from a device other than a modem, it may be asserted to indicate that the device is turned on and ready to function, or it may not be used at all. If unused, DCE Ready should be permanently asserted (logic "0", positive voltage) within the DCE device or by use of a self-connect jumper in the cable. Alternatively, the DTE device may be programmed to ignore this signal.

Pin 20 – DTE Ready (DTR): This signal is asserted (logic "0", positive voltage) by the DTE device when it wishes to open a communications channel. If the DCE device is a modem, the assertion of DTE Ready prepares the modem to be connected to the telephone circuit and, once connected, maintains the connection. When DTE Ready is deasserted (logic "1", negative voltage), the modem is switched to "on-hook" to terminate the connection.

IMPORTANT: If the DCE device is not a modem, it may require DTE Ready to be asserted before the device can be used, or it may ignore DTE Ready altogether. If the DCE device (e.g., a printer) is not responding, confirm that DTE Ready is asserted before you search for other explanations.

Pin 8 – Received Line Signal Detector (CD) (also called carrier detect): This signal is relevant when the DCE device is a modem. It is asserted (logic "0", positive voltage) by the modem when the telephone line is "off-hook," a connection has been established, and an answer tone is being received from the remote modem. The signal is deasserted when no answer tone is being received or when the answer tone is of inadequate quality to meet the local modem's requirements (perhaps due to a noisy channel).

Pin 12 – Secondary Received Line Signal Detector (SCD): This signal is equivalent to the Received Line Signal Detector (pin 8), but refers to the secondary channel.

Pin 22 – Ring Indicator (RI): This signal is relevant when the DCE device is a modem and is asserted (logic "0", positive voltage) when a ringing signal is being received from the telephone line. The assertion time of this signal will approximately equal the duration of the ring signal, and it will be deasserted between rings or when no ringing is present.

Pin 23 – Data Signal Rate Selector: This signal may originate either in the DTE or DCE devices (but not both) and is used to select one of two prearranged baud rates. The asserted condition (logic "0", positive voltage) selects the higher baud rate.

Group 5: Transmitter and Receiver Timing Signals

Pin 15 – Transmitter Signal Element Timing (TC) (also called Transmitter Clock): This signal is relevant only when the DCE device is a modem and is operating with a synchronous protocol. The modem generates this clock signal to control exactly the rate at which data are sent on Transmitted Data (pin 2) from the DTE device to the DCE device. The logic "1" to logic "0" (negative voltage to positive voltage) transition on this line causes a corresponding transition to the next data element on the Transmitted Data line. The modem generates this signal continuously, except when it is performing internal diagnostic functions.

Pin 17 – Receiver Signal Element Timing (RC) (also called Receiver Clock): This signal is similar to TC described above, except that it provides timing information for the DTE receiver.

Pin 24 – Transmitter Signal Element Timing (ETC) (also called External Transmitter Clock): Timing signals are provided by the DTE device for use by a modem. This signal is used only when TC and RC (pins 15 and 17) are not in use. The logic "1" to logic "0" transition (negative voltage to positive voltage) indicates the time-center of the data element. Timing signals will be provided whenever the DTE is turned on, regardless of other signal conditions.

Group 6: Channel Test Signals

Pin 18 – Local Loopback (LL) This signal is generated by the DTE device and is used to place the modem into a test state. When Local Loopback is asserted (logic "0", positive voltage), the modem redirects its modulated output signal, which is normally fed into the telephone line, back into its receive circuitry. This enables data generated by the DTE to be echoed back through the local modem to check the condition of the modem circuitry. The modem asserts its Test Mode signal on Pin 25 to acknowledge that it has been placed in local loopback condition.

Pin 21 – Remote Loopback (RL): This signal is generated by the DTE device and is used to place the remote modem into a test state. When Remote Loopback is asserted (logic "0", positive voltage), the remote modem redirects its received data back to its transmitted data input, thereby remodulating the received data and returning it to its source. When the DTE initiates such a test, transmitted data are passed through the local modem, the telephone line, the remote modem, and back, to exercise the channel and confirm its integrity. The remote modem signals the local modem to assert Test Mode on pin 25 when the remote loopback test is underway.

Pin 25 – Test Mode (TM): This signal is relevant only when the DCE device is a modem. When asserted (logic "0," positive voltage), it indicates that the modem is in a Local Loopback or Remote Loopback condition. Other internal self-test conditions may also cause Test Mode to be asserted and depend on the modem and the network to which it is attached.

Table 3.5   Conventional usage of signal names

	DTE Side		DCE Side	
2	Transmitted Data	$\longrightarrow$	Received Data	2
3	Received Data	$\longleftarrow$	Transmitted Data	3
4	Request Send	$\longrightarrow$	Clear to Send	4
5	Clear to Send	$\longleftarrow$	Request to Send	5
14	Sec. Transmitted Data	$\longrightarrow$	Sec. Received Data	14
16	Sec. Received Data	$\longleftarrow$	Sec. Transmitted Data	16
19	Sec. Request to Send	$\longrightarrow$	Sec. Clear to Send	19
13	Sec. Clear to Send	$\longleftarrow$	Sec. Request to Send	13

IMPORTANT: Signal names that imply a direction, such as Transmit Data and Receive Data, are named from the point of view of the DTE device. If the EIA232 standard was strictly followed, these signals would have the same name for the same pin number on the DCE side as well. Unfortunately, this is not done in practice by most engineers, probably because no one can keep straight which side is DTE and which side is DCE. As a result, direction-sensitive signal names are changed at the DCE side to reflect their *drive direction* at DCE. Table 3.5 gives a list with the conventional usage of signal names:

3.4.3.2 RS-232 DB9 pin configuration and function
Although the RS-232 25-pin configuration with a 25-way D-type connector is very widely used, in many applications the smaller 9-way D-type connector is used. It provides an adequate size and cost benefit. Also, the RS-232 9-pin configuration is quite sufficient in most circumstances because many of the lines available for RS-232 signaling are rarely used. This means that the 9-way connector is able to provide all the required connectivity for most applications.

The pin configuration of DB-9 port is shown in Figure 3.13. The pinout is also shown for the DEC modified modular jack (Figure 3.14). Table 3.6 categorizes the pins according to the function. Table 3.7 defines the signal for the DTE device – looking into the DTE connector. DTE device is often a computer. Table 3.8 shows the RS-232 signal definitions for the DCE device – looking into the DCE connector. DCE device is often a modem.

By using these configurations, it means that an RS-232 cable connects two RS-232 devices, that is, a DTE to a DCE will be wired in a one-to-one configuration, that is, pin 1 to pin 1, etc. When using RS-232, it is

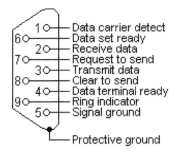

Figure 3.13   RS-232 DB9 pinout.

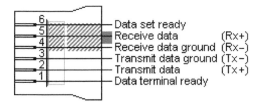

Figure 3.14   DEC MMJ pinout.

Table 3.6   Pin functions for RS-232 DB9

Data	
TXD (pin 3)	Serial Data Output
RXD (pin 2)	Serial Data Input
Handshake	
RTS (pin 7)	Request to Send
CTS (pin 8)	Clear to Send
DSR (pin 6)	Data Set Ready
DCD (pin 1)	Data Carrier Detect
DTR (pin 4)	Data Terminal Ready
Ground	
GND (pin 5)	Ground
Other	
RI (pin 9)	Ring Indicator

essential to ensure that the correct connections are being used. It can be very difficult to fault-find when the wrong connections are made as nothing happens. Checking the voltages are correct and that the correct pins have been connected normally resolves the issues.

Table 3.7   RS-232 DB9 pin definition for the DTE device

RS-232 9-pin signal definition for the DTE device – looking into the DTE connector
DTE RS-232 device is often a PC

Pin no.	Circuit Name	Abbreviation	Source
1	Received line signal detect		
2	Receive data	RXD	
3	Transmit data	TXD	
4	Data terminal ready	DTR	
5	Signal ground		
6	Data set ready	DSR	DCE
7	Request to send	RTS	
8	Clear to send	CTS	
9	Ring indicator		

Table 3.8   RS-232 DB9 pin definition for the DCE device

RS-232 9-pin signal definition for the DCE device – looking into the DCE connector
DCE device is often a modem

Pin No.	Circuit Name	Abbreviation	Source
1	Received line signal detect		
2	Transmit data	TXD	
3	Receive data	RXD	
4	Data terminal ready	DTR	
5	Signal ground		
6	Data set ready	DSR	DCE
7	Clear to send	CTS	
8	Request to send	RTS	
9	Ring indicator		

Sometimes confusion can be caused by defining which is the DTE and which is the DCE. Once this has been correctly done, and the correct pins connected, the system should operate well.

Note: The serial port pin and signal assignments are with respect to the DTE. For example, data are transmitted from the TD pin of the DTE to the RD pin of the DCE.

The pin description for the RS-232 DB9 pins is as follows:

- **DTR** (data terminal ready): When terminal is turned on, it sends out signal DTR to indicate that it is ready for communication.

- **DSR** (data set ready): When DCE is turned on and has gone through the self-test, it asserts DSR to indicate that it is ready to communicate.
- **RTS** (request to send): When the DTE device has byte to transmit, it asserts RTS to signal the modem that it has a byte of data to transmit.
- **CTS** (clear to send): When the modem has room for storing the data it is to receive, it sends out signal CTS to DTE to indicate that it can receive the data now.
- **DCD** (data carrier detect): The modem asserts signal DCD to inform the DTE that a valid carrier has been detected and that contact between it and the other modem is established.
- **RI** (ring indicator): An output from the modem and an input to a PC indicates that the telephone is ringing. It goes on and off in synchronous with the ringing sound.
- **RxD** (received data): The RxD pin is the Data Receive pin. This is the pin where the receiver receives data.
- **TxD** (transmitted data): The TxD pin is the Data Transmit pin. This is the pin through which data are transmitted to the receiver.
- **GND:** Ground pin.

The above signals are divided into groups (Table 3.6): data, hand-shake/control, and ground. In the following, the signals are given in more details.

a. The Data Pins

Most serial port devices support full-duplex communication meaning that they can send and receive data at the same time. Therefore, separate pins are used for transmitting and receiving data. For these devices, the TD, RD, and GND pins are used. However, some types of serial port devices support only one-way or half-duplex communications. For these devices, only the TD and GND pins are used. In the course of explanation, it is assumed that a full-duplex serial port is connected to the DCE. The TD pin carries data transmitted by a DTE to a DCE. The RD pin carries data that are received by a DTE from a DCE.

b. The Control Pins or Handshaking Pins:

The 9-pin serial ports provide several control pins whose functions are to signal the presence of connected devices and control the flow of data. The control pins include RTS and CTS, DTR and DSR, and CD and RI.

The RTS and CTS Pins: The RTS and CTS pins are used to signal whether the devices are ready to send or receive data. This type of data flow control, called hardware handshaking, is used to prevent data loss during transmission. When enabled for both the DTE and DCE, hardware handshaking using RTS and CTS follows these steps:

1. The DTE asserts the RTS pin to instruct the DCE that it is ready to receive data.
2. The DCE asserts the CTS pin indicating that it is clear to send data over the TD pin. If data can no longer be sent, the CTS pin is unasserted.
3. The data are transmitted to the DTE over the TD pin. If data can no longer be accepted, the RTS pin is unasserted by the DTE and the data transmission is stopped.

The DTR and DSR Pins: Many devices use the DSR and DTR pins to signal if they are connected and powered. Signaling the presence of connected devices using DTR and DSR follows these steps:

1. The DTE asserts the DTR pin to request that the DCE connect to the communication line.
2. The DCE asserts the DSR pin to indicate it is connected.
3. DCE unasserts the DSR pin when it is disconnected from the communication line.

The DTR and DSR pins were originally designed to provide an alternative method of hardware handshaking. However, the RTS and CTS pins are usually used in this way and not the DSR and DTR pins. However, you should refer to your device documentation to determine its specific pin behavior.

The CD and RI Pins: The CD and RI pins are typically used to indicate the presence of certain signals during modem–modem connections. CD is used by a modem to signal that it has made a connection with another modem or has detected a carrier tone. CD is asserted when the DCE is receiving a signal of a suitable frequency. CD is unasserted if the DCE is not receiving a suitable signal. RI is used to indicate the presence of an audible ringing signal. RI is asserted when the DCE is receiving a ringing signal. RI is unasserted when the DCE is not receiving a ringing signal (e.g., it is between rings).

Overall Procedure of Communication

Figure 3.15 represents a typical digital transmission system, which will be described here. (Note: This is actually the initialization and handshaking description for a typical UART, the Intel 8251A.)

To start with, it should be mentioned that the signals alongside the arrowheads represent the minimum number of necessary signals for the execution

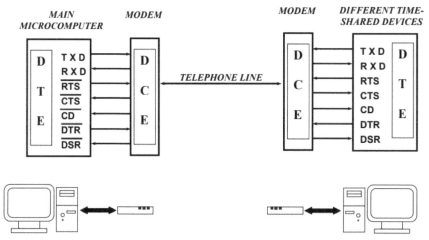

Figure 3.15 A typical digital transmission system.

of a typical communication standard or a protocol, being elaborated later. These signals occur when the main control terminal wants to send some control signal to the end device or if the end device wants to send some data, say an alarm or some process output, to the main controller.

Both the main microcomputer and the end device or the time-shared device can be referred to as terminals.

Whenever a terminal is switched on, it first performs a self-diagnostic test, in which it checks itself and if it finds that its integrity is fully justified it asserts the DTR (data-terminal ready) signal low. As the modem senses it getting low, it understands that the terminal is ready.

The modem then "replies" the terminal by asserting DSR (data-set ready) signal low. Here the direction of the arrows is of prime importance and must be remembered to get the full understandability of the whole procedure.

If the terminal is actually having some valuable data to convey to the end terminal, it will assert the RTS (request-to-send) signal low back to the modem and, in turn, the modem will assert the CD (carrier-detect) signal to the terminal indicating as if now it has justified the connection with the terminal computer.

But it may be possible that the modem may not be fully ready to transmit the actual data to the telephone; this may be because of its buffer saturation and several other reasons. When the modem is fully ready to send the data along the telephone line, it will assert the CTS (Clear-to send) signal back to the terminal.

The terminal then starts sending the serial data to the modem. When the terminal gets exhausted of the data, it asserts the RTS signal low indicating the modem that it has not got any more data to be sent. The modem in turn unasserts its CTS signal and stops transmitting.

The same way initialization and the handshaking processes are executed at the other end. Therefore, it must be noted here that the very important aspect of data communication is the definition of the handshaking signals defined **for transferring serial data to and from the modem.**

3.4.4 Mechanical Interface Characteristics

RS-232 serial connector pin assignment

The third area covered by RS-232 concerns the mechanical interface. In particular, the two connectors that are specified by RS-232 are the 25-pin D-type connector, DB25 connector, and the 9-pin DB9 connector. The DB25 is the minimum connector size that can accommodate all the signals defined in the functional portion of the standard. The pin assignment for this connector is shown in Figure 3.16, and images are shown in Figure 3.17. The connector for DCE equipment is male for the connector housing and female for the connection pins. Likewise, the DTE connector is a female housing with male connection pins. Although RS-232 specifies a 25 position connector, it should be noted that often this connector is not used. This is due to the fact that most applications do not require all of the defined signals and therefore a 25-pin connector is larger than necessary. This being the case, it is very common for other connector types to be used. Perhaps the most popular is the 9 position DB9S connector which is also illustrated in Figure 3.16. Figure 3.16 shows that all the signals needed to let DTE communicate (transmit and receive) with modem are provided by DB9 as well as by DB25. Accordingly, the DB9 connector provides the means to transmit and receive the necessary signals for modem applications.

Considering the DB25 connector configuration and the signal definitions given in Section 3.3.3.1, it is easy to note that the signals cover the two types of lines that are specified in the RS-232 specification: primary channel and secondary channel. The primary channels are normally used and operate at the normal or higher data rates. The secondary channels are for providing control information. If it is used, it will usually send data at a much slower rate than the primary channel. As the secondary lines are rarely used or even implemented on equipment, manufacturers often use these connector pins for other purposes.

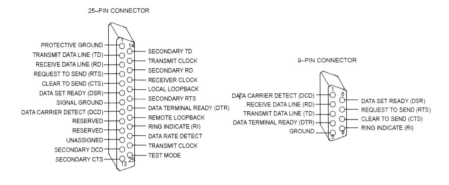

(a)

Pin	Description	Pin	Description
1	Protective ground	14	Secondary Transmitted Data
2	Transmitted Data (TxD)	15	Transmitted signal element time
3	Received Data (RxD)	16	Secondary Receive Data
4	Request to send (RTS)	17	Receive signal element timing
5	Clear to send (CTS)	18	Local Loopback
6	Data Set Ready (DSR)	19	Secondary Receive Data
7	Signal ground (GND)	20	Data Terminal Ready (DTR)
8	Data Carrier Detect (DCC)	21	Signal Quality detector/Remote Loopback
9/10	Reserved for data testing	22	Ring Indicator (RI)
11	Unassigned	23	Data Signal Rate Select
12	Secondary Data Carrier Detect	24	Transmit signal element timing
13	Secondary Clear to Send	25	Test Mode

(b)

Figure 3.16 RS-232 connector pin assignments.

Figure 3.17 Images of D-SUB 25 MALE connector in RS-232-based systems.

In practice, only one serial communication channel with accompanying handshaking is present. Only very few computers have been manufactured where both serial RS-232 channels are implemented. Examples of this are the Sun SparcStation 10 and 20 models and the Dec Alpha Multia. Also, on a number of Telebit modem models, the secondary channel is present. It can be used to query the modem status while the modem is on-line and busy communicating. On personal computers, the smaller DB9 version is more commonly used today.

3.4.4.1 DB9 to DB25 converters and DB25 to DB9 converters

The original pinout for RS-232 was developed for a 25-pin D-sub-connector. Since the introduction of the smaller serial port on the IBM-AT, 9-pin RS-232 connectors are commonly used.

Considering again Figure 3.16, it is easy to note that:

- most of the pins of DB25 are similar to that of a DB9 port.
- in DB25 connector there are **two** TxD and RxD pairs of pins. This means that serial communication through the DB25 connector could take place through two channels simultaneously. Applications that need two channel are limited and the majority of applications need one channel.
- it is so simple to let a microcontroller to communicate with a PC. This takes place through RxD, TxD, and Ground Pins.

The above shows the simplicity of having 9- to 25-pin converters and 25- to 9-pin converters that will help in case of mixed applications.

- **RS-232 DB9 to DB25 converter: DB9 to DB25 adapter**

In mixed applications, a 9- to 25-pin converter can be used to connect connectors of different sizes. Figure 3.18 shows an RS-232 DB9 to DB25 converter.

In this converter, signals on the DB9 DTE side are directly mapped to the DB25 assignments for a DTE device. This configuration can be used to adapt a 9-pin COM connector on the back of a computer to mate with a 25-pin serial DCE device, such as a modem. This adapter may also be in the form of a cable.

- **DB25 to DB9 adapter**

Signals on the DB25 DTE side are directly mapped to the DB9 assignments for a DTE device (Figure 3.19). This configuration is used to adapt a 25-pin COM connector on the back of a computer to mate with a 9-pin serial DCE device, such as a 9-pin serial mouse or modem. This adapter may also be in the form of a cable.

- **DB25 to DB9 adapter (Pin 1 Connected to Shield)**

This adapter has the same wiring as the previous one (Figure 3.18) except that pin 1 is wired to the connector shell (shield). Note that the cable's shield is usually a foil blanket surrounding all conductors running the length of the cable and joining the connector shells. Pin 1 of the EIA232 specification, called out as "shield," may be separate from the earth ground usually associated with the connector shells.

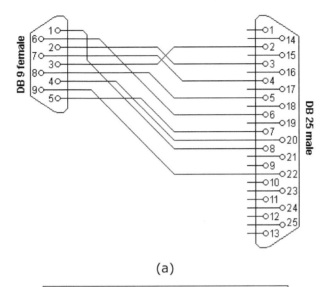

(a)

DB9 - DB25 conversion		
DB9	**DB25**	**Function**
1	8	Data carrier detect
2	3	Receive data
3	2	Transmit data
4	20	Data terminal ready
5	7	Signal ground
6	6	Data set ready
7	4	Request to send
8	5	Clear to send
9	22	Ring indicator

(b)

Figure 3.18   RS-232 DB9 to DB25 converter (adapter).

3.4.4.2 Loopback test plugs

Serial loopback test plugs are used when it is required to test a serial port of a computer. Using the loopback connector, the equipment can be used on its own to check that it can communicate, even if it is only with itself.

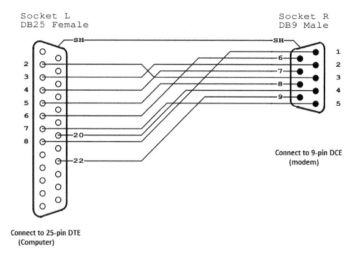

(a)

DB25 – DB9 conversion		
DB25	**DB9**	**Function**
SH	SH	Shield
2	3	Transmit data
3	2	Receive data
4	7	Request to send
5	8	Clear to send
6	6	Data set ready
7	5	Signal ground
8	1	Data carrier detect
20	4	Data Terminal Ready
22	9	Ring indicator

Figure 3.19   DB25 to DB9 adapter.

A loopback test can, for example, verify if the used USB to serial RS-232 adapter has been installed properly and if it can send and receive data as intended. In a troubleshooting situation, this will help determining if a communication problem is caused by the adapter, equipment, or the drivers.

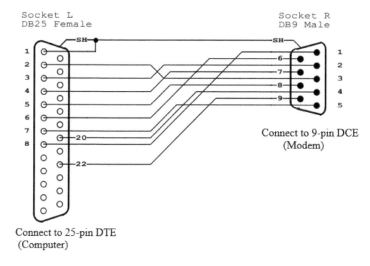

Connect to 25-pin DTE
(Computer)

(a)

DB25 – DB9 conversion		
DB25	**DB9**	**Function**
SH	1 SH	Shield
2	3	Transmit data
3	2	Receive data
4	7	Request to send
5	8	Clear to send
6	6	Data set ready
7	5	Signal ground
8	1	Data carrier detect
20	4	Data Terminal Ready
22	9	Ring indicator

Figure 3.20   DB25 to DB9 adapter with pin 1 connected to shield.

The test loopback connector approach to testing has its limitations as it will not be able to detect speed mismatches or the like as the same issue is likely to be present on transmit and receive. However, it will be able to test the basic functionality and provide a good level of confidence about the ability of the equipment to send and receive and generally check the operation of the serial port.

RS-232 loopback connectors are very easy to make up. They can be made very easily either for DB9 or DB25 way connectors.

The loopback plug connects serial inputs to serial outputs so that the port may be tested. It usually consists of a connector without a cable and includes internal wiring to reroute signals back to the sender. There is more than one way to wire up a serial loopback connector. The connections given below form a commonly used configuration. Although it may not be necessary to incorporate all the handshaking connections for all applications, they have nevertheless been incorporated so that the serial loopback connector should work for all situations.

Next we give three examples of serial loopback test plug, and latter, we introduce many others and a table depicts the pin assignments for most loopback plugs and cables that may be used when testing a system.

Example 3.1: DB9 Serial Loopback Connections
The connections for the serial loopback connector are given in Figure 3.21 and given in Table 3.8 for a DB-9 D-type connector using the common pinout connections.

The DB9 female connector (Figure 3.21) would attach to a DTE device such as a personal computer. When the computer receives data, it will not know whether the signals it receives come from a remote DCE device set to echo characters or from a loopback connector. Use loopback connectors

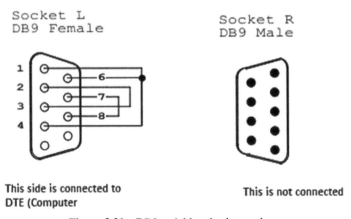

Figure 3.21   DB9 serial loopback test plug.

Table 3.9 DB9 loopback connections

DB9 Connector Serial Loopback Connections			
	Connection 1	Connection 2	Connection 3
Link 1	1	6	4
Link 2	2	3	–
Link 3	7	8	–

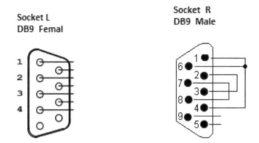

DB9 CONNECTOR SERIAL LOOPBACK CONNECTIONS			
	CONNECTION 1	CONNECTION 2	CONNECTION 3
Link 1	1	6	4
Link 2	2	3	--
Link 3	7	8	--

Figure 3.22 9-Pin male single-port loopback plug wiring diagram.

to confirm proper operation of the computer's serial port. Once confirmed, insert the serial cable you plan to use and attach the loopback to the end of the serial cable to verify the cable.

Example 3.2: 9-Pin Male Single-Port Loopback Plug

Figure 3.22 shows the wiring for male 9-pin RS-232 and RS-423 single-port loopback plugs.

Example 3.3: 9-Pin to 9-Pin Port-to-Port Loopback Cable

Figure 3.23 shows the wiring for 9-pin RS-232 and RS 423 port to 9-pin RS-232 and RS-423 port loopback cables. Both plugs are male.

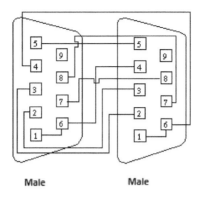

Male   Male

9-pin to 9-pin Port-to-Port Loopback Table	
Connect first connector to:	**Second Connector**
Pins 1 and 6	Pin4
Pin 2	Pin 3
Pin 3	Pin 2
Pin 4	Pins 1 and 6
Pin 5	Pin 5
Pin 7	Pin 8
Pin 8	Pin 7

Figure 3.23   9-Pin to 9-Pin port-to-port loopback cable wiring diagram.

Running the test

Testing occurs in a few steps. Data are sent on the Tx line, and the received information on the Rx input is then compared with the original data. The signal level on the DTR and RTS lines is also controlled by the test software and the attached inputs are read back in the software to see if these signal levels are properly returned. The second RS-232 test plug has the advantage that the ring indicator RI input line can also be tested. This input is used by modems to signal an incoming call to the attached computer.

3.5 Cables

Whenever RS-232 is used, RS-232 cables will be needed to provide the required electrical connection. These cables can take a variety of forms in terms of the physical methods used for construction as well as the number of connections that are incorporated within the overall RS-232 cable.

When constructing a cable, it should be remembered that not all the signals provided within RS-232 standard need to be implemented in a practical

RS-232 data cable. As a result, there are many design varieties for serial data cables. Such design varieties of serial data cables may be seen on the Internet and elsewhere.

In addition to this, further confusion is caused by the fact that RS-232 was designed to connect modems and terminal devices or teleprinters so that telephone lines could be used to transmit data over the telephone system, and serial data cables were required to connect the equipment at the remote end. This intended application for RS-232 gives rise to the terminology associated with these serial data connections, that is, DCE (Data Communications Equipment or Modem) and the DTE (Data Terminal Equipment or teleprinter).

Handshaking is another factor that has to be considered while studying serial cables: Different data systems utilize different levels of handshaking when using RS-232. Many configurations can be used to achieve handshaking.

The above discussion highlights the fact of having large varieties of RS-232 serial cable. This is the subject of this section.

Types of Serial Cables

To make serial communication possible between DTE and DCE, the following types exist:

1. DTE – DCE is called a Straight Cable
2. DTE – DTE is called a Null-Modem Cable
3. DCE – DCE is called a Tail Circuit Cable

The three types are summarized here, and the details of the Null Modem will follow.

3.5.1 Straight-through Cable

The **Straight-through cable**, as the name implies, is a one-to-one connector, that is, a transmit pin of one device is connected to the transmit pin of another device, and a receiver pin of one device is connected to the receiver pin of another device. Apart from connections, the cable length depends upon the wiring capacitance. As per specification, the cable length is nearly 80 ft.

Example 3.4: Three wires straight cable

The simplest form of straight cable is shown in Figure 3.24 which can be used just to send the data without any hard-wired handshaking. Three wires

Straight Cable Connection

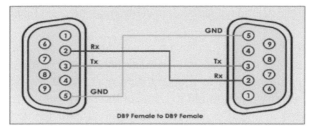

Figure 3.24   Three wires straight cable connection.

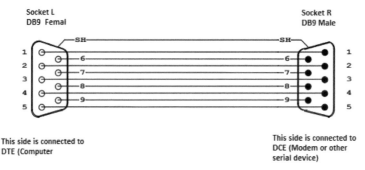

Figure 3.25   DB9 extension straight cable.

Tx, Tr, and GND are used to complete the communication. The transmit pins of the two devices are connected together exactly as the receiving pins. The GRD pins are connected, also, together.

Example 3.5: DB9 Extension: All-line Direct Extension

Figure 3.25 shows a 9-pin DTE-to-DCE serial cable that would result if the RS-232 (EIA232) standard was strictly followed. All 9 pins plus shield are directly extended from DB9 Female to DB9 Male. There are no crossovers or self-connects present. This cable is in use to connect modems, printers, or any device that uses a DB9 connector to a PC's serial port.

This cable may also serve as an extension cable to increase the distance between a computer and serial device. This is limited by the 25 ft separation between devices. If the distance is more than 25 ft, a signal booster can be used.

Example 3.6: DB25 All-line Direct Extension

Figure 3.26 shows a 25-pin DTE-to-DCE serial cable that would result if the EIA232 standard was strictly followed. All 25 pins plus shield are directly

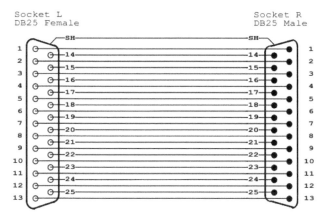

Figure 3.26 25-pin extension straight cable.

extended from DB25 Female to DB25 Male. There are no crossovers or self-connects present. This cable can be used to connect modems, printers, or any serial device that uses a DB25 connector to a PC's serial port.

This cable may also serve as an extension cable to increase the distance between computer and serial device. This can be used up to 25 ft separation between devices. In case if the separation exceeds 25 ft, booster is needed.

Caution: The male end of this cable (right) also fits a PC's parallel printer port. Accordingly, this cable may be used to extend the length of a printer cable, but the user must avoid attaching a serial device to the computer's parallel port. Doing so may cause damage to both devices.

3.5.2 Tail-circuit Cable

We are giving here some examples of tail circuit cable. The pins are connected as given in Figure 3.27.

3.5.3 RS-232 Null Modem Cable

3.5.3.1 Null modem: an introduction

As mentioned before, serial communication with RS-232 is one of the oldest and most widely spread communication methods in the computer world. The way this type of communication can be performed is pretty well defined in standards.

The RS-232 standards show the use of DTE/DCE communication, the way a computer should communicate with a peripheral device like a modem.

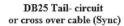

DB25 Tail- circuit
or cross over cable (Sync)

Pin	Pin
1	1
2	3
3	2
4	8
6	20
7	7
8	4
17	24
20	6
24	17

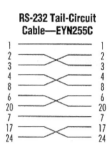

RS-232 Tail-Circuit
Cable—EYN255C

Cross Pinned Cables for Asynchronous Data Transfer

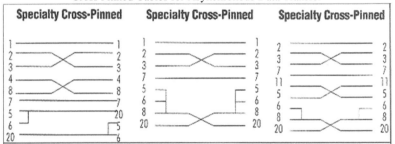

Figure 3.27 Examples of tail-circuit cables.

Figure 3.28 shows how two computers are communicating using DTE/DCE. Figure 3.29 gives more details and the names of the different devices.

As can be seen in the diagram, the PC is the DTE and the modem is the DCE. Communication between each PC and its associated modem is accomplished using the RS-232 standard. Communication between the two modems is accomplished via telecommunication.

It should be noted that although a microcomputer is usually the DTE in RS-232 applications, this is not mandatory according to a strict interpretation of the standard.

Many applications that use serial communication today do not involve modem: null modem communication. A serial null modem configuration with DTE/DTE communication is not so well defined, especially when it comes to flow control. The terminology null modem for the situation where two computers communicate directly is so often used nowadays that most

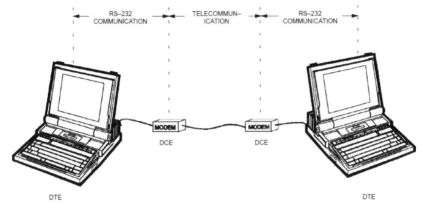

Figure 3.28 Modem communication between two PCs.

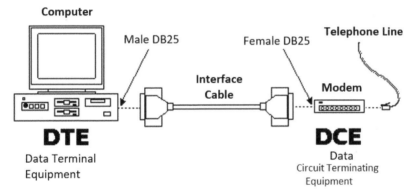

Figure 3.29 Names of devices.

people don't realize anymore the origin of the phrase and that a null modem connection is an exception, not the rule.

In history, practical solutions were developed to let two computers talk with each other using a null modem serial communication line. In most situations, the original modem signal lines are reused to perform some sort of handshaking. Handshaking can increase the maximum allowed communication speed because it gives the computers the ability to control the flow of information. High amounts of incoming data are allowed if the computer is capable to handle it, but not if it is busy performing other tasks. If no flow control is implemented in the null modem connection, communication is only possible at speeds at which it is sure the receiving side can handle the amount information even under worst-case conditions.

3.5.3.2 Original Use of RS-232

The two pins RTS (Request to Send) and CTS (Clear to Send) found in all connectors are used for flow control. With DTE/DCE communication (i.e., a computer communicating with a modem device), RTS is an output on the DTE and input on the DCE. CTS is the answering signal coming from the DCE.

Before sending a character, the DTE asks permission by setting its RTS output. No information will be sent until the DCE grants permission using the CTS line. If the DCE cannot handle new requests, the CTS signal will go low, a simple but useful mechanism allowing flow control in one direction. The assumption is that the DTE can always handle incoming information faster than the DCE can send it. In the past, this was true. Modem speeds of 300 baud were common and 1200 baud was seen as a high-speed connection.

For further control of the information flow, both devices have the ability to signal their status to the other side. For this purpose, the DTR data terminal ready and DSR data set ready signals are present. The DTE uses the DTR signal to signal that it is ready to accept information, whereas the DCE uses the DSR signal for the same purpose. Using these signals involves not a small protocol of requesting and answering as with the RTS/CTS handshaking. These signals are in one direction only.

The last flow control signal present in DTE/DCE communication is the CD carrier detect. It is not used directly for flow control, but mainly an indication of the ability of the modem device to communicate with its counterpart. This signal indicates the existence of a communication link between two modem devices.

RS-232 null modem cables

As mentioned above, the easiest way to connect two PCs is using an RS-232 null modem cable. The only problem in this solution is the large variety of RS-232 null modem cables available. For simple connections, a three-line RS-232 cable connecting the signal ground and receive and transmit lines is sufficient. Depending of the software used, some sort of handshaking may however be necessary. It is possible, as a way of selecting the proper null modem, to use one or another of the available RS-232 null modem selection tables. Table 3.10 is an example of the RS-232 null modem selection table. As an example of using the selection table, for a Windows 95/98/ME Direct Cable Connection, the RS-232 null modem cable with loop back handshaking is a good choice.

Table 3.10 Null modem layout selection table (source: https://www.lammertbies.nl/comm/info/RS-232_null_modem.html#conc)

	Cable Without Handshaking	Loopback Handshaking	Partial Handshaking	Full Handshaking
Software flow control only	+ + +	+ +	+	+
DTE/DCE compatiblehardware flow controlat low speeds	–	+ + +	+ +	–
DTE/DCE compatible hardware flow control at high speeds	–	+	+ + +	–
High speed communication using special software	–	–	+ +	+ + +

+ + + Recommended cable; + + Good alternative; + Works, but not recommended; – Does not work.

RS-232 null modem cables with handshaking can be defined in numerous ways, with loopback handshaking to each PC or complete handshaking between the two systems. Before discussing null modem cables with handshaking, we introduce at first the principle of RS-232 handshaking.

Comments related to the use of the table:

- The right null modem cable to choose mainly depends on the application and the software that will be used. As a general guide line, I would advise to follow as Table 3.10.
- As general guide line, the null modem cable with partial handshaking works in most cases. If you are developing software which must work with all kinds of cables, it is best to use software flow control only and ignore all modem control inputs.

3.6 RS-232 Handshaking

The method used by RS-232 for communication allows for a simple connection of three lines: TX, RX, and ground. For the data to be transmitted, both sides have to be clocking the data at the same baud rate. Although this method is sufficient for most applications, it is limited in being able to respond to problems such as the receiver getting overloaded. This is where

RS232 handshaking signals

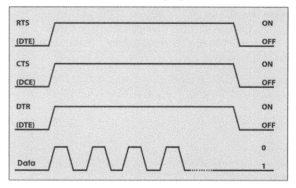

Figure 3.30   RS-232 handshaking signals.

serial handshaking (data flow control) can help. To start the data flow, the handshaking (the control signals) needs to indicate that data are ready to be sent and received across equipment. In other words, the equipment at either end of the link is ready to send the data and ready to receive the data.

In general, it is possible to identify two uses for handshaking:

1. To start the data flow
2. To guarantee no loss of data while communicating

A. How handshaking works to start data flow?

The handshaking exchange to start the data flow is quite straightforward and can be seen as a number of distinct stages (Figure 3.30):

1. The data terminal equipment (DTE) puts the RTS line into the "On" state. In this state, no data is transmitted.
2. The data communication equipment (DCE) puts the CTS line into the "On" state.
3. The DTE puts the DTR line into the "On" state.
4. The DTR line remains in the "On" state while data are being transmitted.

At the end of the transmission, DTR and RTS are pulled to the OFF state and then the DCE pulls the CTS line to the OFF state. This series of handshake controls was devised to allow the DTE to request control of the communication link from the related modem and then to let the modem inform the terminal equipment that the control has been acquired. In this way, the communication will only take place when both ends of the link are ready.

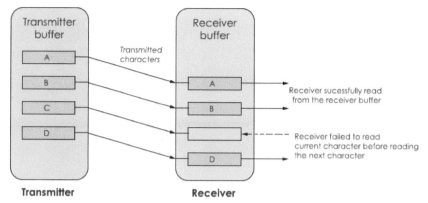

Figure 3.31 Data flow.

B. Use of handshaking to guarantee no loss in data

To send and receive the information without loss of data, it is necessary to maintain robust communication between the transmitter and receiver. To do that, buffer is used. Buffer is a temporary storage location which allows the transmitter and receiver to store the data until the information is processed by each other at different speeds.

In Figure 3.31, the transmitter and receiver have their own buffer. The transmit buffer holds the characters to be sent to the receiver, while the receive buffer holds the characters received from the transmitter. If the transmitter sends data at a higher speed, the receiver may fail to receive. In this case, character "C" is missed by the receiver. To avoid this, handshaking is used. Handshaking allows the transmitter and receiver device to agree before the communication is going to start.

As will be discussed in Section 3.6.1.2, the connectors DB9 and DB25 are used for handshaking purpose, see Figure 3.32. When no handshaking is performed, only the TxD (Transmitter) and RxD are cross-coupled. Other pins, RTS, CTS, DSR, and DTR are connected in loopback fashion.

To use the handshaking technique, RTS and CTS are cross-coupled. Also, DTR and DSR are also connected in cross mode.

Important note

Before discussing handshaking in some detail, it is important to note that:

- Handshake lines ensure a computer won't transmit data if the receiving computer is not ready.

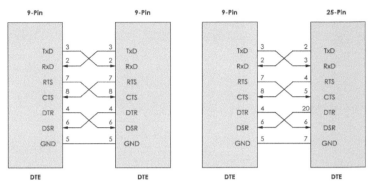

Figure 3.32   Handshaking (hardware handshaking).

- It is crucial to note that you can tie handshake lines to fixed voltages. This ensures the handshake lines will remain operational consistently.
- Furthermore, RS-232 handshaking is not always necessary. And even though your RS-232 monitoring hardware may be equipped with inputs and outputs that can be used for handshaking, there is no guarantee that handshaking is needed to optimize the value of your hardware.
- Determining whether the application needs RS-232 handshaking may be difficult. It needs carful study from the designer to decide about that.

3.6.1 Forms of RS-232 Handshaking: RS-232 Flow Control

There are three popular forms of handshaking with RS-232: Software Handshaking, Hardware Handshaking, and XModem. In the following, the three forms are discussed.

3.6.1.1 Software Handshaking

Both software flow control and hardware flow control need software to perform the handshaking task. This makes the term software flow control somewhat misleading. What is meant is that with hardware flow control, additional lines are present in the communication cable which signal handshaking conditions. With software flow control, which can be either XON-XOFF flow control or EXT/ACK flow control, bytes are sent to the sender using the standard communication lines.

Using hardware flow control implies that more lines must be present between the sender and the receiver, leading to a thicker and more expensive cable. Therefore, software flow control is a good alternative if it is not needed to gain maximum performance in communications. Software flow control makes use of the data channel between the two devices which reduces the bandwidth. The reduction in bandwidth is in most cases, however, not so astonishing that it is a reason to not use it.

When implementing RS-232 software handshaking, there are, as mentioned before, two commonly used methods: XON/XOFF and EXT/ACK.

The lines necessary are still the simple three line set of TX, RX, and ground since the control characters are sent over the transmission line like regular data. There is a function, for example, the function SetXMode allows the user to enable or disable the use of two control characters, XON and XOFF. These characters are sent by the receiver of the data to pause the transmitter during communication.

a. Software Handshaking Using XON-XOFF

In this case, the two bytes that have been predefined in the ASCII character set to be used with software flow control are named XOFF and XON, because they can stop and restart transmitting. The byte value of XOFF is 19 or hex 13; it can be simulated by pressing Ctrl-S on the keyboard. XON has the value 17 or hex 11 assigned which is equivalent to Ctrl-Q.

The XON/XOFF operates by sending the control characters along the data line from the equipment receiving the data. The "X" in the name means "transmitter," so XON and XOFF are commands for switching a transmitter on or off, respectively. When XON is received at the transmitting end, data transmission starts.

Using software flow control is easy. If sending of characters must be postponed, the character XOFF is sent on the line, and to restart the communication again, XON is used. Sending the XOFF character only stops the communication in the direction of the device which issued the XOFF.

As an example, assume that the transmitter begins to transmit data at a high baud rate. During the transmission, the receiver finds that the input buffer is becoming full due to the CPU being busy with other duties. To temporarily pause the transmission, the receiver sends XOFF until the input buffer has been emptied. Once the receiver is ready for more data, it sends XON to resume communication. LabWindows will send XOFF when its input buffer becomes half full. In addition, in case the XOFF transmission was corrupted, LabWindows will also transmit XOFF when the buffer has reached 75% and 90% capacity. Obviously, the transmitter must also be following this protocol for it to succeed.

This method has a few disadvantages. The biggest drawback to this method is also the most important fact to keep in mind: decimal 17 and 19 are now off limits for data values. In ASCII transmissions, this typically does not matter since these values are non-character values; however, if the data are being transmitted via binary, it is very likely that these values could be transmitted as data and the transmission would fail.

Another disadvantage is overhead cost: using bytes on the communication channel takes up some bandwidth. One other reason is more severe. Handshaking is mostly used to prevent an overrun of the receiver buffer, the buffer in memory used to store the recently received bytes. If an overrun occurs, this affects the way new coming characters on the communication channel are handled. In the worst case where software has been designed badly, these characters are thrown away without checking them. If such a character is XOFF or XON, the flow of communication can be severely damaged. The sender will continuously supply new information if the XOFF is lost, or never send new information if no XON was received.

This also holds for communication lines where signal quality is bad. What happens if the XOFF or XON message is not received clearly because of noise on the line? Special precaution is also necessary that the information sent does not contain the XON or XOFF characters as information bytes.

Therefore, serial communication using software flow control is only acceptable when communication speeds are not too high, and the probability that buffer overruns or data damage occur are minimal.

b. Software Handshaking Using EXT/ACK

A second method is called EXT/ACK. Using this method, the data are separated into blocks, and after each block has been sent, the control code ETX is transmitted to show the end of this block of text.

Once the data have been accepted, and there is sufficient space in the input buffer, the ACK or acknowledgement control code is sent. Once this has been received, the next block of data is sent.

Other codes

Other codes are also used within what is effectively software control of an RS-232 communication link. These codes are used for a variety of purposes from indicating tabs, form feeds as well as providing audible "bell" warnings. These codes are naturally focused on some of the older teletypes that used two wire RS-232 links.

Although not all of the codes are used these days because many are intended for use by the older teletype equipment, the same processes are valid for modern equipment. This set of codes are known as control codes, which even gives rise to the "CTRL" key on today's keyboards. Additionally, the idea of software handshaking, although it was not so widely used in the early days, is now well established and is usually the preferred method of implementing control of communications systems in view of the flexibility offered by processors.

3.6.1.2 Hardware handshaking: hardware flow control

The second method of handshaking is to use actual hardware lines. Hardware flow control is superior compared to software flow control using the XON and XOFF characters. The main problem is that an extra investment is needed. Extra lines are necessary in the communication cable to carry the handshaking information. To explain the meaning of extra lines, we must remember that the method used by RS-232 for communication allows for a simple connection of three lines: TX, RX, and ground. Any line above that is an extra line.

Hardware flow control is sometimes referred to as RTS/CTS flow control. This term mentions the extra input and output used on the serial device to perform this type of handshaking. RTS/CTS in its original outlook is used for handshaking between a computer and a device connected to it such as a modem.

First, the computer sets its RTS (Request to Send) line to signal the device that some information is present. The device checks if there is room to receive the information, and if so, it sets the CTS (Clear to Send) line to start the transfer. When using a null modem connection, this is somewhat different. There are two ways to handle this type of handshaking in that situation.

The first one is where the RTS of each side is connected with the CTS side of the other. In that way, the communication protocol differs somewhat from the original one. The RTS output of computer A signals computer B that A is capable of receiving information, rather than a request for sending information as in the original configuration. This type of communication can be performed with a null modem cable for full handshaking. Although using this cable is not completely compatible with the original way hardware flow control was designed, if software is properly designed for it, it can achieve the highest possible speed because no overhead is present for requesting on the RTS line and answering on the CTS line.

In the second situation of null modem communication with hardware flow control, the software side looks quite similar to the original use of the handshaking lines. The CTS and RTS lines of one device are connected directly to each other. This means that the request to send query answers itself. As soon as the RTS output is set, the CTS input will detect a high logical value indicating that sending of information is allowed. This implies that information will always be sent as soon as sending is requested by a device if no further checking is present. To prevent this from happening, two other pins on the connector are used, the data set ready DSR and the data terminal ready DTR. These two lines indicate if the device attached is working properly and willing to accept data. When these lines are cross-connected (as in most null modem cables), flow control can be performed using these lines. A DTR output is set, if that computer accepts incoming characters. Handshaking using the four extra lines,CTS, RTS, DSR, and DTR, besides the receive line Rx, the transmit line Tx and the ground is shown in Figure 3.32 before. Sometimes this configuration is called "six signal lines" handshake (Rx, Tx, CTS, RTS, DSR, and DTR).

Eight Signal Lines Handshaking

In the latest version of RS-232 E standard, the handshaking is redefined where CTS (clear to send) is no longer a response to RTS but it indicates permission from DCE for the DTE devices. In the similar way, RTS indicates permission from the DTE for the DCE to send data. RTS and CTS are controlled by DTE and DCE and independent of each other. Table 3.11 gives the signals in use in this case.

To give more details on how handshaking takes place, the handshaking system with eight signal lines is explained (Figure 3.33).

When Data Carrier Detect is off, it indicates to the local terminal that remote DTE has not switched on its RTS, and local terminal can gain control over the line. When this circuit is on locally, it indicates to the local terminal

Table 3.11 Signals in use in case of eight lines handshaking

Pin number	Description
1	Protective Ground
2	Transmit Data (TxD)
3	Receive Data (RxD)
4	Request to send (RTS)
5	Clear to Send (CTS)
6	Data Set Ready (DSR)
8	Data Carrier Detect (DCD)
20	Data Terminal Ready (DTR)
7	Signal Ground (GND)

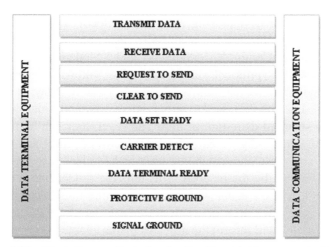

Figure 3.33 Block diagram displaying details of handshaking system in RS-232.

that remote modem has received a RTS ON condition from its terminal and remote DTE is in control over the carrier line. RXD stands for receive data from modem to DTE. TXD transmits data from DTE to modem. The DTR data terminal ready pin is generally on when terminal is ready to establish communication channel through its modem. But when DTR does not want to accept calls from remote terminal, the circuit is off. Both the modems switch on their Data Set Ready circuit on when communication path has been established between two sites. Now, when the terminal is ready to transmit, it switches the Request to Send circuit on indicating local modem that it is ready to send data. This request gets passed on to remote modem. The RTS controls the direction of data transmission. Once terminal is ready to transmit local modem switches on the CTS circuit to indicate that it is ready to receive the data from DTE. It also gains control over the telephone line. Next when the modem receives the call, the Ring Indicator switches on/off informing DTE that a call is coming to indicate remote modem is requesting dial-up. This is a simple handshaking system with eight signal lines.

3.6.1.3 XModem handshaking

The last mode discussed here is the XModem file transfer protocol. This protocol is very common in modem communication. Although it is often used for modem communication, the XModem protocol can be used directly between other devices if they both follow the protocol. In LabWindows, the actual implementation of XModem is hidden from the user. As long as the PC is connected to another device using XModem protocol, the LabWindows' XModem functions can be used to transfer files from one site to another. The functions are XModemConfig, XModemSend, and XModemReceive.

XModem uses a protocol based on the following parameters: start_of_data, end_of_trans, neg_ack, ack, wait_delay, start_delay, max_tries, packet_size. These parameters need to be agreed upon by both sides. Standard XModem has a standard definition of these; however, they can be modified through the XModemConfig function in LabWindows to meet any require-ment. The way that these parameters are used in XModem is by having the neg_ack character sent by the receiver. This tells the sender that it is ready to receive data. It will try again with start_delay time in-between each try until either it reaches max_tries or receives start_of_data from the sender. If it reaches max_tries it will inform the user that it was unable to communicate with the sender. If it does receive start_of_data from the sender, it will read the packet of information that follows. This packet contains the packet number, the complement of the packet number as an error check, the actual data packet of packet_size bytes, and a checksum on the data for more error checking. After reading the data, the receiver will call wait_delay and then send ack back to the sender. If the sender does not receive ack, it will resend the data packet max_tries or until it receives ack. If it never receives the ack, it informs the user that it has failed to transfer the file.

Note: Since the data must be sent in packets of packet_size bytes, when the last packet is sent, if there is not enough data to fill the packet, the data packet is padded with ASCII NUL (0) bytes. This can cause the received file to be larger than the original. It is also important to remember not to use XON/XOFF with the XModem protocol since the packet number from the XModem transfer is very likely to increment to the XON/OFF control character values, which would cause a breakdown in communication.

3.6.1.4 Null modems and handshaking
There are many null modems with handshaking. In this section, we introduce six examples, four with DB9 connectors and 2 with DB25 connectors. More are given in the following examples.

Example 3.7: Null modem without handshaking
The simplest null modem is that without handshaking. In this configuration (Figure 3.34), only the data lines (Tx and Rx) and signal ground are cross connected, and no handshaking lines are used. Also all other pins have no connection.

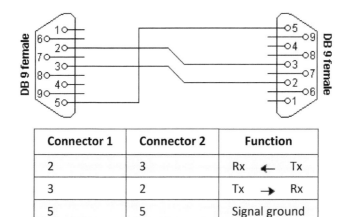

Connector 1	Connector 2	Function
2	3	Rx ← Tx
3	2	Tx → Rx
5	5	Signal ground

Figure 3.34 Simple null modem without handshaking.

In all null modems, the three wires Tx, Rx, and ground are in use. Using the three wires may arise some problems in some circumstances: if either of the two devices checks the DSR or CD inputs. These signals normally define the ability of the other side to communicate. As they are not connected, their signal level will never go high. This might cause a problem.

The same holds for the RTS/CTS handshaking sequence. If the software on both sides is well structured, the RTS output is set high and then a waiting cycle is started until a ready signal is received on the CTS line. This causes the software to hang because no physical connection is present to either CTS line to make this possible. The only type of communication which is allowed on such a null modem line is data-only traffic on the cross-connected Rx/Tx lines.

This limits the use of this null modem cable. For example, this null modem cable can be used when communicating with devices which do not have modem control signals like electronic measuring equipment, etc. Communication links like those present in the Norton Commander program can also use this null modem cable.

It is important to note that with this simple null modem cable, no hardware flow control can be implemented. The only way to perform flow control is with software flow control using the XOFF and XON characters.

Example 3.8: Null modem with loop back handshaking

The simple null modem cable without handshaking shows incompatibilities with common software. The main problem with this cable is that there is a

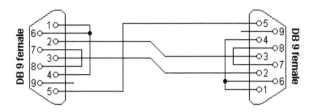

Connector 1	Connector 2	Function
2	3	Rx ← Tx
3	2	Tx → Rx
5	5	Signal ground
1 + 4 + 6	-	DTR → CD + DSR
-	1 + 4 + 6	DTR → CD + DSR
7 + 8	-	RTS → CTS
-	7 + 8	RTS → CTS

Figure 3.35 Null modem with loopback handshaking.

possibility for the software to hang if it checks the modem signal lines in a proper way. With this null modem cable, good written programs will perform worse than badly written programs.

To overcome this problem and still be able to use a cheap null modem communication cable with only three lines in it, a fake null modem cable layout has been defined. The null modem cable with loopback handshaking resulted from this, see Figure 3.35.

The main purpose of this null modem cable is to let well-defined software think there is handshaking available, with a null modem cable which has no provisions for it.

Compatibility issues

Consider first the DSR signal (pin 6). This input indicates that the other side is ready to start communicating. In the layout, the line is linked back to the DTR output (pin 4). This means that the software doesn't see the ready signal of the other device, but its own. The same holds for the CD input (pin 1). The assumption is that if software has been written to check the DSR line to test communication availability, it will probably also set the DTR output to indicate its own state. This is true for at least 99% of all serial communication software. This implies that at least 99% of all serial communication software is capable of faking its own DSR check with this null modem cable.

The same trick is used with the CTS input. In the original use, RTS is set, and then CTS is checked before starting the communication. By setting the RTS output (pin 7), the CTS input on the same connector (pin 8) is receiving clearance immediately. There is no possibility of a software hangup because of dangling RTS requests.

Other issues to consider

The null modem cable with loopback handshaking is often advised as the best low-cost available null modem cable. It is better at this stage to compare this cable with the simple null modem cable discussed in Example 3.4. The simple null modem cable without handshaking has the disadvantage that it does not permit proper written software to communicate with it. Software which is aware of the lack of handshaking signals can however use it without problems.

On the other hand, the null modem cable with loop back handshaking can be used with more software, but it has no functional enhancements over the simple cable! There is no way both devices can control data flow, other than by using XON/XOFF handshaking. If the software is designed for using hardware flow control, it seems to work with this null modem cable, but on unpredictable moments, data loss may occur. This means that the null modem cable allows communication as long as no flow control is needed, but when data speeds reach the limit the receivers can handle, communication may stop immediately without an assignable reason. Therefore, although this null modem cable is cheap and easy to make, use it with care! Despite these warnings, this type of null modem cable has been used successfully between Windows 95/98/ME computers with a Direct Cable Connection.

Example 3.9: Null modem with partial handshaking

The simple null modem cable and the null modem cable with loopback handshaking are useful, but have no provisions for hardware flow control. If it is absolutely necessary that hardware flow control is used, the null modem with partial handshaking can be an alternative (Figure 3.36).

Compatibility issues

This null modem cable is the best of two worlds. There is the possibility of hardware flow control without being incompatible with the original way flow

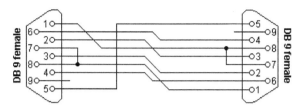

Connector 1	Connector 2	Function
1	7 + 8	$RTS_2 \rightarrow CTS_2 + CD_1$
2	3	Rx $\leftarrow$ Tx
3	2	Tx $\rightarrow$ Rx
4	6	DTR $\rightarrow$ DSR
5	5	Signal ground
6	4	DSR $\leftarrow$ DTR
7 + 8	1	$RTS_1 \rightarrow CTS_1 + CD_2$

Figure 3.36   Null modem with partial handshaking.

control was used with DTE/DCE communication. Let us first consider the RTS/CTS flow control lines present on pins 7 and 8. As with the loopback null modem cable, these signals are not connected to the other device, but directly looped back on the same connector. This means that RTS/CTS flow control is allowed to be used in the software, but it has no functional meaning. Only when the software at the other side checks the CD signal at pin 1, the RTS information will reach the other device. This would however be only the case in specifically developed software which uses the CD input for this purpose.

More important, however, is the cross-connection of the DSR (pin 6) and DTR (pin 4) lines. By cross connecting these lines, their original function is simulated pretty well. The DTR output is used to signal the other device that communication is possible. This information is read on the DSR input, the same input used for this purpose with modem communication. Because of this cross connection, the DTR output line can be used for simple flow control. Incoming data are allowed when the output is set, and blocked if the output is not set.

Software using only the RTS/CTS protocol for flow control cannot take advantage of the partial handshaking null modem cable. Most software, however, will also check the DSR line and in that case – when using the

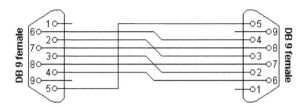

Connector 1	Connector 2	Function
2	3	Rx ← Tx
3	2	Tx → Rx
4	6	DTR → DSR
5	5	Signal ground
6	4	DSR ← DTR
7	8	RTS → CTS
8	7	CTS ← RTS

Figure 3.37   Null modem with full handshaking.

null modem cable with partial handshaking – the best possible hardware flow control can be achieved which is still compatible with the original use with modems.

Example 3.10: Null modem with full handshaking

The most expensive null modem cable is the null modem cable suitable for full handshaking. In this null modem cable, seven wires are present. Only the ring indicator RI and carrier detect CD signal are not linked. The cable is shown in Figure 3.37.

Compatibility issues

The null modem cable with full handshaking does not permit the older way of flow control to take place. The main incompatibility is the cross-connection of the RTS and CTS pins. Originally, these pins are used for a question/answer type of flow control. When the full handshaking null modem cable is used, there is no request anymore. The lines are purely used for telling the other side if communication is possible.

The main advantage of this cable is that there are two signaling lines in each direction. Both RTS and DTR outputs can be used to send flow control information to the other device. This makes it possible to achieve very high

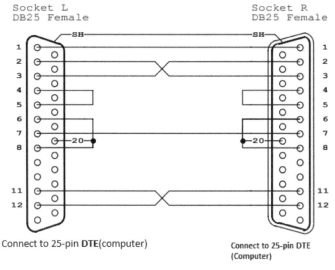

Connect to 25-pin **DTE**(computer)

Connect to 25-pin DTE (Computer)

Connector 1	Connector 2	Function		
Sh	Sh			
1	1			
2	3	Rx	←	Tx
3	2	Tx	→	Rx
4 + 5	-	DTR	→	DSR
-	4+5			
6+8+20L	-	DSR	←	DTR
7	7	RTS	→	RTS
11	12	CTS	←	RTS
12	11	RTS → CTS		
-	6+8+20			

Figure 3.38 DB25 null modem.

communication speeds with this type of null modem cable, provided that the software has been designed for it. Because of the high possible connection speed, this null modem cable can be used with Interlink to connect two MS-DOS PCs.

This is the type of cable Microsoft recommends for the direct cable connection in their knowledge base article. For the DB9 connector, they also added a connection of DTR to CD on each connector but they didn't define this connection for the DB25 connector version and they also didn't

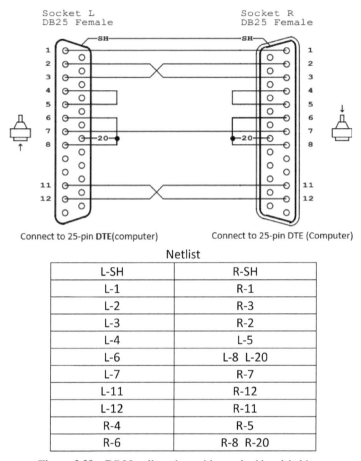

Figure 3.39 DB25 null modem with standard handshaking.

mention the CD input in the descriptive text, so it is safe to leave the CD input disconnected.

Example 3.11: DB25 Null Modem (no handshaking)

This female-to-female null modem cable can be used in any application where it is needed to connect two DTE devices (e.g., two computers). A male-to-male equivalent of this cable would be used to connect two DCE devices.

Note that Pins 11 and 12 are not necessary for this null modem cable to work. As is often the case, the manufacturer of equipment that uses this cable had a proprietary application in mind.

This cable employs NO handshaking lines between devices. The handshake signals on each side are artificially made to appear asserted by the use of self-connects on each side of the cable (e.g., between pins 4 and 5). Without hardware handshaking, you risk buffer overflow at one or both ends of the transmission unless STX and ETX commands are inserted in the dataflow by software.

Example 3.12: DB25 Null Modem (standard handshaking)

This female-to-female cable can be used in any application where it is needed to connect two DTE devices (e.g., two computers). A male-to-male equivalent of this cable would be used to connect two DCE devices.

The cable shown in Figure 3.39 is intended for EIA232 asynchronous communications (most PC-based systems). If you are using synchronous communications, the null modem will have additional connections for timing signals not shown here.

NOTE: Not all null modem cables connect handshaking lines the same way. Refer to the manual for your equipment if you experience problems. In this cable, the DTE Ready (pin 20) on one side asserts the DCE Ready (pin 6) and the Request to Send (pin 5) on the other side.

4

RS-232 Implementation, Applications, and Limitations

4.1 Introduction

This chapter continues the discussion about RS-232. The chapter introduces the following topics:

- Practical RS-232 Implementation
- RS-232 Applications
- RS-232 Application Limitations
- Advantages and Disadvantages of RS-232
- Difference between RS-232 and UART

4.2 Practical RS-232 Implementation

4.2.1 Voltage Level Conversion

RS-232, as mentioned in Chapter 2, is an interface between UART/USART and external devices which are normally ICs using TTL and LVTTL logic families. Such logic families do not operate using RS-232 voltage levels. RS-232, as interface between UART and external ICs, must guarantee the compatibility between voltages on the two sides. In many cases, this matter needs to use "level conversion" circuits. Level conversion is performed by special RS-232 ICs. These ICs typically have line drivers that generate the voltage levels required by RS-232 and line receivers that can receive RS-232 voltage levels without being damaged. These line drivers and receivers typically invert the signal as well since logic 1 is represented by a low voltage level for RS-232 communication and likewise logic 0 is represented by a high voltage level. Figure 4.1 illustrates the function of an RS-232 line driver/receiver in a typical modem application. In this particular example, the

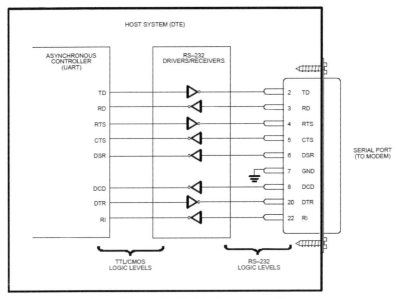

Figure 4.1   Typical RS-232 modem application.

signals necessary for serial communication are generated and received by the Universal Asynchronous Receiver/Transmitter (UART).

The RS-232 line driver/receiver IC performs the level translation necessary between the CMOS/TTL and RS-232 interface.

The UART just mentioned performs the "overhead" tasks necessary for asynchronous serial communication.

For example, the asynchronous nature of this type of communication usually requires that start and stop bits be initiated by the host system to indicate to the peripheral system when communication will start and stop. Parity bits are also often employed to ensure that the data sent have not been corrupted. The UART usually generates the start, stop, and parity bits when transmitting data and can detect communication errors upon receiving data. The UART also functions as the intermediary between byte-wide (parallel) and bit-wide (serial) communication; it converts a byte of data into a serial bit stream for transmitting and converts a serial bit stream into a byte of data when receiving.

Now that an elementary explanation of the TTL/CMOS to RS-232 interface has been provided we can consider some "real world" RS-232 applications. It has already been noted that RS-232 applications rarely follow the RS-232 standard precisely. Perhaps the most significant reason this is true is due to the fact that many of the defined signals are not necessary for most applications.

As such, the unnecessary signals are omitted. Many applications, such as a modem, require only nine signals (two data signals, six control signals, and ground).

Other applications may require only five signals (two for data, two for handshaking, and ground), while others may require only data signals with no handshake control.

We will begin our investigation of "real-world" implementations by first considering the typical modem application.

4.2.2 RS-232 in Modem Applications

Modem applications are one of the most popular uses for the RS-232 standard. Figure 3.28, given before, illustrates a typical modem application utilizing the RS-232 interface standard.

As can be seen in the figure, the PC is the DTE and the modem is the DCE. Communication between each PC and its associated modem is accomplished using the RS-232 standard. Communication between the two modems is accomplished via telecommunication.

It should be noted that although a microcomputer is usually the DTE in RS-232 applications, this is not mandatory according to a strict interpretation of the standard.

Many modem applications require only nine signals (including ground). Although some designers choose to use a 25-pin connector, it is not necessary since there are only nine interface signals between the DTE and DCE. With this in mind, many have chosen to use 9- or 15-pin connectors (see Figure 3.14 for 9-pin connector pin assignment). The "basic nine" signals used in modem communication are illustrated in Figure 3.16. Note that with respect to the DTE, three RS-232 drivers and five receivers are necessary. The functionality of these signals is described below. Note that for the following signal descriptions, "ON" refers to a high RS-232 voltage level (+5 to +15 V) and "OFF" refers to a low RS-232 voltage level (−5 to −15 V). Keep in mind that a high RS-232 voltage level actually represents logic 0 and a low RS-232 voltage level refers to logic 1.

Transmitted Data (TD): One of the two separate data signals. This signal is generated by the DTE and received by the DCE.

Received Data (RD): One of the two separate data signals. This signal is generated by the DCE and received by the DTE.

Request to Send (RTS): When the host system (DTE) is ready to transmit data to the peripheral system (DCE), RTS is turned ON. In simplex and duplex systems, this condition maintains the DCE in receive mode. In half-duplex systems, this condition maintains the DCE in receive mode and disables transmit mode. The OFF condition maintains the DCE in transmit mode. After RTS is asserted, the DCE must assert CTS before communication can commence.

Clear to Send (CTS): CTS is used along with RTS to provide handshaking between the DTE and the DCE. After the DCE sees an asserted RTS, it turns CTS ON when it is ready to begin communication.

Data Set Ready (DSR): This signal is turned on by the DCE to indicate that it is connected to the telecommunication line.

Data Carrier Detect (DCD): This signal is turned ON when the DCE is receiving a signal from a remote DCE which meets its suitable signal criteria. This signal remains ON as long as a suitable carrier signal can be detected.

Data Terminal Ready (DTR): DTR indicates the readiness of the DTE. This signal is turned ON by the DTE when it is ready to transmit or receive data from the DCE. DTR must be ON before the DCE can assert DSR.

Ring Indicator (RI): RI, when asserted, indicates that a ringing signal is being received on the communication channel. The signals described above form the basis for modem communication. Perhaps the best way to understand how these signals interact is to give a brief step-by-step example of a modem interfacing with a PC. The following steps describe a transaction in which a remote modem calls a local modem:

1. The local PC monitors the RI (Ring Indicator) signal via software.
2. When the remote modem wants to communicate with the local modem, it generates an RI signal. This signal is transferred by the local modem to the local PC.
3. The local PC responds to the RI signal by asserting the DTR (Data Terminal Ready) signal when it is ready to communicate.
4. After recognizing the asserted DTR signal, the modem responds by asserting DSR (Data Set Ready) after it is connected to the communication line. DSR indicates to the PC that the modem is ready to exchange further control signals with the DTE to commence communication. When DSR is asserted, the PC begins monitoring DCD for indication that data are being sent over the communication line.

5. The modem asserts DCD (Data Carrier Detect) after it has received a carrier signal from the remote modem that meets the suitable signal criteria.
6. At this point, data transfer can begin. If the local modem has full-duplex capability, the CTS (Clear to Send) and RTS (Request to Send) signals are held in the asserted state. If the modem has only half-duplex capability, CTS and RTS provide the handshaking necessary for controlling the direction of the data flow. Data are transferred over the RD and TD signals.
7. When the transfer of data has been completed, the PC disables the DTR signal. The modem follows by inhibiting the DSR and DCD signals. At this point, the PC and modem are in the original state described in step number 1.

4.2.3 RS-232 in Minimal Handshake Applications

Even though the modem application discussed above is simplified from the RS-232 standard in terms of the number of signals needed, it is still more complex than the requirements of many systems. For many applications, two data lines and two handshake control lines are all that is necessary to establish and control communication between a host system and a peripheral system.

For example, an environmental control system may need to interface with a thermostat using a half-duplex communication scheme. At times, the control systems may desire to read the temperature from the thermostat and at other times may need to load temperature trip points to the thermostat. In this type of simple application, five signals may be all that is necessary (two for data, two for handshake control, and one for ground).

Figure 4.2 illustrates a simple half-duplex communication interface. As can be seen in this diagram, data are transferred over the TD (Transmit Data) and RD (Receive Data) pins, and handshake control is provided by the RTS (Ready to Send) and CTS (Clear to Send) pins.

- RTS is driven by the DTE to control the direction of data.
- When it is asserted, the DTE is placed in transmit mode.
- When RTS is inhibited, the DTE is placed in receive mode. CTS, which is generated by the DCE, controls the flow of data. When asserted, data can flow. However, when CTS is inhibited, the transfer of data is interrupted.
- The transmission of data is halted until CTS is reasserted.

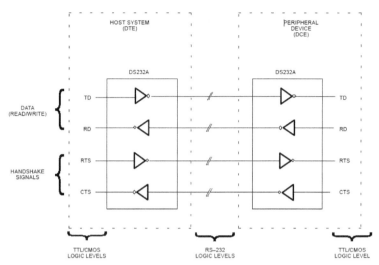

Figure 4.2 Half-duplex communication scheme.

4.3 RS-232 Applications

The RS-232 standard has come a long way since its initial release in 1962. Since then the standard has seen a number of revisions, but more importantly, RS-232 has been used in an ever increasing number of applications. Originally it was devised as a method of connecting telephone modems to teleprinters or teletypes. This enabled messages to be sent along telephone lines – the use of computers was still some way off.

As computers started to be used, links to printers were required. The RS-232 standard provided an ideal method of connection, and therefore, it started to be used in a rather different way. However its use really started to take off when personal computers were first introduced. Here the RS-232 standard provided an ideal method of linking the PC to the printer.

The RS-232 standard provided an ideal method of linking many other remote items to computers and data recorders. As a result, RS-232 became an industry standard, used in a host of applications that were never conceived when it was first launched in 1962.

The RS-232 standard has a variety of updates and minor variations as a result of being taken on by a number of different bodies. For the most part, these different standards are interchangeable and RS-232 and V24, for example, are exactly the same for most applications.

The standard has been in use for over half a century, and in that time, it has provided reliable and effective communications over a serial link.

Applications in modern computers

In the book PC 97 *Hardware Design Guide*, Microsoft deprecated support for the RS-232 compatible serial port of the original IBM PC design. Today, RS-232 has mostly been replaced in personal computers by USB for local communications. Advantages compared to RS-232 are that USB is faster, uses lower voltages, and has connectors that are simpler to connect and use. Disadvantages of USB compared to RS-232 are that USB is far less immune to electromagnetic interference (EMI) and that maximum cable length is much shorter (15 m for RS-232 v.s. 3–5 m for USB depending on USB speed used).

In fields such as laboratory automation or surveying, RS-232 devices may continue to be used. Some types of programmable logic controllers, variable frequency drivers, servo drivers, and computerized numerical control equipment are programmable via RS-232. Computer manufacturers have responded to this demand by re-introducing the DE-9M connector on their computers or by making adapters available.

RS-232 ports are also commonly used to communicate to headless systems such as servers, where no monitor or keyboard is installed, during boot when operating system is not running yet and therefore no network connection is possible. A computer with an RS-232 serial port can communicate with the serial port of an embedded system (such as a router) as an alternative to monitoring over Ethernet.

In general, RS-232 communication is used in different applications. Some of them are listed as follows:

- Teletypewriter devices
- Demodulator applications
- PC COM port interfacing
- In embedded systems for debugging
- Modems and printers
- Handheld equipment
- CNC controllers, Software debuggers, etc.
- Barcode scanners and Point of Sales (POS) terminals.

4.4 RS-232 Application Limitations

As mentioned earlier, the RS-232 standard was first introduced in 1962. In the more than three decades since, the electronics industry has changed immensely, and therefore, there are some limitations in the RS-232 standard.

One limitation, the fact that over 20 signals have been defined by the standard, has already been addressed – simply do not use all of the signals or the 25-pin connector if they are not necessary.

Other limitations in the standard are not necessarily as easy to correct, however.

4.4.1 Generation of RS-232 Voltage Levels

We start with the limitation that is related to the generation of RS-232 voltage levels. In Section 3.4.2, while discussing the electrical characteristics of RS-232 electrical characteristics, we mentioned that RS-232 does not use the conventional 0 and 5 V levels implemented in TTL and CMOS designs. In Section 4.2.1, we showed the need for using Voltage Level Conversion to solve this problem.

Drivers have to supply $+5$ to $+15$ V for logic 0 and -5 to -15 V for logic 1. This means that extra power supplies are needed to drive the RS-232 voltage levels.

Typically, a $+12$ V and a -12 V power supply are used to drive the RS-232 outputs. This is a great inconvenience for systems that have no other requirements for these power supplies. With this in mind, some RS-232 manufacturers, for example, Dallas Semiconductor, their RS-232 products have on-chip charge-pump circuits that generate the necessary voltage levels for RS-232 communication. The first charge pump essentially doubles the standard $+5$ V power supply to provide the voltage level necessary for driving logic 0. A second charge pump inverts this voltage and provides the voltage level necessary for driving logic 1. These two charge pumps allow the RS-232 interface products to operate from a single $+5$ V supply.

In summary, the large voltage swings and requirement for positive and negative supplies increases power consumption of the interface and complicates power supply design. The voltage swing requirement also limits the upper speed of a compatible interface.

4.4.2 Maximum Data Rate

Another limitation in the RS-232 standard is the maximum data rate. The standard defines a maximum data rate of 20 kbps. This is unnecessarily slow for many of today's applications. Some RS-232 manufacturers guarantee higher rate. For example, RS-232 products manufactured by Dallas Semiconductor guarantee up to 250 kbps and typically can communicate up to 350 kbps. While providing a communication rate at this frequency, the devices still maintain a maximum 30 V/ms maximum slew rate to reduce the likelihood of cross-talk between adjacent signals.

4.4.3 Maximum Cable Length

The next limitation to discuss concerning RS-232 communication is cable length. The RS-232 standard, actually, does not define a maximum cable length, but instead defines the maximum capacitance that a compliant drive circuit must tolerate. (Note: The standard included, once, cable length specification but it has been replaced by the maximum capacitance.) The standard is allowing a maximum load capacitance specification of 2500 pF. To determine the total length of cable allowed, one must determine the total line capacitance. Figure 4.3 shows a simple approximation for the total line capacitance of a conductor. As can be seen in the diagram, the total capacitance is approximated by the sum of the mutual capacitance between the signal conductors and the conductor to shield capacitance (or stray capacitance in the case of unshielded cable).

As an example, let's assume that the user has decided to use non-shielded cable when interconnecting the equipment. The cable mutual capacitance (Cm) of the cable is found in the cable's specifications to be 20 pF per foot. If we assume that the input capacitance of the receiver is 20 pF, this leaves the user with 2480 pF for the interconnecting cable. From the equation in Figure 4.3, the total capacitance per foot is found to be 30 pF. Dividing 2480 pF by 30 pF reveals that the maximum cable length is approximately 80 ft. If a longer cable length is required, the user would need to find a cable with a smaller mutual capacitance.

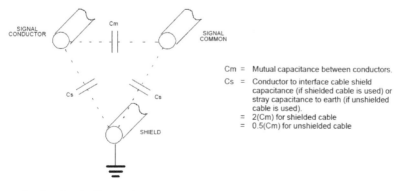

Cc = Cm + Cs = Total line capacitance per unit length

Figure 4.3   Interface cable capacitive model per unit length.

Another factor that limits the cable length is the use of common earth. Use of a common earth or ground limits the length of the cable as the ground will degrade over distance.

Cable Length and Communication Speed

The cable length limitations that are based on the maximum capacitance mentioned in the standard allow the maximum communication speed to occur. The maximum length can be increased by reducing the communication speed. If speed is reduced by a factor of 2 or 4, the maximum length increases dramatically. Texas Instruments has done some practical experiments years ago at different baud rates to test the maximum allowed cable lengths. (Keep in mind that the RS-232 standard was originally developed for 20 Kbps). By halving the maximum communication speed, the allowed cable length increases by a factor of 10.

RS-232 cable length according to Texas Instruments	
Baud Rate	Maximum Cable Length (ft)
19200	50
9600	500
4800	1000
2400	3000

4.4.4 Other RS-232 Limitations

- Single-ended signaling referred to a common signal ground that limits the noise immunity and transmission distance.
- Multi-drop connection among more than two devices is not defined. While multi-drop "work-around" has been devised, they have limitations in speed and compatibility.
- The standard does not address the possibility of connecting a DTE directly to a DTE, or a DCE to a DCE. Null modem cables can be used to achieve these connections, but these are not defined by the standard, and some such cables use different connections than others.
- The definitions of the two ends of the link are asymmetric. This makes the assignment of the role of a newly developed device problematic; the designer must decide on either a DTE-like or DCE-like interface and which connector pin assignments to use.

- The handshaking and control lines of the interface are intended for the setup and takedown of a dial-up communication circuit; in particular, the use of handshake lines for flow control is not reliably implemented in many devices.
- No method is specified for sending power to a device. While a small amount of current can be extracted from the DTR and RTS lines, this is only suitable for low power devices such as a computer mouse. This limits the usefulness as every device must have its own independent power supply.
- The 25-pin D-sub connector recommended in the standard is large compared to current practice.

4.5 Advantages and Disadvantages of RS-232

4.5.1 Advantages of RS-232

The **advantages of RS-232** make it as a standard serial interface for system-to-system communication and also for the following benefits:

- Simple protocol design.
- RS-232 interface is supported in many compatible legacy devices due to its simplicity. It is widely used for point-to-point connection between DTE and DCE devices.
- Hardware overhead is lesser than parallel communication.
- Recommended standard for short distance applications.
- It also supports long distances of about 50 ft (for low baud rates) and with error correction capabilities.
- Compatible with DTE and DCE communication.
- Low cost protocol for development.
- RS-232 is immune to noise due to use of ± 5 V or higher for binary logic 0 and logic 1.
- Converters or adaptors are available at cheaper rates for conversion from RS-232 to RS-485/USB/Ethernet etc.

4.5.2 Disadvantages of RS-232

Following are the **disadvantages of RS-232**:

- It is suitable for system-to-system communications. It is not suitable for chip-to-chip or chip-to-sensor device communications.
- It doesn't support full-duplex communication.

- It is a single-ended protocol which shifts the ground potential.
- It supports lower speed for long distances. Higher speed (i.e., 115200 baud) can be achieved for short distances only.
- RS-232 interface requires separate transceiver chips which will add cost to the system. It is used for single master and single slave configuration and not for single master-multiple slaves mode.
- It is unbalanced transmission.
- Longer cable length introduces cross talk during serial communication. Hence, this protocol is restricted for long distance communication.

4.6 Difference between RS-232 and UART

The main difference between RS-232 and UART is that RS-232 is a half-duplex communication protocol, whereas UART is a full duplex communication protocol.

Microcontrollers don't tolerate RS-232 voltages and may be damaged. To avoid this, UART (Universal Asynchronous Transmitter Receiver) is used. It sends and receives the data in serial form. To do the level conversion of voltages, RS-232 driver IC such as MAX232 is used between the UART and serial port.

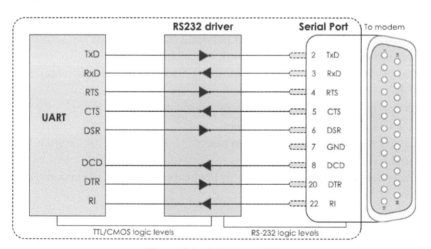

Figure 4.4   RS-232–UART.

5

RS-232 Variants RS-422/RS-423, RS-449, RS-485, and EIA-530

5.1 Introduction: RS-232 Variants – Background

During the late 1970s, the EIA began developing two new serial data standards to replace RS-232. RS-232 had a number of issues that limited its performance and practicality. Among these was the relatively large voltages used for signaling, $+5$ and -5 V for mark and space. To supply these, a $+12$ V power supply was typically required, which made it somewhat difficult to implement in a market that was rapidly being dominated by the transistor–transistor logic (TTL) circuitry and even lower-voltage CMOS implementations. These high voltages and unbalanced communications also resulted in relatively short cable lengths, nominally set to a maximum of 50 ft (15 m), although in practice they could be somewhat longer if running at slower speeds.

The reason for the large voltages was due to ground voltages. RS-232 included both a protective ground and a single ground in the standard, but did not define how these were to be implemented. It was often the case that the protective ground was left unconnected, and the signal ground was connected at both ends. As a result, if there was a slight difference in ground potential at the two ends of the cable, the voltage in the signal ground pin might not be zero, and large signal voltages were needed to provide a positive signal in this case.

To address this problem, the new RS-242 and RS-243 standards used well-defined grounding that was always based on the sender's reference and made the signal only 400 mV above or below this reference. In the case of RS-422, for instance, every signal had a second pin operating at the opposite voltage, thereby balancing the voltages and always providing a positive signal. When this process was starting, the decision was made to unbundle the mechanical aspects of the standard from the electrical, with the former becoming the RS-449 standards track.

The primary difference between RS-422 and RS-423 was that the former had a return line for every signal, while the later had a single shared signal ground. This meant that RS-422 had double the number of signal wires. Along with other changes, the number of connections began to grow to the point where even RS-423, which was functionally similar to RS-232, no longer fit in a DB25 connector. This led to the use of the larger DC-37, but even that did not have enough pins to support RS-422, so this was "solved" by adding the additional ground wires to a separate DE-9 connector. This resulted in a "horrendous number of wires" and the conclusion in 1983 that its "success... remains to be seen."

RS-485 was introduced as an improvement over RS-422. RS-485 increases the number of devices that can be connected from 10 to 32 and defines the electrical characteristics necessary to ensure adequate signal voltages under maximum load.

The RS-449 serial data standard was also introduced as an enhancement to RS-232. It was aimed at providing serial data transmission at speeds up to 2 Mbps while still being able to maintain compatibility with RS-232.

RS-449 standard was rarely used, although it could be found on some network communication equipment. EIA-449-1 was rescinded in January 1986 and superseded by EIA/TIA-530-A, and the final version of EIA-449-1 was withdrawn in September 2002. The most widespread use of RS-422/423, the early Apple Macintosh computers, used a simple 9-pin DIN connector and for inter-machine links used only three-wire connectors.

In summary, in search for serial interfaces that can avoid some of the limitations of RS-232, it is possible to identify the following serial interfaces which are similar to RS-232:

- RS-422 (a high-speed system similar to RS-232 but with differential signaling)
- RS-423 (a high-speed system similar to RS-422 but with unbalanced signaling)
- RS-449 (a functional and mechanical interface that used RS-422 and RS-423 signals – it never caught on like RS-232 and was withdrawn by the EIA)
- RS-485 (a descendant of RS-422 that can be used as a bus in multidrop configurations)

- EIA-530 (a high-speed system using RS-422 or RS-423 electrical properties in an EIA-232 pinout configuration, thus combining the best of both; supersedes RS-449)
- MIL-STD-188 (a system like RS-232 but with better impedance and rise time control)
- EIA/TIA-561 8 Position Non-Synchronous Interface Between Data Terminal Equipment and Data Circuit Terminating Equipment Employing Serial Binary Data Interchange
- EIA/TIA-562 Electrical Characteristics for an Unbalanced Digital Interface (low-voltage version of EIA/TIA-232)
- TIA-574 (standardizes the 9-pin D-subminiature connector pinout for use with EIA-232 electrical signaling, as originated on the IBM PC/AT)

In this section, the first five standards are introduced.

5.1.1 Introduction to RS-422 and RS-423 (EIA Recommended Standard 422 and 423)

RS-422 and RS-423 were designed, specifically, to overcome the distance and speed limitations of RS-232. Although they are similar to the more advanced RS-232C, they can accommodate higher baud rates, longer cable lengths, and multiple receivers. RS-422 and RS-423 are using different techniques to overcome the distance and speed limitation. RS-423 achieved that by introducing an adjustable slew rate. RS-422 is able to achieve these improvements through using differential or balanced transmission techniques. RS-422 uses both differential transmitters and receivers, which means that it is much more resilient to common mode interference, a key issue with long lines.

It may be mentioned here to avoid any ambiguity in understanding the RS-422 and the RS-423 standards that the standard RS-423 is an advanced counterpart of RS-422 which has been designed to tolerate the ground voltage differences between the sender and the receiver for the more advanced version of RS-232, that is, the RS-232C.

5.1.1.1 RS-422
What is RS-422?

RS-422, also known as TIA/EIA RS-422-A Standard or X.27 standard, is the serial connection historically used on Apple Macintosh computers. RS-422, originated by the Electronic Industries Alliance (EIA), specifies the electrical characteristics of a digital signaling circuit. RS-422 uses a differential electrical signal, as opposed to unbalanced signals referenced to ground

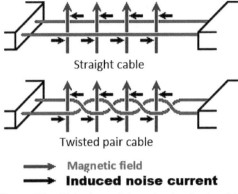

Figure 5.1   Noise in straight and twisted pair cable.

with the RS-232. RS-422 provides for data transmission, using balanced, or differential, signaling, with unidirectional/non-reversible, terminated or non-terminated transmission lines, point to point, or multi-drop. In contrast to EIA-485, RS-422/V.11 does not allow multiple drivers but only multiple receivers. Differential transmission uses two lines each for transmit and receive signals which results in greater noise immunity and longer distances as compared to the RS-232. Twisting the lines helps to reduce the noise. The noise currents induced by an external source are reversed in every twist. Instead of amplifying each other as in a straight line, the reversed noise currents reduce each other's influence. Figure 5.1 explains this in more detail. Differential signaling can transmit data at rates as high as 10 Mbit/s, or may be sent on cables as long as 1500 m. (Note: More discussion about twisted pair and differential signals are given while discussing RS-485). Some systems directly interconnect using RS-422 signals, or RS-422 converters may be used to extend the range of RS-232 connections. The RS-422 standard only defines signal levels; other properties of a serial interface, such as electrical connectors and pin wiring, are part of the RS-449 and RS-530 standards. The mentioned advantages make RS-422 a better fit for industrial applications.

Revision B of RS-422, published in May 1994, was reaffirmed by the Telecommunications Industry Association in 2005.

5.1.1.2 RS-422 characteristics

Several key advantages offered by this standard include the differential receiver, a differential driver and data rates as high as 10 Megabits per second at 12 m (40 ft). Since the signal quality degrades with cable length,

Table 5.1 RS-422 characteristics

	RS-422
Standard	TIA/EIA-422
Physical Media	Twisted Pair
Network topology/ cabling	Point-to-point (single ended), Multi-dropped
Maximum devices	10 (1 driver and 10 receivers)
Communication modes	Half duplex, full duplex
Maximum distance	1500 m (4900 ft) at 100 Kbps
Mode of operation	Differential
Maximum binary rate	100 kbit/s–10 Mbit/s
Voltage levels	−6 V to +6 V (maximum differential voltage)
Mark (1)	Negative voltages
Space (0)	Positive voltages
Signaling	Balanced
Available signals	Tx+, Tx−, Rx+, Rx− (full duplex)
Driver output current capability	150 Ma
Connector types	Not specified

the maximum data rate decreases as cable length increases. The maximum cable length is not specified in the standard, but guidance is given in its annex. (This annex is not a formal part of the standard, but is included for information purposes only.) Limitations on line length and data rate vary with the parameters of the cable length, balance, and termination, as well as the individual installation. In Table 5.1, it is given a maximum length of 1500 m, but this is with a termination and with applications that can tolerate greater timing and amplitude distortion.

RS-422 specifies the electrical characteristics of a single balanced signal. The standard was written to be referenced by other standards that specify the complete DTE/DCE interface for applications which require a balanced voltage circuit to transmit data. These other standards would define protocols, connectors, pin assignments, and functions. Standards such as EIA-530 (DB-25 connector) and EIA-449 (DC-37 connector) use RS-422 electrical signals. Some RS-422 devices have 4 screw terminals for pairs of wire, with one pair used for data in each direction.

RS-422 cannot implement a true multi-point communications network such as with EIA-485 since there can be only one driver on each pair of wires. However, one driver can fan-out to up to 10 receivers.

RS-422 can interoperate with interfaces designed to MIL-STD-188-114B, but they are not identical. RS-422 uses a nominal 0 to 5 V signal, while MIL-STD-188-114B uses a signal symmetric about 0 V. However, the tolerance for common mode voltage in both specifications allows them to interoperate. Care must be taken with the termination network.

EIA-423 is a similar specification for unbalanced signaling (RS-428).

When used in relation to communications wiring, RS-422 wiring refers to cable made of 2 sets of twisted pair, often with each pair being shielded, and a ground wire. While a double pair cable may be practical for many RS-422 applications, the RS-422 specification only defines one signal path and does not assign any function to it. Any complete cable assembly with connectors should be labeled with the specification that defined the signal function and mechanical layout of the connector, such as RS-449.

Table 5.1 gives the summary for RS-422 characteristics.

5.1.1.3 Network topology with RS-422

There are more differences between RS-422 and RS-232 than the maximum data speed and cable length. RS-232 was specifically defined as an interface between computers, printers, and terminals with modems. The modem would translate the communication signals to protocol usable for long distance communication, where long distance could also mean a device on the other side of the control room or building. RS-422 allows the direct connection of intelligent devices, without the need of modems. Furthermore, where the RS-232 line driver is only designed to serve one receiver, a RS-422 line driver can serve up to 10 receivers in parallel. This allows one central control unit to send commands in parallel to up to 10 slave devices. Unfortunately, those slave devices cannot send information back over a shared interface line. RS-422 allows a multi-drop network topology, rather than a multi-point network where all nodes are considered equal and every node has send and receive capabilities over the same line. RS-485 can be used in any application that needs a multi-point communication network rather than multi-drop. In this case the design will have a maximum of 32 parallel send and 32 receive units parallel on one communication channel.

5.1.1.4 RS-422 applications

Although RS-232 has been the most popular standard for serial data transmission, the higher speeds offered by RS-422 are a distinct advantage, and as a result, it is being used more widely. In addition to this, it can be used almost interchangeably with RS-232, and this provides a significant advantage. RS-422 is a common solution for RS-232 extenders. They consist of RS-232 ports on either end of an RS-422 connection.

Historically, the most widespread use of RS-422 was on the early Macintosh computers. This was implemented in a multi-pin connector that had enough pins to support the majority of the common RS-232 pins; the first model used a 9-pin D connector, but this was quickly replaced by a mini-DIN-8 connector. The ports could be put into either RS-232 or RS-422 mode, which changed the behavior of some of the pins while turning others on or off completely. These connectors were used both to support RS-232 devices like modems, as well as AppleTalk networking, RS-422 printers, and other peripherals. Two such ports were part of every Mac until they were replaced, along with ADB ports, by Universal Serial Bus on the iMac in 1998.

RS-422 and RS-423 fit well in process control applications in which instructions are sent out to many actuators or responders. Ground voltage differences can occur in electrically noisy environments where heavy electrical machinery is operating.

Broadcast automation systems and post-production linear editing facilities use RS-422A to remotely control the players/recorders located in the central apparatus room. In most cases, the Sony 9-pin connection is used, which makes use of a standard DE-9 connector. This is a de facto industry standard connector for RS-422 used by many manufacturers.

5.2 RS-423

RS-423 serial information

Introduction to RS-423

The RS-423 standard is one of the lesser known serial communication standards. Its older brother RS-232 is widely known because serial ports with this interface are present on almost all computer systems. RS-422 and RS-485 are differential which makes them useful in applications where noise immunity is an issue, like in industrial applications. The single-ended RS-423 standard sits somewhere in-between these standards as an enhancement of RS-232 with longer cable lengths and higher allowed data rates.

Although RS-423 is currently not widely implemented, it has seen a broad usage in the late eighties of the previous century because of its backward compatibility with RS-232. Hewlett Packard shipped their computers with a serial interface capable of communicating on both RS-232 and RS-423 levels, and Digital Equipment Corporation used the RS-423 signal levels on their DEC Connect MMJ serial interface standard. Because RS-232 and RS-423 are both single-ended serial communication standards with comparable signal levels, it was possible to use these RS-423 interfaces in both pure RS-423 applications and mixed RS-232/RS-423 situations.

Single-ended variable slew rate

Although RS-423 is a single-ended serial communication interface just like RS-232, the standard allows higher data rates and cable lengths. This is achieved by introducing an adjustable slew rate. With RS-232, the slew rate is fixed at a speed of 30 V/μs. RS-423 has a variable slew rate which can be set dependent on cable length and data rate. This adds noise immunity to the interface. On longer cable lengths where noise can be an issue, a lower maximum data rate can be set and the slew rate is changed accordingly. RS-422 and RS-485, which are both differential serial interfaces, do not need such a system of noise immunity.

The maximum allowed cable length with RS-423 is 1200 m, just as with RS-422 and RS-485. The maximum data rate is 100 Kbps which is only five times faster than the default RS-232 and much slower than possible with RS-422 and RS-485. Detailed information on the interface characteristics of RS-423 can be found in Table 5.6 in which serial interface comparison is given.

5.3 RS-449

The RS-449 serial data standard was intended as an enhancement to RS-232. It was aimed at providing serial data transmission at speeds up to 2 Mbps while still being able to maintain compatibility with RS-232.

Although never applied on personal computers, this interface was found on some network communication equipment. The RS-449 standard has now been discontinued and may also be seen in some references as EIA-449, TIA-449, and ISO 4902.

The **RS-449** specification, also known as **EIA-449** or **TIA-449**, defines the functional and mechanical characteristics of the interface between data terminal equipment, typically a computer, and data communications equipment, typically a modem or terminal server. It was an effort to replace RS-232C, offering much higher performance and longer cable lengths, but emerged as an unwieldy system requiring both DC-37 and DE-9 connectors. The effort was eventually abandoned in favor of RS-530, which used a single DB-25 connector. The full title of the standard is *"EIA-449 General Purpose 37-Position and 9-Position Interface for Data Terminal Equipment and Data Circuit-Terminating Equipment Employing Serial Binary Data Interchange"*.

RS-449 interface

One of the ways in which the RS-449 data communications standard is able to send at high speeds without stray noise causing interference is to use a differential form of signaling, exactly as RS-422. Earlier data communication standards such as RS-232 used signaling that was referenced to earth, and while this was easier to implement and cheaper to cable, it introduced limitations into the system.

What Is Slew Rate

What is slew rate?

Slew rate effect on a square wave: red = desired output, green = actual output.

In electronics, **slew rate** is defined as the change of voltage or current, or any other electrical quantity, per unit of time. Slew rate is usually expressed in V/μs.

Electronic circuits may specify minimum or maximum limits on the slew rates for their inputs or outputs, with these limits only valid under some set of given conditions (e.g., output loading). When given for the output of a circuit, such as an amplifier, the slew rate specification guarantees that the speed of the output signal transition will be at least the given minimum, or at most the given maximum. When applied to the input of a circuit, it instead indicates that the external driving circuitry needs to meet those limits in order to guarantee the correct operation of the receiving device. If these limits are violated, some error might occur and correct operation is no longer guaranteed. For example, when the input to a digital circuit is driven too slowly, the digital input value registered by the circuit may oscillate between 0 and 1 during the signal transition. In other cases, a *maximum* slew rate is specified in order to limit the high frequency content present in the signal, thereby preventing such undesirable effects as ringing or radiated EMI.

As discussed before, by using twisted wire pairs for the data lines, any unwanted noise will be picked up by both wires together. As the RS-449 receivers use a differential input, and they are not referenced to ground, any noise that is picked up does not affect the input. This means that higher levels of noise can be tolerated without any degradation to the performance to the data communications system.

For the RS-449 interface, 10 additional circuit functions have been provided when compared to RS-232. Additionally, three of the original interchange circuits have been abandoned.

In order to minimize any confusion that could easily occur, the circuit abbreviations have been changed. In addition to this, the RS-449 interface requires the use of 37-way D-type connectors and 9-way D-type connectors, the latter being necessary when use is made of the secondary channel interchange circuits.

RS-449 Primary connector pinout and interface connections

The RS-449 primary connector, which is the one used as standard, uses a 37-way D-type connector. The pinout and connections are given in the Table 5.2.

Within the RS-449 interface, a number of differential connections are defined. In the pinout table above, they are labeled as either "A and B" or "+" and "−". When setting up a connection, it is necessary to ensure that the correct polarities are used. As twisted pairs are used for the A and B connections, it is often possible to mix them. If this happens, the interface will not work.

RS-449 auxiliary connector

A second connector is defined for use when the secondary channel interchange circuits are needed. This connector uses a 9-way D-type connector (Table 5.3).

RS-449 secondary connector

The RS-449 data communications interface is an interface standard that is able to provide data communications with speeds of up to 2 Mbps. Retaining some similarities to RS-232, it is a more comprehensive interface capable of greater speeds and operation with greater levels of data integrity.

Table 5.2   RS-449 primary connector pinout and connections

Pin	Signal Name	Description	Pin	Signal Name	Description
1		Shield	19	SG	Signal Ground
2	SI	Signal Rate Indicator	20	RC	Receive Common
3	n/a	Unused	21	n/a	Unused
4	SD−	Send Data (A)	22	SD+	Send Data (B)
5	ST−	Send Timing (A)	23	ST+	Send Timing (B)
6	RD−	Receive Data (A)	24	RD+	Receive Data (B)
7	RS−	Request To Send (A)	25	RS+	Request To Send (B)
8	RT−	Receive Timing (A)	26	RT+	Receive Timing (B)
9	CS−	Clear To Send (A)	27	CS+	Clear To Send (B)
10	LL	Local Loopback	28	IS	Terminal In Service
11	DM−	Data Mode (A)	29	DM+	Data Mode (B)
12	TR−	Terminal Ready (A)	30	TR+	Terminal Ready (B)
13	RR−	Receiver Ready (A)	31	RR+	Receiver Ready (B)
14	RL	Remote Loopback	32	SS	Select Standby
15	IC	Incoming Call	33	SQ	Signal Quality
16	SF/SR+	Signal Freq./Sig. Rate Select.	34	NS	New Signal
17	TT−	Terminal Timing (A)	35	TT+	Terminal Timing (B)
18	TM−	Test Mode (A)	36	SB	Standby Indicator
			37	SC	Send Common

Table 5.3   RS-449 auxiliary connector

Pin	Signal Name	Description
1		Shield
2	SRR	Secondary Receive Ready
3	SSD	Secondary Send Data
4	SRD	Secondary Receive Data
5	SG	Signal Ground
6	RC	Receive Common
7	SRS	Secondary Request to Send
8	SCS	Secondary Clear to Send
9	SC	Send Common

5.4 RS-485

What Is RS-485?

RS-232, as detailed before, is an interface to connect one DTE, data terminal equipment, to one DCE, data communication equipment, at a maximum speed of 20 Kbps with a maximum cable length of 50 ft. This was sufficient in the old days where almost all computer equipment were connected using modems, but soon after people started to look for interfaces capable of one or more of the following:

- Connect DTE's directly without the need of modems
- Connect several DTE's in a network structure
- Ability to communicate over longer distances
- Ability to communicate at faster communication rates

RS-485 is the most versatile communication standard in the standard series defined by the EIA, as it performs well on all the four points. That is why RS-485 is currently a widely used communication interface in data acquisition and control applications where multiple nodes communicate with each other.

RS-485 (also known as EIA-485, TIA-485(-A) Standard) is an improvement over RS-422, because it increases the number of devices from 10 to 32 and defines the electrical characteristics necessary to ensure adequate signal voltages under maximum load. With this enhanced multi-drop capability, it is possible to create networks of devices connected to a single RS-485 serial port. The noise immunity and multi-drop capability make RS-485 the serial connection of choice in industrial applications requiring many distributed devices networked to a PC or other controller for data collection, HMI, or other operations. RS-485 is a superset of RS-422; thus, all RS-422 devices may be controlled by RS-485. RS-485 is able to provide a data rate of 10 Mbps at distance s up to 50 ft. RS-485 hardware may also be used for serial communication with up to 4000 ft of cable with a lower speed of 100 Kbps. RS-485 uses, as RS-422, differential signals that results in longer distance and higher bit rate.

Although RS-485 was never intended for domestic use, it found many applications where remote data acquisition was required and also widely used communication interface in control applications where multiple nodes communicate with each other.

Table 5.4 summarizes the characteristics of RS-485.

Table 5.4   RS-485 characteristics

TIA-485-A (Revision of EIA-485)	
Standard	ANSI/TIA/EIA-485-A-1998 Approved: March 3, 1998 Reaffirmed: March 28, 2003
Physical media	Balanced interconnecting cable
Network topology	Point-to-point, multi-dropped, multi-point
Maximum devices	At least 32 unit loads
Maximum distance	Not specified
Mode of operation	Different receiver levels: binary 1 (OFF) (Voa–Vob < -200 mV) binary 0 (ON) (Voa–Vob $> +200$ mV)
Available signals	A, B, C
Connector types	Not specified

The characteristics of RS-485 make it useful in industrial control systems and similar applications.

5.4.1 RS-485 Specification Overview

Table 5.5 provides the highlight details behind RS-485.

Differential signals with RS-485: Longer distances and higher bit rates

One of the main problems with RS-232 is the lack of immunity for noise on the signal lines. The transmitter and receiver compare the voltages of the data and handshake lines with one common zero line. Shifts in the ground level can have disastrous effects. Therefore, the trigger level of the RS-232 interface is set relatively high at ±3 V. Noise is easily picked up and limits both the maximum distance and communication speed. With RS-485 on the contrary, there is no such thing as a common zero as a signal reference. Several volts difference in the ground level of the RS-485 transmitter and receiver does not cause any problems. The RS-485 signals are floating and each signal is transmitted over a Sig+ line and a Sig− line. The RS-485 receiver compares the voltage difference between both lines, instead of the absolute voltage level on a signal line. This works well and prevents the existence of ground loops,

Table 5.5 RS-485 specification

RS-485 Highlight Specifications	
Attribute	Specification
Standard	ANSI/TIA/EIA-485-A-1998 Approved: March 3, 1998 Reaffirmed: March 28, 2003
Cabling: Network topology	Multi-drop, Multi-point, Point-to-point
Number of devices	32 transmitters32 receivers
Communications modes	Half duplex
Maximum distance	4000 ft @ 100 Kbps
Maximum data rate	10 Mbps @ 50 ft
Signaling	Balanced interconnecting cable
Mark (data = 1) condition	1.5 V to 5 V (B greater than A)
Space (data = 0) condition	1.5 V to 5 V (A greater than B)
Driver output current capability	250 mA
Connector type	Not defined

a common source of communication problems. The best results are achieved if the Sig+ and Sig− lines are twisted. Figure 5.1 given before explains why.

Going back to Figure 5.1, noise is generated by magnetic fields from the environment. As mentioned before, the figure shows the magnetic field lines and the noise current in the RS-485 data lines that is the result of that magnetic field. In the straight cable, all noise current is flowing in the same direction, practically generating a looping current just like in an ordinary transformer. When the cable is twisted, it is easy to note that in some parts of the signal lines, the direction of the noise current is the opposite from the current in other parts of the cable. Because of this, the resulting noise current is many factors lower than with an ordinary straight cable. Shielding – which is a common method to prevent noise in RS-232 lines – tries to keep hostile magnetic fields away from the signal lines. Twisted pairs in RS-485 communication however add immunity which is a much better way to fight noise. The magnetic fields are allowed to pass, but do no harm. If high noise immunity is needed, often a combination of twisting and shielding is used as for example in STP, shielded twisted pair and FTP, foiled twisted pair networking cables. Differential signals and twisting allow RS-485 to communicate over much longer communication distances than achievable with RS-232. With RS-485, communication distances of 1200 m are possible.

Differential signal lines also allow higher bit rates than possible with non-differential connections. Therefore, RS-485 can overcome the practical communication speed limit of RS-232. Currently, RS-485 drivers are produced that can achieve a bit rate of 35 Mbps.

5.4.2 Characteristics of RS-485 Compared to RS-232, RS-422, and RS-423

Table 5.6 gives a comparison between RS-232, RS-422, RS-423, and RS-485.

From the information given in Table 5.6, it is possible to get the following conclusions:

1. The speed of the differential interfaces RS-422 and RS-485 is far superior to the single-ended versions RS-232 and RS-423.
2. There is a maximum slew rate defined for both RS-232 and RS-423. This has been done to avoid reflections of signals. The maximum slew rate also limits the maximum communication speed on the line. For both other interfaces – RS-422 and RS-485 – the slew rate is indefinite. To avoid reflections on longer cables, it is necessary to use appropriate termination resistors.
3. The maximum allowed voltage levels for all interfaces are in the same range, but that the signal level is lower for the faster interfaces. Because

Table 5.6 Characteristics of RS-232, RS-422, RS-423, and RS-485

Characteristics of RS-232, RS-422, RS-423, and RS-485				
Differential	No	No	Yes	Yes
Max number of drivers	1	1	1	32
Max number of receivers	1	10	10	32
Modes of operation	Half duplex Full duplex	Half duplex	Half duplex	Half duplex
Network topology	Point-to-point	Multidrop	Multidrop	Multipoint
Max distance (acc. standard)	15 m	1200 m	1200 m	1200 m
Max speed at 12 m	20 kbs	100 kbs	10 Mbs	35 Mbs
Max speed at 1200 m	(1 kbs)	1 kbs	100 kbs	100 kbs
Max slew rate	30 V/μs	Adjustable	n/a	n/a
Receiver input resistance	3.7 kΩ	$\geq$4 kΩ	$\geq$4 kΩ	$\geq$12 kΩ
Driver load impedance	3.7 kΩ	$\geq$450Ω	100 Ω	54 Ω
Receiver input sensitivity	$\pm$3 V	$\pm$200 mV	$\pm$200 mV	$\pm$200 mV
Receiver input range	$\pm$15 V	$\pm$12 V	$\pm$10 V	-7.12 V
Max driver output voltage	$\pm$25 V	$\pm$6 V	$\pm$6 V	-7.12 V
Min driver output voltage (with load)	$\pm$5 V	$\pm$3.6 V	$\pm$2.0 V	$\pm$1.5 V

of this, RS-485 and the others can be used in situations with a severe ground level shift of several volts, where at the same time high bit rates are possible because the transition between logical 0 and logical 1 is only a few hundred millivolts.

4. RS-232 is the only interface capable of full-duplex communication. This is, because on the other interfaces, the communication channel is shared by multiple receivers and – in the case of RS-485 – by multiple senders. RS-232 has a separate communication line for transmitting and receiving which – with a well written protocol – allows higher effective data rates at the same bit rate than the other interfaces. The request and acknowledge data needed in most protocols does not consume bandwidth on the primary data channel of RS-232.

5. RS-485 has the same distance and data rate specifications as RS-422 and uses differential signaling but, unlike RS-422, allows multiple drivers on the same bus. As depicted in Figure 5.2, each node on the bus can include both a driver and receiver forming a multi-point star network. Each driver at each node remains in a disabled high impedance state until called upon to transmit. This is different than drivers made for RS-422 where there is only one driver and it is always enabled and cannot be disabled.

5.4.3 Network topology with RS-485

Network topology is probably the reason why RS-485 is now the favorite of the four mentioned interfaces in data acquisition and control applications. RS-485 is the only of the interfaces capable of internetworking multiple transmitters and receivers in the same network. When using the default RS-485 receivers with an input resistance of 12 kΩ, it is possible to connect 32 devices to the network. Currently available high-resistance RS-485 inputs allow this number to be expanded to 256. RS-485 repeaters are also available which make it possible to increase the number of nodes to several thousands, spanning multiple kilometers.

Figure 5.2 shows the general network topology of RS-485. N nodes are connected in a multipoint RS-485 network. For higher speeds and longer lines, the termination resistances are necessary on both ends of the line to eliminate reflections. Use 100 Ω resistors on both ends. The RS-485 network must be designed as one line with multiple drops, not as a star. Although total cable length maybe shorter in a star configuration, adequate termination is not possible anymore and signal quality may degrade significantly.

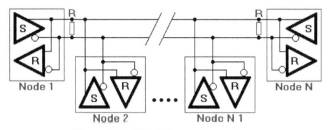

Figure 5.2   RS-485 network topology.

5.4.4 RS-485 Functionality

This section discusses how does RS-485 function in practice. Default, all the senders on the RS-485 bus are in tri-state with high impedance. In most higher-level protocols, one of the nodes is defined as a master that sends queries or commands over the RS-485 bus. All other nodes receive these data. Depending of the information in the sent data, zero or more nodes on the line respond to the master. In this situation, bandwidth can be used for almost 100%. There are other implementations of RS-485 networks where every node can start a data session on its own. This is comparable with the way Ethernet networks function. Because there is a chance of data collision with this implementation, theory tells us that in this case, only 37% of the bandwidth will be effectively used. With such an implementation of a RS-485 network, it is necessary that there is error detection implemented in the higher level protocol to detect the data corruption and resend the information at a later time.

There is no need for the senders to explicitly turn the RS-485 driver on or off. RS-485 drivers automatically return to their high impedance tri-state within a few microseconds after the data have been sent. Therefore, it is not needed to have delays between the data packets on the RS-485 bus.

RS-485 is used as the electrical layer for many well-known interface standards, including Profibus and Modbus. Therefore, RS-485 will be in use for many years in the future.

5.4.5 RS-485 Applications

Often RS-485 links are used for simple networks, and they may be connected in a 2 or 4 wire mode. In a typical application, several address able devices may be linked to a single controlled (PC), and in this way, a single line may be used for communication. It is also possible to convert between RS-485 and

RS-232 using simple interface converters that may include optical isolation between the two circuits as well as surge suppression for any electrical 'spikes' that may be picked up.

Using RS-485, it is possible to construct a multi-point data communications network. The standard specifies that up to 32 drivers or transmitters along with 32 receivers can be used on a system. This means that there can be 32 nodes capable to both transmit and receive. This can be extended further by using "automatic" repeaters and high-impedance drivers / receivers. In this way, it is possible to have hundreds of nodes on a network. In addition to this, RS-485 extends the common mode range for both drivers and receivers in the "tri-state" mode and with power off. Also, RS-485 drivers are able to withstand "data collisions" (bus contention) problems and bus fault conditions.

As RS-485 networks become larger, the problem of data collisions becomes greater. This can be solved, at least in part by ensuring the hardware units (converters, repeaters, micro-processor controls) are designed to remain in a receive mode until they are ready to transmit data.

Another approach is to design a 'single master' system. Here the master initiates a communications request to a "slave node" by addressing that unit. The hardware detects the start-bit of the transmission and thereby enables the transmitter. Once the requested data are sent, the hardware reverts back into a receive mode.

Advantages of RS-485
- Among all of the asynchronous standards mentioned above, this standard offers the maximum data rate.
- Apart from that special hardware for avoiding bus contention and,
- A higher receiver input impedance with lower Driver load impedances is its other asset.

5.5 EIA-530

Interface Standards

Currently known as **TIA-530-A**, but often called **EIA-530**, or **RS-530**, is a balanced serial interface standard that generally uses a 25-pin connector, originally created by the Telecommunications Industry Association (TIA).

The standard is finalized in 1987 (revision A finalized in 1992), and the specification defines the cable between the DTE and DCE devices. It is to be used in conjunction with EIA-422 and EIA-423, which define the electrical signaling characteristics. Because TIA-530 calls for the more common 25-pin connector, it displaced the similar EIA-449, which also uses EIA-422/423, but a larger 37-pin connector.

Two types of **interchange circuits** ("signals" or "leads") between the DCE and DTE are defined in TIA-530: **Category I**, which uses the balanced characteristics of EIA-422, and **Category II**, which is the unbalanced EIA-423. Most of the interchange circuits are Category I, with the exception of Local Loopback (pin 18), Remote Loopback (pin 21), and Test Mode (pin 25) being Category II.

TIA-530 originally used Category I circuits for what is commonly called "Data Set Ready" (DCE Ready, pins 6 and 22) and "Data Terminal Ready" (DTE Ready, pins 20 and 23). Revision A changed these interchange circuits to Category II (para 4.3.6 and 4.3.7 of the standard) and added a "Ring Indicator" on pin 22. Pin 23 is grounded in TIA-530-A.

Confusion between the revisions has led to many incorrect wiring diagrams of this interface and most manufacturers still adhere to the original TIA-530 standard. Care should be taken to ensure devices are of the same standard before connecting to avoid complications.

The majority of the signals conform to the RS-422 standard and for the majority of requirements requiring RS-422 signaling, the RS-530 cable is suitable. Some of the link management controls signals are implemented using V.10 (RS-423) single-ended interfaces, and a variant of this standard called RS-530A / EIA-530A also uses V.10 for the DTR signal. Note that the EIA standards have effectively replaced the RS standards and have now been themselves superseded by TIA standards.

Interface Characteristics
RS-530 is a differential communications interface with some single-ended link management signals, typically limited to a maximum throughput of 10 Mbps. Communications over distances exceeding 1000 m is possible at low bit rates, the actual performance being mostly dependent on cable specification. Separate clock lines are used for receiving and transmitting data.

Interface Applications
EIA-530 interfaces are commonly found on communications equipment in some parts of the world where high throughput and/or long distances

are required. The interface also offers good noise immunity enabling reliable communications in environments where there are high levels of EMI (electromagnetic interference).

Applications include high-speed connections between satellite modems and host computer systems.

Pin Configuration

EIA-530, or RS-530, is a balanced serial interface standard that generally uses a 25-pin connector. The R-S530 is not an actual interface, but a generic connector specification. The connector pinning can be used to support RS-422, RS-423, V.35, and X.21 to name the most popular ones (Table 5.7).

RS-530 is just like RS-422 and uses a differential signaling on a DB25 – RS-232 format; EIA-530 Transmit (and the other signals) use a twisted pair of wires (TD+ & TD-) instead of TD and a ground reference as in RS-232 or V.24. This interface is used for HIGH SPEED synchronous protocols. Using a differential signaling allows for higher speeds over long cabling. This standard is applicable for use at data signaling rates in the range from 20,000 to a nominal upper limit of 2,000,000 bits per second. Equipment complying with this standard, however, need not operate over this entire data signaling rate range. They may be designed to operate over a narrower range as appropriate for the specific application.

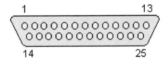

25 pin D-SUB male connector at the DTE (Computer).

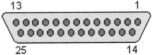

25 pin D-SUB female connector at the DCE (Modem)

Figure 5.3   Pin configuration of DTE and DCE, RS-530.

Table 5.7 Pin configuration of EIA-530 interface

Pin	Name	Dir.	Signal	Pin	Name	Dir.	Signal
1		—	Shield	14		OUT	Transmitted Data Return
2	TxD	OUT	Transmitted Data	15		IN	Transmit Signal Element Timing
3	RxD	IN	Received Data	16		IN	Received Data Return
4	RTS	OUT	Request to Send	17		IN	Rec. Sig. Element Timer Return
5	CTS	IN	Clear to Send	18	LL	OUT	Local Loopback
6	DSR	IN	DCE Ready	19		OUT	Request to Send
7	SGND	—	Signal Ground	20	DTR	OUT	DTE Ready
8	DCD	IN	Received Line Signal Detector	21	RL	OUT	Remote Loopback
9		IN	Receiver Signal Element Timer	22		IN	DCE Ready
10		IN	Received Line Signal Detector	23		OUT	DTE Ready
11		OUT	Transmit Signal Element Timing	24		OUT	Trans. Sig Element Timing Return
12		IN	Transmit Signal Element Timing	25		IN	Test Mode
13		IN	Clear to Send				

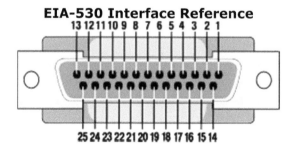

References

[1] "Interface Between Data Terminal Equipment and Data Circuit-Terminating Equipment Employing Serial Binary Data Interchange," TIA/EIA-232-F Standards, Electronics Industries Association Engineering Department.

[2] "Electrical Characteristics of Balanced Digital Interface Circuits," TIA/EIA-422-B Standards, Electronics Industries Association Engineering Department.

[3] "Standard for Electrical Characteristics of Generators and Receivers for Use in Balanced Digital Multipoint Systems," TIA/EIA-485-A Standards, Electronics Industries Association Engineering Department.

[4] "The I^2C Specification," Version 2.1, Philips Semiconductors.

[5] Aleaf, Abdul, "Microwire Serial Interface," Application Note AN-452, National Semiconductor.

[6] Goldie, John, "Summary of Well Known Interface Standards," Application Note AN-216, National Semiconductor.

[7] Nelson, Todd, "The Practical Limits of RS-485," Application Note AN-979, National Semiconductor.

[8] Wilson, Michael R., "TIA/EIA-422-B Overview," Application Note AN-1031, National Semiconductor.

[9] Goldie, John, "Ten Ways to Bulletproof RS-485 Interfaces," Application Note AN-1057, National Semiconductor.

6

Serial Peripheral Interface (SPI)

In the first chapter of the book, we introduced serial communication and the most famous protocols: UART, SPI, I2C, CAN, etc. As a fact, the three most common multi-wire serial data transmission formats that have been in use for decades are I2C, UART, and SPI. Chapter 2 introduced the first serial communication protocols: UART/USART. This chapter introduces to the reader another serial communication protocol: **Serial Peripheral Interface (SPI)**. As a case study, the SPI of (AVR) microcontrollers will be discussed.

6.1 Serial Peripheral Interface (SPI): Introduction

Serial Peripheral Interface, which is commonly known as S-P-I or "spy," is one of the most popular interface specifications used in embedded systems. Since its introduction in the late 1980s by Motorola, the SPI protocol has been widely used for short distance communication in embedded systems. It has been accepted as a *de facto* standard, it is available in almost all architectures, including 8051, x86, ARM, PIC, AVR, MSP etc., and is thus widely used. This means that there should not be any portability issues and the user can connect devices of two different architectures together as well.

SPI is a synchronous serial data transfer protocol where two or more serial devices are connected to each other in full-duplex mode at a very high speed. SPI uses clock signal for synchronization. The clock signal is provided by the master: Only the master device can control the clock line. The clock signal controls when data can change and when it is valid for reading. As a result of having clock signal, the SPI clock can vary without disrupting the data. The data rate will simply change along with the changes in the clock rate. This makes SPI ideal when the microcontroller is being clocked imprecisely, such as by a RC oscillator.

Through the SPI protocol, devices communicate with each other using master–slave architecture. It, for example, provides a simple and low-cost interface between a microcontroller (master) and its peripherals (slaves). Although multiple slave devices can be supported by SPI, the number of master devices is limited to one. The *Master* device is the one which initiates the connection and controls it. The master device originates the frame for reading and writing. Once the connection is initiated, the *Master* and one or more *Slave(s)* can communicate by transmitting and/or receiving data. SPI as a full-duplex connection, the *Master* can send data to *Slave(s)* and the *Slave(s)* can also send the data to the *Master* at the same time.

Multiple slave devices are supported through selection with individual slave select (SS) lines. SPI interface bus is commonly used for interfacing microprocessor or microcontroller with memory like EEPROM, RTC (Real Time Clock), ADC (Analog to Digital Converters), DAC (Digital to Analog Converters), displays like LCDs, Audio ICs, sensors like temperature and pressure, Secure Digital Cards, memory cards like MMC or SD Cards, 2.4 GHz wireless transmitter/receivers or even other microcontrollers.

Sometimes SPI is called a *four-wire* serial bus, contrasting with three-, two-, and one-wire serial buses. The SPI may be accurately described as a synchronous serial interface, but it is different from the Synchronous Serial Interface (SSI) protocol, which is also a four-wire synchronous serial communication protocol. SSI Protocol employs differential signaling and provides only a single simplex communication channel.

SPI is not standardized; accordingly, it is possible to encounter situations where either the Most Significant Bit (MSB) or the Least Significant Bit (LSB) is transferred first. For this reason, it is recommended to check the datasheet for the SPI device selected for the application, and the designer has to set up the data-handling routines accordingly.

In full-duplex mode, data rates over 1 Mbps can be achieved – this is one of the main advantages of the SPI bus. Compared to I2C, SPI also supports using simple hardware interfacing and provides a higher throughput.

One unique benefit of SPI is the fact that data can be transferred without interruption. Any number of bits can be sent or received in a continuous stream. With I2C and UART, data are sent in packets, limited to a specific number of bits. Start and stop conditions define the beginning and end of each packet, so the data are interrupted during transmission.

However, the SPI protocol also has some drawbacks – the lack of error-checking mechanism and slave acknowledgment feature are some of the major disadvantages (see the latter for advantages and disadvantages of SPI).

Many protocols are developed from SPI aiming to enhance the performance of SPI. Such systems are discussed in Section 6.6.

6.2 SPI Bus Transaction: SPI Operation

SPI is known as four-wire serial bus: it consists of four signals (Figure 6.1):

- Master out slave in (MOSI)
- Master in slave out (MISO)
- Serial clock (SCK),
- Slave select (SS) or chip select (CS)

The SPI operation is based upon shift registers. Every device, whether *Master* or *Slave,* has an 8-bit shift register inside it. The size of the shift register could be more than 8-bit as well (like 10-bit, 12-bit, etc.), but it should be the same for both *Master* and *Slave*, and the protocol should support it.

6.2.1 Hardware Setup

The hardware requirement for implementing SPI is very simple when compared to UART and I2C. Consider a Master and a single Slave connected using SPI bus. Figure 6.2 explains the hardware setup: The *Master* and *Slave* are connected in such a way that the two shift registers form an inter-device circular buffer.

There is an 8-bit shift register inside each of the *Master* and *Slave* devices. These shift registers operate in Serial-In/Serial-Out (SISO) mode. The output of the *Master's* shift register, MOSI, is connected to the input of the *Slave's* shift register; and the output of the *Slave's* shift register, MISO, is connected to the input of *Master's* shift register. This makes the connection operate, as mentioned before, like a circular/ring buffer (Figure 6.2).

Figure 6.1 SPI interface: Single master to single slave.

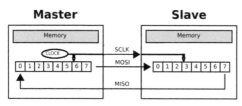

Figure 6.2 A typical hardware setup using two shift registers to form an inter-chip circular buffer.

As a synchronous serial data transfer protocol, there must be a clock to synchronize the data transfer between the master and the slave. Since the *Master* is responsible for initiating and controlling the connection, the clock source, SCK, of the *Master* device is to be used to synchronize the data transfer. The clock source is shown in Figure 6.2, inside the *Master and it is used to* control the operation of *both* the shift registers.

6.2.2 Data Transfer Operation

Before explain the data transfer operation, it is important to mention here the following facts:

- SPI is a Data Exchange protocol. As data are being clocked out, new data are also being clocked in. Data are always "exchanged" between devices. No device can just be a "transmitter" or just a "receiver" in SPI. This fact means that as the master sends data to the slave, the slave must respond by transmitting something to the master. This can be a garbage if there is no specific data to be transmitted by the slave.
- When one "transmits" data, the incoming data must be read before attempting to transmit again. If the incoming data are not read, then the data will be lost and the SPI module may become disabled as a result. Always read the data after a transfer has taken place, even if the data have no use in your application.
- These data exchanges are controlled by the clock line, SCK, which is controlled by the master device.

Mentioning the above facts and to understand the transfer of the data between the Master and slave, consider Figure 6.3. Both, *Master* and *Slave*, place the data (byte) they wish to transfer in their respective shift registers before the communication starts. Let the data in the *Master's* shift register be A7 through A0 (MSB through LSB) whereas the data in the *Slave's* shift register is B7 through B0 (MSB through LSB). The *Master* switches the SS/CS pin to a low voltage state, which activates the slave and then generates 8 clock pulses. After each clock pulse, one bit of information is transfer from Master to Slave and vice versa.

Figure 6.3(a) represents the initial state before any clock pulse arrives. As the master generates the 8 clock pulses, the contents of the Master's shift register are transferred to the Slave's shift register and vice versa (Figure 6.3(b) to 6.3(e))

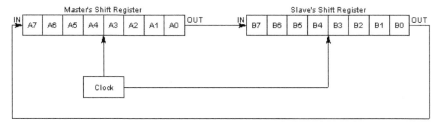

(a) Initial state

Master generates the first clock pulse:

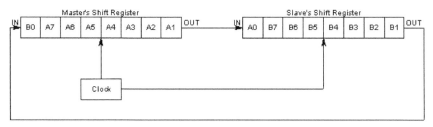

(b) Master generates the first clock

Master generates the second clock pulse:

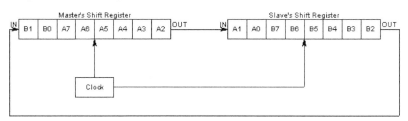

(c) Status after the master generating the second pulse

Master generates the seventh clock pulse:

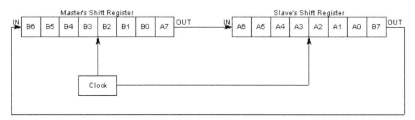

(d) Status after the master generating the sevens pulse

Figure 6.3

Master generates the last clock pulse:

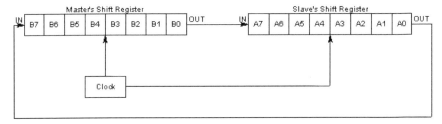

(e) The status after generating the last clock pulse, the eighth clock

Figure 6.3   Data transfer between master and slave.

As soon as the first clock pulse arrives, the shift registers come into operation and the data in the registers is shifted by one bit towards the right. This evicts bit A0 from *Master* and bit B0 from *Slave*. Since the *Master* and *Slave* are connected to form a ring/circular buffer, the evicted bit occupies the MSB position of the other device. Which means, bit A0 gets evicted from *Master* and occupies MSB position in *Slave's* shift register; whereas bit B0 gets evicted from *Slave* and occupies MSB position in *Master's* shift register. This first cycle is shown in Figure 6.3(b), while Figure 6.3(c) shows the second cycle, etc. At the end of the clock pulses, the Master has completely received B, and the Slave has received A.

As seen from Figure 6.3, during each SPI clock cycle, the following takes place, which represents a full-duplex transmission:

- *Master* sends a bit to the MOSI line; *Slave* reads it from the same line.
- *Slave* sends a bit to the MISO line; *Master* reads it from the same line.

6.3 SPI Bus Interface

From the above discussion, it is easy to describe the needed connection and the interface between *Master* and *Slave*.

The *Master* and *Slave* are connected by means of four wires. Each of these wires carries a particular signal defined by the SPI bus protocol. These four signals/wires are follows:

1. **MOSI (Master Out Slave In):** This is the wire/signal which goes from the output of *Master's* shift register to the input of the *Slave's* shift register. Hence, MOSI pins on both the master and slave are connected together. Master In/Slave Out or MISO is the data generated by Slave and must be transmitted to Master

2. **MISO (Master In Slave Out):** This is the wire/signal which goes from the output of *Slave's* shift register to the input of the *Master's* shift register.

   MISO pins on both the master and slave are ties together. Even though the Signal in MISO is produced by the Slave, the line is controlled by the Master. The Master generates a clock signal at SCLK and is supplied to the clock input of the slave. Chip Select (CS) or Slave Select (SS) is used to select a particular slave by the master.

   Since the clock is generated by the Master, the flow of data is controlled by the master. For every clock cycle, one bit of data is transmitted from master to slave and one bit of data is transmitted from slave to master.

   This process happens simultaneously and after 8 clock cycles, a byte of data is transmitted in both directions and hence, SPI is a full-duplex communication.

   If the data have to be transmitted by only one device, then, as mentioned before, the other device has to send something (even garbage or junk data), and it is up to the device whether the transmitted data are actual data or not.

   This means that for every bit transmitted by one device, the other device has to send one bit data, that is, the Master simultaneously transmits data on MOSI line and receive data from slave on MISO line.

   If the slave wants to transmit the data, the master has to generate the clock signal accordingly by knowing when the slave wants to send the data in advance.

3. **SCK/SCLK (Serial Clock):** This is the output of the clock generator for *Master* and clock input for *Slave*.

   The clock signal synchronizes the output of data bits from the master to the sampling of bits by the slave. One bit of data is transferred in each clock cycle, the speed of data transfer is determined by the frequency of the clock signal. SPI communication is always initiated by the master since the master configures and generates the clock signal.

   The clock signal in SPI can be modified using the properties of clock polarity and clock phase. These two properties work together to define when the bits are output and when they are sampled. Clock polarity can be set by the master to allow for bits to be output and sampled on either the rising or falling edge of the clock cycle. Clock phase can be set for output and sampling to occur on either the first edge or second edge of the clock cycle, regardless of whether it is rising or falling. This is discussed in the Section 6.4.

4. **SS' (Slave Select):** This is discussed in the next section: "Multiple Slaves."

The MOSI, SCK, and SS' signals are directed from *Master* to *Slave,* whereas the MISO signal is directed from *Slave* to *Master.* Consider Figure 6.1 which represents interface having single *Master* and single *Slave.* It is possible to note that during each SPI clock cycle, a full duplex transmission occurs as follows:

- *Master* sends a bit to the MOSI line; *Slave* reads it from the same line.
- *Slave* sends a bit to the MISO line; *Master* reads it from the same line.

Note: The pin names MOSI, MISO, SCLK, and SS' are in common use now. However, in the past and also some microcontrollers currently, alternative pin naming conventions were sometimes used, and so SPI port pin names for older IC products may differ from those depicted here:

Serial Clock: The following can be used:

- SCLK: SCK

Master Output/Slave Input (MOSI): The following alternatives are in use:

- SIMO, MTSR – correspond to MOSI on both master and slave devices, connects to each other
- SDI, DI, DIN, SI – on slave devices; connects to MOSI on master, or to below connections
- SDO, DO, DOUT, SO – on master devices; connects to MOSI on slave, or to above connections

Master Input/Slave Output (MISO): The following alternatives exist:

- SOMI, MRST – correspond to MISO on both master and slave devices, connects to each other
- SDO, DO, DOUT, SO – on slave devices; connects to MISO on master, or to below connections
- SDI, DI, DIN, SI – on master devices; connects to MISO on slave, or to above connections

Slave Select:

- SS: $\overline{SS}$, SSEL, CS, $\overline{CS}$, CE, nSS, /SS, SS#

In other words, MOSI (or SDO on a master) connects to MOSI (or SDI on a slave). MISO (or SDI on a master) connects to MISO (or SDO on a slave). Slave Select is the same functionality as chip select and is used instead of an addressing concept. Pin names are always capitalized as in Slave Select, Serial Clock, and Master Output Slave Input.

SPI Configuration: Multiple Slaves – Slave Select (SS') Signal

As mentioned earlier, SPI can be used to connect one *Master* to multiple *Slaves* as well. Having multiple *Masters* is also possible, but it does nothing but increase the complexity due to clock synchronization issues. Multiple Masters is, in general, very rare. Having multiple *Slaves* is where the Slave Select (SS') signal comes into effect.

SS' (which means SS complemented) signal is in active low configuration, that is, to select a particular *Slave*, we need to provide a LOW signal level to SS' of the *Slave*.

Two configurations are possible to connect the Master with the Slaves:

* Independent slave configuration also called Parallel Configuration
* Daisy Chain Configuration, also known as Cascaded Configuration

a. Independent slave configuration or Parallel Configuration

This configuration is shown in Figure 6.4.

In the independent slave configuration, all the *Slaves* share the same MOSI, MISO, and SCK signals. Even though multiple slaves are connected to the master in the SPI bus, only one slave will be active at any time. The SS'

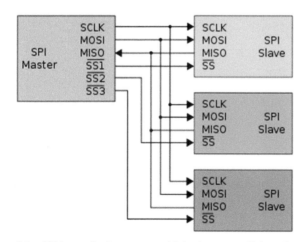

Figure 6.4 SPI bus – single master multiple slaves: parallel configuration.

signal is responsible for choosing a particular *Slave*. The *Slave* gets enabled only when its input SS' signal goes LOW.

A pull-up resistor between power source and chip select line is highly recommended for each independent device to reduce cross-talk between devices. This configuration is the one normally used in SPI. Since the MISO pins of the slaves are connected together, they are required to be tri-state pins (high, low, or high-impedance).

b. Daisy chain configuration

Some products that implement SPI may be connected in a daisy chain configuration (Figure 6.5). In this configuration, all the Slaves are selected at a time, and the output of one *Slave* goes to the input of another *Slave*, and so on. The SPI port of each slave is designed to send out during the second group of clock pulses an exact copy of the data it received during the first group of clock pulses. The whole chain acts as a communication shift register; daisy chaining is often done with shift registers to provide a bank of inputs or outputs through SPI. Each slave copies input to output in the next clock cycle until active low SS line goes high. Such a feature only requires a single SS line from the master, rather than a separate SS line for each slave.

Some of the applications that can potentially interoperate with SPI that require a daisy chain configuration include SGPIO, JTAG, and Two Wire Interface.

To show how daisy chain works consider Figure 6.6 and consider the case in which the master transmits 3 bytes of data in to the SPI bus. First, the 1st

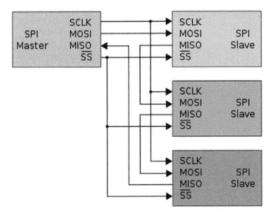

Figure 6.5   Daisy-chained SPI bus: master and cooperative slaves.

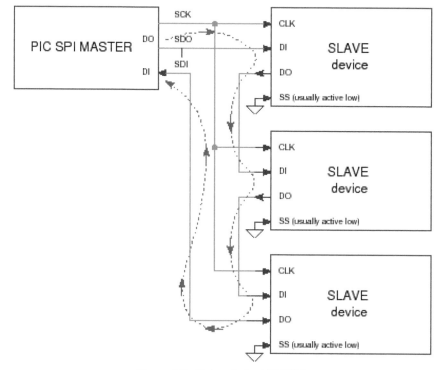

Figure 6.6 Daisy-chained PIC SPI.

byte of data is shifted to slave 1. When the 2nd byte of data reaches slave 1, the first byte is pushed into slave 2.

Finally, when the 3rd byte of data arrives into the first slave, the 1st byte of data is shifted to slave 3 and the second byte of data is shifted to slave 2.

If the master wants to retrieve information from the slaves, it has to send 3 bytes of junk data to the slaves so that the information in the slaves comes to the master.

6.4 Clock Polarity and Phase

Because, as mentioned before, there is no official specification, what exactly SPI is and what not, it is necessary to consult the data sheets of the components one wants to use. Important are the permitted clock frequencies and the type of valid transitions. There are no general rules for transitions where data should be latched.

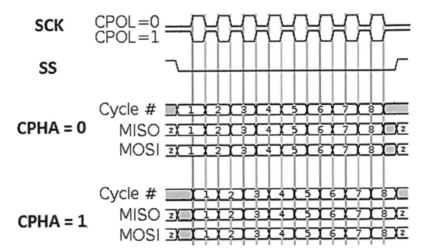

Figure 6.7   A timing diagram showing clock polarity and phase. Red lines denote clock leading edges, and blue lines, trailing edges.

Keeping synchronization in mind, *Master's* role does not end with simply generating clock pulses at a particular frequency (usually within the range of 10 kHz to 100 MHz). In fact, *Master* and *Slave* should agree on a particular synchronization protocol as well, or else everything will go wrong and data will get lost. This is where the concept of clock polarity (CPOL) and clock phase (CPHA) comes in.

The timing diagram is shown in Figure 6.7. The timing applies to both the master and the slave device.

- **CPOL (Clock Polarity):** This determines the base value of the clock, that is, the value of the clock when SPI bus is idle.

    - When CPOL = 0, base value of clock is zero, that is, SCK is LOW when idle.
    - When CPOL = 1, base value of clock is one, that is, SCK is HIGH when idle.

- **CPHA (Clock Phase):** This determines the clock transition at which data will be sampled/captured.

    - When CPHA = 0, data are sampled at clock's rising/leading edge.
    - When CPHA = 1, data are sampled at clock's falling/trailing edge.

Figure 6.8 shows the sampling edge and toggling edge during the four modes.

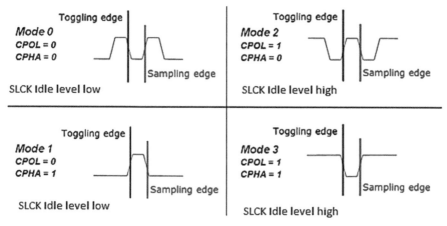

Figure 6.8   Sampling edge and toggling edge during the four modes.

Table 6.1   SPI modes

SPI Mode	Clock Polarity (CPOL/CKP)	Clock Phase (CPHA)	Clock Edge (CKE/NCPHA)
0	0	0	1
1	0	1	0
2	1	0	1
3	1	1	0

Table 6.2   CPOL and CPHA functionality

	Leading Edge	Trilling Edge	SPI mode
CPOL = 0, CPHA =0	Sample (Rising)	Setup (Falling)	0
CPOL = 0, CPHA =1	Setup (Rising)	Sample (Falling)	1
CPOL = 1, CPHA =0	Sample (Falling)	Setup (Rising)	2
CPOL = 1, CPHA =1	Setup (Falling)	Sample (Rising)	3

Modes and Mode numbers

The combinations of polarity and phases are often referred to as modes which are commonly numbered according to the following convention (Table 6.1), with CPOL as the high-order bit and CPHA as the low-order bit:

Table 6.2 shows the functionality of the four modes in case of ATmega32. The different modes will be discussed again while discussing how to program the SPI of the AVR.

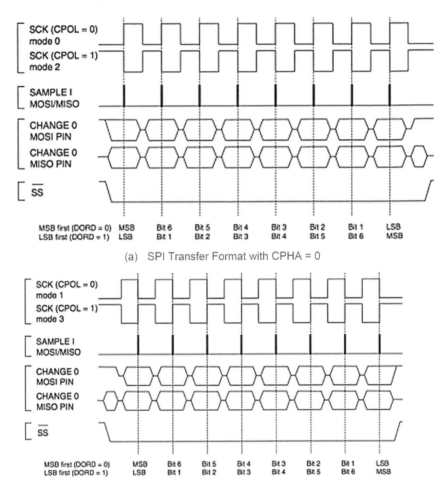

(a)   SPI Transfer Format with CPHA = 0

(b)   SPI Transfer Format with CPHA = 1

Figure 6.9   Example of the effect of CPOL and CPHA.

From Tables 6.1 and 6.2 and Figure 6.7, the four modes are summarized as follows:

Mode 0:

Mode 0 occurs when Clock Polarity is LOW and Clock Phase is 0 (CPOL = 0 and CPHA = 0). During Mode 0, data transmission occurs during rising edge of the clock.

Mode 1:
Mode 1 occurs when Clock Polarity is LOW and Clock Phase is 1 (CPOL = 0 and CPHA = 1). During Mode 1, data transmission occurs during falling edge of the clock.

Mode 2:
Mode 2 occurs when Clock Polarity is HIGH and Clock Phase is 0 (CPOL = 1 and CPHA = 0). During Mode 2, data transmission occurs during rising edge of the clock.

Mode 3:
Mode 3 occurs when Clock Polarity is HIGH and Clock Phase is 1 (CPOL = 1 and CPHA = 1). During Mode 3, data transmission occurs during rising edge of the clock.

Each transaction begins when the slave-select line is driven to logic low (slave select is typically an active-low signal). The exact relationship between the slave-select, data, and clock lines depends on how the clock polarity (CPOL) and clock phase (CPHA) are configured.

With non-inverted clock polarity (i.e., the clock is at logic low when slave select transitions to logic low):

- Mode 0: Clock phase is configured such that data are sampled on the rising edge of the clock pulse and shifted out on the falling edge of the clock pulse. This corresponds to the first blue clock trace in the above diagram. Note that data must be available before the first rising edge of the clock.
- Mode 1: Clock phase is configured such that data is sampled on the falling edge of the clock pulse and shifted out on the rising edge of the clock pulse. This corresponds to the second blue clock trace in the above diagram.

With inverted clock polarity (i.e., the clock is at logic high when slave select transitions to logic low):

- Mode 2: Clock phase is configured such that data are sampled on the falling edge of the clock pulse and shifted out on the rising edge of the clock pulse. This corresponds to the first orange clock trace in the above diagram. Note that data must be available before the first falling edge of the clock.
- Mode 3: Clock phase is configured such that data are sampled on the rising edge of the clock pulse and shifted out on the falling edge of the

clock pulse. This corresponds to the second orange clock trace in the above diagram.

Valid communications

Some slave devices are designed to ignore any SPI communications in which the number of clock pulses is greater than specified. Others do not care, ignoring extra inputs and continuing to shift the same output bit. It is common for different devices to use SPI communications with different lengths, as, for example, when SPI is used to access the scan chain of a digital IC by issuing a command word of one size (perhaps 32 bits) and then getting a response of a different size (perhaps 153 bits, one for each pin in that scan chain).

About SPI protocol

A master–slave pair must use the same set of parameters – SCLK frequency, CPOL, and CPHA for a communication to be possible. If multiple slaves are used, that are fixed in different configurations, the master will have to reconfigure itself each time it needs to communicate with a different slave.

This is basically all what is defined for the SPI protocol. SPI does not define any maximum data rate, not any particular addressing scheme; it does not have an acknowledgement mechanism to confirm receipt of data and does not offer any flow control. Actually, the SPI master has no knowledge of whether a slave exists, unless "something" additional is done outside the SPI protocol. For example, a simple codec won't need more than SPI, while a command-response type of control would need a higher-level protocol built on top of the SPI interface. SPI does not care about the physical interface characteristics like the I/O voltages and standard used between the devices. Initially, most SPI implementation used a non-continuous clock and byte-by-byte scheme. But many variants of the protocol now exist that use a continuous clock signal and an arbitrary transfer length.

Interrupts

SPI devices sometimes use another signal line to send an interrupt signal to a host CPU. Examples include pen-down interrupts from touchscreen sensors, thermal limit alerts from temperature sensors, alarms issued by real time clock chips, SDIO, and headset jack insertions from the sound codec in a cell phone. Interrupts are not covered by the SPI standard; their usage is neither forbidden nor specified by the standard.

Bit banging

In computer engineering and electrical engineering, **bit banging** is slang for various techniques for data transmission in which software is used to generate and process signals instead of dedicated hardware. Software directly sets and samples the state of pins on a microcontroller and is responsible for all parameters of the signal: timing, levels, synchronization, etc. In contrast to bit banging, dedicated hardware (such as a modem, UART, or shift register) handles these parameters and provides a (buffered) data interface in other systems, so software is not required to perform signal demodulation. Bit banging can be implemented at very low cost and is used in embedded systems.

Bit banging allows the same device to use different protocols with minimal or no hardware changes required. In many cases, bit banging is made possible because more recent hardware operates much more quickly than hardware did when standard communications protocols were created.

In the following, two programs are given showing sending a byte on an SPI bus.

Program 1: Example of bit banging the master protocol

Below is an example of bit banging the SPI protocol as an SPI master with CPOL=0, CPHA=0, and eight bits per transfer. The example is written in the C programming language. Because this is CPOL=0, the clock must be pulled low before the chip select is activated. The chip select line must be activated, which normally means being toggled low, for the peripheral before the start of the transfer, and then deactivated afterward. Most peripherals allow or require several transfers while the select line is low; this routine might be called several times before deselecting the chip.

```
/*
 * Simultaneously transmit and receive a byte on the SPI.
 *
 * Polarity and phase are assumed to be both 0, i.e.:
 *   - input data is captured on rising edge of SCLK.
 *   - output data is propagated on falling edge of SCLK.
 *
 * Returns the received byte.
 */
uint8_t SPI_transfer_byte(uint8_t byte_out)
{
    uint8_t byte_in = 0;
```

```
    uint8_t bit;

    for (bit = 0x80; bit; bit >>= 1) {
        /* Shift-out a bit to the MOSI line */
        write_MOSI((byte_out & bit) ? HIGH : LOW);

        /* Delay for at least the peer's setup time */
        delay(SPI_SCLK_LOW_TIME);

        /* Pull the clock line high */
        write_SCLK(HIGH);

        /* Shift-in a bit from the MISO line */
        if (read_MISO() == HIGH)
            byte_in | = bit;

        /* Delay for at least the peer's hold time */
        delay(SPI_SCLK_HIGH_TIME);

        /* Pull the clock line low */
        write_SCLK(LOW);

    }

    return byte_in;
}
```

Program 2: Another program for sending a byte on an SPI bus

```
// transmit byte serially, MSB first

void send_8bit_serial_data(unsigned char data)

{
    int i;

    // select device (active low)
    output_low(SD_CS);

    // send bits 7..0
    for (i = 0; i < 8; i++)
    {
        // consider leftmost bit
        // set line high if bit is 1, low if bit is 0
        if (data & 0x80)
            output_high(SD_DI);
        else
            output_low(SD_DI);
        // pulse clock to indicate that bit value should be read
        output_low(SD_CLK);
        output_high(SD_CLK);
```

```
        // shift byte left so next bit will be leftmost
        data <<= 1;

    }

    // deselect device
    output_high(SD_CS);

}
```

6.5 SPI Pros and Cons

Advantages

SPI uses 4 pins for communications while the other communication protocols available, for example, on AVR use lesser number of pins like 2 or 3. Then why does one use SPI? Here are some of the advantages of SPI:

- Full-duplex communication in the default version of this protocol
- Push-pull drivers (as opposed to open drain) provide good signal integrity and high speed
- Higher throughput than I2C or SMBus. Not limited to any maximum clock speed, enabling potentially high speed
- Complete protocol flexibility for the bits transferred
  - Not limited to 8-bit words
  - Arbitrary choice of message size, content, and purpose
- Extremely simple hardware interfacing
  - Typically lower power requirements than I2C or SMBus due to less circuitry (including pull up resistors)
  - No arbitration or associated failure modes
  - Slaves use the master's clock and do not need precision oscillators
  - Slaves do not need a unique address – unlike I2C or GPIB or SCSI
  - Transceivers are not needed
- Uses only four pins on IC packages, and wires in board layouts or connectors, much fewer than parallel interfaces
- At most one unique bus signal per device (chip select); all others are shared
- Signals are unidirectional allowing for easy galvanic isolation.
- Simple software implementation

Disadvantages

- Requires more pins on IC packages than I2C, even in the *three-wire* variant
- No in-band addressing; out-of-band chip select signals are required on shared buses
- No hardware flow control by the slave (but the master can delay the next clock edge to slow the transfer rate)
- No hardware slave acknowledgment (the master could be transmitting to nowhere and not know it)
- Typically supports only one master device (depends on device's hardware implementation)
- No error-checking protocol is defined
- Without a formal standard, validating conformance is not possible
- Only handles short distances compared to RS-232, RS-485, or CAN-bus. (Its distance can be extended with the use of transceivers like RS-422.)
- Many existing variations, making it difficult to find development tools like host adapters that support those variations
- SPI does not support hot swapping (dynamically adding nodes).
- Interrupts must either be implemented with out-of-band signals or be faked by using periodic polling similarly to USB 1.1 and 2.0.
- Some variants like dual SPI, quad SPI, and three-wire serial buses defined below are half-duplex.

6.6 SPI Variants

SPI Bus 3-Wire and Multi-IO Configurations

In addition to the standard 4-wire configuration, the SPI interface has been extended to include a variety of IO standards including 3-wire for reduced pin count and dual or quad I/O for higher throughput.

In 3-wire mode, MOSI and MISO lines are combined to a single bidirectional data line as shown in Figure 6.10. Transactions are half-duplex to allow for bidirectional communication. Reducing the number of data lines and operating in half-duplex mode also decreases maximum possible throughput; many 3-wire devices have low performance requirements and are instead designed with low pin count in mind.

Multi I/O variants such as dual I/O and quad I/O add additional data lines to the standard for increased throughput. Components that utilize multi

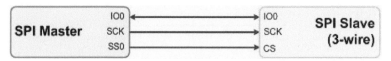

Figure 6.10 3-wire SPI configuration with one slave.

I/O modes can rival the read speed of parallel devices while still offering reduced pin counts. This performance increase enables random access and direct program execution from flash memory (execute-in-place).

Some of such SPI variants are introduced here.

6.6.1 Intelligent SPI Controllers

A **Queued Serial Peripheral Interface** (**QSPI**) is a type of SPI controller that uses a data queue to transfer data across the SPI bus. It has a wrap-around mode allowing continuous transfers to and from the queue with only intermittent attention from the CPU. Consequently, the peripherals appear to the CPU as memory –mapped parallel devices. This feature is useful in applications such as control of an A/D Converter. Other programmable features in QSPI are chip selects and transfer length/delay.

SPI controllers from different vendors support different feature sets; such DMA queues are not uncommon, although they may be associated with separate DMA engines rather than the SPI controller itself, such as used by **Multichannel Buffered Serial Port** (**MCBSP**). Most SPI master controllers integrate support for up to four chip selects, although some require chip selects to be managed separately through GPIO lines.

6.6.2 Microwire

Microwire, often spelled μ**Wire**, is essentially a predecessor of SPI and a trademark of National Semiconductor. It's a strict subset of SPI: half-duplex and using SPI mode 0. Microwire chips tend to need slower clock rates than newer SPI versions, perhaps 2 MHz versus 20 MHz. Some Microwire chips also support a three-wire mode.

Microwire/Plus

Microwire/Plus is an enhancement of Microwire and features full-duplex communication and support for SPI modes 0 and 1. There was no specified improvement in serial clock speed.

6.6.3 Three-wire Serial Buses

As mentioned, one variant of SPI uses single bidirectional data line (slave out-/slave in, called SISO) instead of two unidirectional ones (MOSI and MISO). This variant is restricted to a half-duplex mode. It tends to be used for lower performance parts, such as small EEPROMs used only during system startup and certain sensors, and Microwire. Few SPI master controllers support this mode, although it can often be easily bit-banged in software.

6.6.4 Dual SPI

Because the full-duplex nature of SPI is rarely used, an extension uses both data pins in a half-duplex configuration to send two bits per clock cycle. Typically a command byte is sent requesting a response in dual mode, after which the MOSI line becomes SIO0 (serial I/O 0) and carries even bits, while the MISO line becomes SIO1 and carries odd bits. Data are still transmitted MSbit first, but SIO1 carries bits 7, 5, 3, and 1 of each byte, while SIO0 carries bits 6, 4, 2, and 0.

This is particularly popular among SPI ROMs, which have to send a large amount of data, and comes in two variants:

- Dual read commands accept the send and address from the master in single mode, and return the data in dual mode.
- Dual I/O commands send the command in single mode, then send the address and return data in dual mode.

6.6.5 Quad SPI (Quad I/O SPI)

While dual SPI re-uses the existing serial I/O lines, quad SPI adds two more I/O lines (SIO2 and SIO3) and sends 4 data bits per clock cycle. Again, it is requested by special commands, which enable quad mode after the command itself is sent in single mode.

SQI Type 1: Commands sent on single line but addresses and data sent on four lines.

SQI Type 2: Commands and addresses sent on a single line but data sent/received on four lines.

Quad I/O devices can, for example, offer up to 4 times the performance of a standard 4-wire SPI interface when communicating with a high speed device. Figure 6.11 shows an example of a single-quad IO slave configuration.

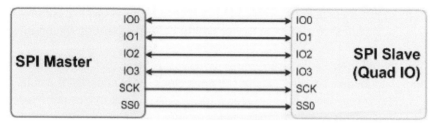

Figure 6.11   Quad IO SPI configuration with one slave.

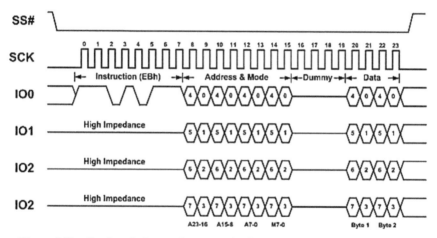

Figure 6.12   Quad mode fast read sequence for Spansion S25FL016K or equivalent.

Quad IO is gaining popularity with flash memories for its increased performance. Instead of using a single output and single input interface, Quad IO utilizes 4 separate half-duplex data lines for both transmitting and receiving data for up to four times the performance of standard 4-wire SPI.

Quad IO SPI Transaction

To show how to achieve transactions in case of Quad I/O, we consider a practical case: Case of using Spansion S25FL016K serial NOR flash device.

Figure 6.12 shows an example read command for a Spansion S25FL016K serial NOR flash device. To read from the device, a fast read command (EBh) is first sent by the master on the first IO line, while all others are tri-stated. Next, the host sends the address; since the interface now has 4 bidirectional data lines, it can utilize these to send a complete 24-bit address along with 8 mode bits in just 8 clock cycles. The address is then followed with 2 dummy bytes (4 clock cycles) to allow the device additional time to set up the initial address.

After the address cycle and dummy bytes have been sent by the host, the component begins sending data bytes; each clock cycle consists of a data nibble spread across the 4 IO lines, for a total of two clock cycles per byte of data. Compare this to the 16 clock cycles required when using simple read transaction, and it is easy to see why quad mode is gaining popularity for high speed flash memory applications. To create this sequence in the SPI Exerciser command language, we would use the example code:

```
4m    // Start in 4-wire mode

   sson    // Activate slave select
   wt EB   // Write instruction EBh

   qm    // Switch to quad mode
   wt AA AA AA 00 // Write 3-byte address and 8 read mode bits

   wt 55 55  // Write 2 dummy bytes

   rd 2    // Read two data bytes

ssoff    // Deactivate slave select
```

Note that we are changing from 4-wire mode to quad mode in the middle of the transaction. In quad mode, the software automatically distributes the data bytes among the IO lines using the same bit pattern depicted in Figure 6.12.

6.6.6 QPI/SQI

Further extending quad SPI, some devices support a "quad everything" mode where *all* communication takes place over 4 data lines, including commands. This is variously called "QPI" (not to be confused with Intel QuickPath Interconnect) or "serial quad I/O" (SQI).

This requires programming a configuration bit in the device and requires care after reset to establish communication.

Note: The **Intel QuickPath Interconnect (QPI)** is a point-to-point processor interconnect developed by Intel which replaced the front-side bus (FSB) in Xeon, Itanium, and certain desktop platforms starting in 2008. It increased the scalability and available bandwidth. Prior to the name's announcement, Intel referred to it as **Common System Interface (CSI)**.

6.6.7 Double Data Rate

In addition to using multiple lines for I/O, some devices increase the transfer rate by using double data rate transmission.

6.6.8 Intel Enhanced Serial Peripheral Interface Bus (eSPI)

Intel has developed a successor to its Low Pin Count (LPC) bus that it calls the Enhanced Serial Peripheral Interface Bus, or eSPI for short. Intel aims to allow the reduction in the number of pins required on motherboards compared to systems using LPC, have more available throughput than LPC, reduce the working voltage to 1.8 V to facilitate smaller chip manufacturing processes, allow eSPI peripherals to share SPI flash devices with the host (the LPC bus did not allow firmware hubs to be used by LPC peripherals), tunnel previous out-of-band pins through the eSPI bus, and allow system designers to trade off cost and performance.

The eSPI bus can either be shared with SPI devices to save pins or be separate from the SPI bus to allow more performance, especially when eSPI devices need to use SPI flash devices.

This standard defines an Alert# signal that is used by an eSPI slave to request service from the master. In a performance-oriented design or a design with only one eSPI slave, each eSPI slave will have its Alert# pin connected to an Alert# pin on the eSPI master that is dedicated to each slave, allowing the eSPI master to grant low-latency service because the eSPI master will know which eSPI slave needs service and will not need to poll all of the slaves to determine which device needs service. In a budget design with more than one eSPI slave, all of the Alert# pins of the slaves are connected to one Alert# pin on the eSPI master in a wired OR connection, which will require the master to poll all the slaves to determine which ones need service when the Alert# signal is pulled low by one or more peripherals that need service. Only after all of the devices are serviced, will the Alert# signal be pulled high due to none of the eSPI slaves needing service and therefore pulling the Alert# signal low.

This standard allows designers to use 1-bit, 2-bit, or 4-bit communications at speeds from 20 to 66 MHz to further allow designers to trade off performance and cost.

All communications that were out-of-band of the LPC bus like general-purpose input/output (GPIO) and System Management Bus (SMBus) are tunneled through the eSPI bus via virtual wire cycles and out-of-band message cycles, respectively, in order to remove those pins from motherboard designs using eSPI.

This standard supports standard memory cycles with lengths of 1 byte to 4 kilobytes of data, short memory cycles with lengths of 1, 2, or 4 bytes that have much less overhead compared to standard memory cycles, and I/O cycles with lengths of 1, 2, or 4 bytes of data which are low overhead as well. This significantly reduces overhead compared to the LPC bus, where all cycles except for the 128-byte firmware hub read cycle spends more than one-half of all of the bus's throughput and time in overhead. The standard memory cycle allows a length of anywhere from 1 byte to 4 kilobytes in order to allow its larger overhead to be amortized over a large transaction. eSPI slaves are allowed to initiate bus master versions of all of the memory cycles. Bus master I/O cycles, which were introduced by the LPC bus specification, and ISA-style DMA including the 32-bit variant introduced by the LPC bus specification, are not present in eSPI. Therefore, bus master memory cycles are the only allowed DMA in this standard.

eSPI slaves are allowed to use the eSPI master as a proxy to perform flash operations on a standard SPI flash memory slave on behalf of the requesting eSPI slave.

64-bit memory addressing is also added, but is only permitted when there is no equivalent 32-bit address.

The Intel Z170 chipset can be configured to implement either this bus or a variant of the LPC bus that is missing its ISA-style DMA capability and is underclocked to 24 MHz instead of the standard 33 MHz.

6.7 Standards

The SPI bus is a *de facto* standard. However, the lack of a formal standard is reflected in a wide variety of protocol options. Different word sizes are common. Every device defines its own protocol, including whether it supports commands at all. Some devices are transmit-only; others are receive-only. Chip selects are sometimes active-high rather than active-low. Some protocols send the least significant bit first.

Some devices even have minor variances from the CPOL/CPHA modes described above. Sending data from slave to master may use the opposite clock edge as master to slave. Devices often require extra clock idle time before the first clock or after the last one, or between a command and its response. Some devices have two clocks, one to read data, and another to transmit it into the device. Many of the read clocks run from the chip select line.

Some devices require an additional flow control signal from slave to master, indicating when data are ready. This leads to a 5-wire protocol instead of the usual 4. Such a *ready* or *enable* signal is often active-low and needs to be enabled at key points such as after commands or between words. Without such a signal, data transfer rates may need to be slowed down significantly, or protocols may need to have dummy bytes inserted, to accommodate the worst case for the slave response time. Examples include initiating an ADC conversion, addressing the right page of flash memory, and processing enough of a command that device firmware can load the first word of the response. (Many SPI masters do not support that signal directly, and instead rely on fixed delays.)

Many SPI chips only support messages that are multiples of 8 bits. Such chips cannot interoperate with the JTAG or SGPIO protocols, or any other protocol that requires messages that are not multiples of 8 bits.

There are also hardware-level differences. Some chips combine MOSI and MISO into a single data line (SI/SO); this is sometimes called 'three-wire' signaling (in contrast to normal "four-wire" SPI). Another variation of SPI removes the chip select line, managing protocol state machine entry/exit using other methods. Anyone needing an external connector for SPI defines their own: UEXT, JTAG connector, Secure Digital card socket, etc. Signal levels depend entirely on the chips involved.

SafeSPI is an industry standard for SPI in automotive applications. Its main focus is the transmission of sensor data between different devices

6.8 Applications

The full-duplex capability makes SPI very simple and efficient for single master/single slave applications. Some devices use the full-duplex mode to implement an efficient, swift data stream for applications such as digital audio, digital signal processing, or telecommunications channels, but most off-the-shelf chips stick to half-duplex request/response protocols.

SPI is used to talk to a variety of peripherals, such as:

1. Wired transmission of data (although the first preference is mostly USART, but SPI *can* be used when we are using multiple slave or master systems, as addressing is much simpler in SPI).
2. Wireless transmissions through ZigBee, 2.4GHz, etc.

3. Communicate with FLASH and EEPROM memory. This gives the possibility of programming some of the microcontrollers, for example, AVR chips
4. Control devices: audio codecs, digital potentiometers, DAC
5. It is also used to talk to various peripherals – like sensors, memory devices, real-time clocks, communication protocols like Ethernet, ADC, video game controllers, etc.
6. Interface with LCDs and SD cards
7. Read data from a real-time clock.
8. Any MMC or SD card (including SDIO variant)

For high-performance systems, FPGAs sometimes use SPI to interface as a slave to a host, as a master to sensors, or for flash memory used to bootstrap if they are SRAM-based.

Although there are some similarities between the SPI bus and the JTAG (IEEE 1149.1-2013) protocol, they are not interchangeable. The SPI bus is intended for high speed, on board initialization of device peripherals, while the JTAG protocol is intended to provide reliable test access to the I/O pins from an off board controller with less precise signal delay and skew parameters. While not strictly a level sensitive interface, the JTAG protocol supports the recovery of both setup and hold violations between JTAG devices by reducing the clock rate or changing the clock's duty cycles. Consequently, the JTAG interface is not intended to support extremely high data rates.

SGPIO is essentially another (incompatible) application stack for SPI designed for particular backplane management activities. SGPIO uses 3-bit messages.

6.9 Case Study 1: The SPI of the AVR

In this section, the Serial Peripheral Interface (SPI) of AVR is considered as case study of the implementation of SPI in microcontroller.

The SPI of the AVR

The SPI of AVRs is one of the simplest peripherals to program. As the AVR has an 8-bit architecture, so the SPI of AVR is also 8-bit. In fact, usually the SPI bus is of 8-bit width. It is available on PORTB on all of the ICs, whether 28 pin (Figure 6.13) or 40 pin (Figure 6.14).

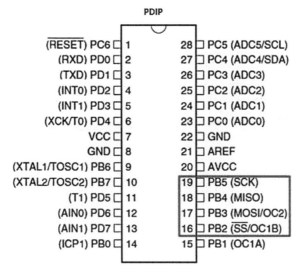

Figure 6.13   SPI pins on 28 pin ATmega8.

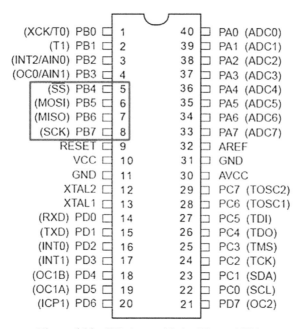

Figure 6.14   SPI pins on 40 pin ATmega16/32.

6.9.1 Register Descriptions

The AVR contains the following three registers that deal with SPI:

1. **SPCR (SPI Control Register)**: This register is basically the master register, that is, it contains the bits to initialize SPI and control it.
2. **SPSR (SPI Status Register)**: This is the status register. This register is used to read the status of the bus lines.
3. **SPDR (SPI Data Register):** The SPI Data Register is the read/write register where the actual data transfer takes place.

The SPI Control Register (SPCR), Figure 6.15

This register controls the SPI. It contains the bits that enable SPI, set up clock speed, configure master/slave, etc. Following are the bits in the SPCR Register.

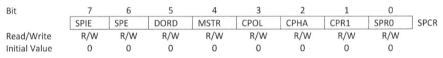

Bit	7	6	5	4	3	2	1	0	
	SPIE	SPE	DORD	MSTR	CPOL	CPHA	CPR1	SPR0	SPCR
Read/Write	R/W	R/W	R/W	R/W	R/W	R/W	R/W	R/W	
Initial Value	0	0	0	0	0	0	0	0	

Figure 6.15 SPCR register.

Bit 7: SPIE (SPI Interrupt Enable)

The SPI Interrupt Enable bit is used to enable interrupts in the SPI. Note that global interrupts must be enabled to use the interrupt functions. Set this bit to "1" to enable interrupts.

Bit 6: SPE (SPI Enable)

The SPI Enable bit is used to enable SPI as a whole. When this bit is set to 1, the SPI is enabled or else it is disabled. When SPI is enabled, the normal I/O functions of the pins are overridden.

Bit 5: DORD (Data Order)

DORD stands for Data ORDer. Set this bit to 1 if you want to transmit LSB first, else set it to 0, in which case it sends out MSB first.

Bit 4: MSTR (Master/Slave Select)

This bit is used to configure the device as *Master* or as *Slave*. When this bit is set to 1, the SPI is in *Master* mode (i.e., clock will be generated by the particular device), else when it is set to 0, the device is in SPI *Slave* mode.

Bit 3: CPOL (Clock Polarity)

This bit selects the clock polarity when the bus is idle. Set this bit to 1 to ensure that SCK is HIGH when the bus is idle, otherwise set it to 0 so that SCK is LOW in case of idle bus.

This means that when CPOL = 0, then the leading edge of SCK is the rising edge of the clock. When CPOL = 1, then the leading edge of SCK will actually be the falling edge of the clock. The CPOL functionally is shown in Figure 6.16.

CPOL	Leading Edge	Trailing Edge
0	Rising	Falling
1	Falling	Rising

Figure 6.16 CPOL functionality.

Bit 2: CPHA (Clock Phase)

This bit determines when the data needs to be sampled. Set this bit to 1 to sample data at the leading (first) edge of SCK, otherwise set it to 0 to sample data at the trailing (second) edge of SCK. CPHA functionality is given in Figure 6.17.

CPHA	Leading Edge	Trailing Edge
0	Sample	Setup
1	Setup	Sample

Figure 6.17 CPHA functionality.

Bit 1,0: SPR1, SPR0 (SPI Clock Rate Select)

These bits, along with the SPI2X bit in the SPSR register (discussed next), are used to choose the oscillator frequency divider, wherein the f_{OSC} stands for internal clock, or the frequency of the crystal in case of an external oscillator.

Table 6.3 gives a detailed description of the frequency divider.

The SPI Status Register (SPSR)

The SPI Status Register, Figure 6.18, is the register that contains information about the status of the SPI bus and also interrupts flag. The following are the bits in the SPSR register.

Bit 7: SPIF (SPI Interrupt Flag)

The SPI Interrupt Flag is set whenever a serial transfer is complete. An interrupt is also generated if SPIE bit (bit 7 in SPCR) is enabled and global

Table 6.3   Frequency divider

SPI2X	SPR1	SPR0	SCK Frequency
0	0	0	$f_{osc}/4$
0	0	1	$f_{osc}/16$
0	1	0	$f_{osc}/64$
0	1	1	$f_{osc}/128$
1	0	0	$f_{osc}/2$
1	0	1	$f_{osc}/8$
1	1	0	$f_{osc}/32$
1	1	1	$f_{osc}/64$

Bit	7	6	5	4	3	2	1	0	
	SPIF	WCOL	-	-	-	-	-	SPI2X	SPSR
Read/Write	R	R	R	R	R	R	R	R/W	
Initial Value	0	0	0	0	0	0	0	0	

Figure 6.18   SPSR register.

interrupts are enabled. This flag is cleared when the corresponding ISR is executed.

Bit 6: WCOL (Write Collision Flag)

The Write COLlision flag is set when data are written on the SPI Data Register (SPDR, discussed next) when there is an impending transfer or the data lines are busy.

This flag can be cleared by first reading the SPI Data Register when the WCOL is set. Usually, if we give the commands of data transfer properly, this error does not occur. This will be discussed latter.

Bit 5:1

These are reserved bits.

Bit 0: SPI2x (SPI Double Speed Mode)

The SPI double speed mode bit reduces the frequency divider from 4x to 2x, hence doubling the speed. Usually this bit is not needed, unless we need very specific transfer speeds, or very high transfer speeds. Set this bit to 1 to enable SPI Double Speed Mode. This bit is used in conjunction with the SPR1:0 bits of SPCR Register.

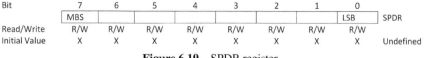

Bit	7	6	5	4	3	2	1	0	
	MBS							LSB	SPDR
Read/Write	R/W	R/W	R/W	R/W	R/W	R/W	R/W	R/W	
Initial Value	X	X	X	X	X	X	X	X	Undefined

Figure 6.19 SPDR register.

The SPI Data Register (SPDR)

The SPI Data register, Figure 6.19, is an 8-bit read/write register. This is the register from where we read the incoming data and write the data to which we want to transmit.

The 7th bit is obviously, the Most Significant Bit (MSB), while the 0th bit is the Least Significant Bit (LSB).

Now we can relate it to bit 5 of SPCR – the DORD bit. When DORD is set to 1, then LSB, that is, the 0th bit of the SPDR is transmitted first and vice versa.

AVR SPI Data Modes

The SPI of AVR offers the same 4 data modes for data communication, wiz SPI Mode 0, 1, 2, and 3, given in Table 6.1. As shown in the table, the only difference in these modes being the clock edge at which data is sampled. This is based upon the selection of CPOL and CPHA bits.

6.9.2 SPI Coded

This section explains how to Code SPI. The datasheets of ATmega 16 are used to write the codes.

Enabling SPI on Master

```
// Initialize SPI Master Device (with SPI interrupt)
void spi_init_master (void)
{
    // Set MOSI, SCK as Output
    DDRB=(1<<5) | (1<<3);

    // Enable SPI, Set as Master
    // Prescaler: Fosc/16, Enable Interrupts
    //The MOSI, SCK pins are as per ATMega8
    SPCR=(1<<SPE) | (1<<MSTR) | (1<<SPR0) | (1<<SPIE);

    // Enable Global Interrupts
    sei();
}
```

In the SPI Control Register (SPCR), the SPE bit is set to 1 to enable SPI of AVR. To set the microcontroller as Master, the MSTR bit in the SPCR is also set to 1. To enable the SPI transfer/receive complete interrupt, the SPIE is set to 1.

In case you don't wish to use the SPI interrupt, do not set the SPIE bit to 1 and do not enable the global interrupts. This will make it look somewhat like this:

```
// Initialize SPI Master Device (without interrupt)
void spi_init_master (void)
{
    // Set MOSI, SCK as Output
    DDRB = (1<<5) | (1<<3);

    // Enable SPI, Set as Master
    //Prescaler: Fosc/16, Enable Interrupts
    SPCR = (1<<SPE) | (1<<MSTR) | (1<<SPR0);
}
```

When a microcontroller is set as Master, the Clock prescaler is also to be set using the SPRx bits.

Example showing how to set the SPI registers:

Here is a code snippet to generate a data transfer between a Master and a Slave. Both Master and Slave are configured to send the MSB first and to use SPI mode 3. The clock frequency of the Master is fosc/16. The Master will send the data 0xAA, and the Slave will send the data 0x55.

Master code:

```
SPI_Init:
sbi DDRB,DDB5      // Set MOSI as output
sbi DDRB,DDB7      // Set SCK as output.
sbi DDRB,DDB4      // Set SS' as output.
ldi r16,01011101b  // Set SPI as a Master, with interrupt disabled,
out SPCR,r16       // MSB first, SPI mode 3 and clock frequency
                   // fosc/16.


SPI_Send:
ldi r16,0xAA
out SPDR,r16       // Initiate data transfer.
Wait:
sbis SPSR,SPIF     // Wait for transmission to complete.
rjmp Wait
in SPDR,r16        // The received data is placed in r16.
```

Slave code:

```
SPI_Init:
sbi DDRB,DDB6      // Set MISO as an output
ldi r16,01001100b  // Set SPI as a Slave, with interrupt disabled,
out SPCR,r16       // MSB first and SPI mode 3.
ldi r16,0x55
out SPDR,r16       // Send 0x55 on Master request.

SPI_Receive:
sbis SPSR,SPIF
rjmp SPI_Receive   // Wait for reception to complete.
in r16,SPDR        // The received data is placed in r16.
```

Next is a program for enabling SPI on slave.

Enabling SPI on Slave

```
// Initialize SPI Slave Device
void spi_init_slave (void)
{
    DDRB = (1<<6);     //MISO as OUTPUT
    SPCR = (1<<SPE);   //Enable SPI
}
```

For setting a microcontroller as a slave, one just needs to set the SPE Bit in the SPCR to 1 and direct the MISO pin (PB4 in case of ATmega16A) as OUTPUT.

Sending and Receiving Data

```
//Function to send and receive data for both master and slave
unsigned char spi_tranceiver (unsigned char data)
{
    // Load data into the buffer
    SPDR = data;

    //Wait until transmission complete
    while(!(SPSR & (1<<SPIF) ));

    // Return received data
    return(SPDR);
}
```

The codes for sending and receiving data are same for both the slave as well as the master. To send data, load the data into the SPI Data Register

(SPDR), and then, wait until the SPIF flag is set. When the SPIF flag is set, the data to be transmitted are already transmitted and are replaced by the received data. So, simply return the value of the SPI Data Register (SPDR) to receive data. We use the return type as unsigned char because it occupies 8 bits and its value is in the range 0–255.

6.9.3 Complete Application for using SPI of AVR

As application of using SPI of AVR, consider using SPI AVR to:

- send some data from *Master* to *Slave*.
- *Slave* in return sends an acknowledgement (ACK) data back to the *Master*.
- *Master* should check for this ACK in order to confirm that the data transmission has completed.

This is a typical example of full-duplex communication. While the *Master* sends the data to the *Slave*, it receives the ACK from the *Slave* simultaneously. The example and the related codes are part of AVR gallery.

Methodology

We would use the primary microcontroller (ATmega8 in this case) as the *Master* device and a secondary microcontroller (ATmega16 in this case) as the *Slave* device. A counter increments in the *Master* device, which is being sent to the *Slave* device. The *Master* then checks whether the received data are the same as ACK or not (ACK is set as 0x7E in this case). If the received data are the same as ACK, it implies that data have been successfully sent and received by the *Master* device. Thus, the *Master* blinks an LED connected to it as many number of times as the value of the counter which was sent to the *Slave*. If the *Master* does not receive the ACK correctly, it blinks the LED for a very long time, thus notifying of a possible error.

On the other hand, *Slave* waits for data to be received from the *Master*. As soon as data transmission begins (from *Master* to *Slave*, the *Slave* sends ACK (which is 0x7E in this case) to the *Master*. The *Slave* then displays the received data in an LCD.

Hardware Connections

Hardware connections are simple. Both the MOSI pins are connected together, MISO pins are connected together and the SCK pins are also connected together. The SS' pin of the slave is grounded, whereas that of the master is left unconnected. To demonstrate the operation of the slave and

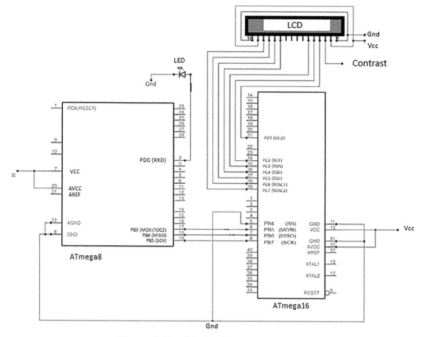

Figure 6.20 SPI hardware connections.

the interrupt, LCD and LED are connected. Figure 6.20 gives the complete circuit diagram which is a standard for such example.

Full Code

The codes for the *Master* and *Slave* are given below. The codes are well commented, so it should be easy to understand what is going on in the code. The reader can also find the code in the AVR code gallery.

Master Code

```
#ifndef F_CPU
#define F_CPU 16000000UL
#endif

#include <avr/io.h>
#include <util/delay.h>
#include <avr/interrupt.h>

#define ACK 0x7E
#define LONG_TIME 10000
```

```
//Initialize SPI Master Device
void spi_init_master (void)
{
    DDRB = (1<<5) | (1<<3);                 //Set MOSI, SCK as Output
    SPCR = (1<<SPE) | (1<<MSTR) | (1<<SPR0); //Enable SPI, Set as Master
                                    //Prescaler:    Fosc/16,     Enable
Interrupts
}

//Function to send and receive data
unsigned char spi_tranceiver (unsigned char data)
{
    SPDR = data;                        //Load data into the buffer
    while(!(SPSR & (1<<SPIF) ));         //Wait until transmission complete
    return(SPDR);                       //Return received data
}

//Function to blink LED
void led_blink (uint16_t i)
{
    //Blink LED "i" number of times
    for (; i>0; --i)
    {
        PORTD | =(1<<0);
        _delay_ms(100);
        PORTD=(0<<0);
        _delay_ms(100);
    }
}

//Main
int main(void)
{
    spi_init_master();              //Initialize SPI Master
    DDRD  | = 0x01;                  //PD0 as Output

    unsigned char data;             //Received data stored here
    uint8_t x = 0;                  //Counter value which is sent

    while(1)
    {
        data = 0x00;                //Reset ACK in "data"
        data = spi_tranceiver(++x);  //Send "x", receive ACK in "data"
        if(data == ACK) {           //Check condition
            //If received data is the same as ACK, blink LED "x" number of
times
            led_blink(x);
        }
        else {
            //If received data is not ACK, then blink LED for a long time
                so as to determine error
            led_blink(LONG_TIME);
        }
        _delay_ms(\ref{GrindEQ__500_});                    //Wait
    }
```

Slave Code

```
#ifndef F_CPU
#define F_CPU 16000000UL
#endif

#include <avr/io.h>
#include <avr/interrupt.h>
#include <util/delay.h>
#include "lcd.h"

#define ACK 0x7E

void spi_init_slave (void)
{
    DDRB=(1<<6);                        //MISO as OUTPUT
    SPCR=(1<<SPE);                      //Enable SPI
}

//Function to send and receive data
unsigned char spi_tranceiver (unsigned char data)
{
    SPDR = data;                        //Load data into buffer
    while(!(SPSR & (1<<SPIF) ));         //Wait until transmission
complete
    return(SPDR);                       //Return received data
}

int main(void)
{
    lcd_init(LCD_DISP_ON_CURSOR_BLINK);    //Initialize LCD
    spi_init_slave();                      //Initialize slave SPI
    unsigned char data, buffer[10];
    DDRA  = 0x00;                          //Initialize PORTA as INPUT
    PORTA = 0xFF;                          //Enable Pull-Up Resistors
    while(1)
    {
        lcd_clrscr();                      //LCD Clear screen
        lcd_home();                        //LCD move cursor to home
        lcd_puts("Testing");
        lcd_gotoxy(0,1);
        data = spi_tranceiver(ACK);        //Receive data, send ACK
        itoa(data, buffer, 10);               //Convert  integer  into
string
        lcd_puts(buffer);                  //Display received data
        _delay_ms(20);                     //Wait
    }
}
```

6.10 Development Tools

When developing or troubleshooting systems using SPI, visibility at the level
of hardware signals can be important.

Host adapters

There are a number of USB hardware solutions to provide computers, running Linux, Mac, or Windows, SPI master, or slave capabilities. Many of them also provide scripting or programming capabilities (Visual Basic, C/C++, VHDL, etc.).

An SPI host adapter lets the user play the role of a master on an SPI bus directly from a PC. They are used for embedded systems, chips (FPGA, ASIC, and SoC) and peripheral testing, programming, and debugging.

The key parameters of SPI adapters are the maximum supported frequency for the serial interface, command-to-command latency, and the maximum length for SPI commands. It is possible to find SPI adapters on the market today that support up to 100 MHz serial interfaces, with virtually unlimited access length.

SPI protocol being a de facto standard, some SPI host adapters also have the ability of supporting other protocols beyond the traditional 4-wire SPI (e.g., support of quad-SPI protocol or other custom serial protocol that derive from SPI).

Protocol analyzers

SPI protocol analyzers are tools which sample an SPI bus and decode the electrical signals to provide a higher-level view of the data being transmitted on a specific bus.

Oscilloscopes

Most oscilloscope vendors offer oscilloscope-based triggering and protocol decoding for SPI. Most support 2-, 3-, and 4-wire SPI. The triggering and decoding capability is typically offered as an optional extra. SPI signals can be accessed via analog oscilloscope channels or with digital MSO channels.

Logic analyzers

When developing or troubleshooting the SPI bus, examination of hardware signals can be very important. Logic analyzers are tools which collect, analyze, decode, and store signals, so people can view the high-speed waveforms at their leisure. Logic analyzers display time-stamps of each signal level change, which can help find protocol problems. Most logic analyzers have the capability to decode bus signals into high-level protocol data and show ASCII data.

6.11 Synchronous Serial Interface (SSI)

Synchronous Serial Interface (SSI) is a widely used serial interface standard for industrial applications between a master (e.g., controller) and a slave (e.g., sensor). It is widely used for absolute position and rotary measuring systems. It enables position and angular information to be transmitted digitally, absolutely and without bus overhead. As a result, it is especially well-suited for applications in which reliability and signal robustness are required in an industrial environment.

SSI is based on RS-422 standards and has a high protocol efficiency in addition to its implementation over various hardware platforms, making it very popular among sensor manufacturers. SSI was originally developed by Max Stegmann GmbH in 1984 for transmitting the position data of absolute encoders – for this reason, some servo/drive equipment manufacturers refer to their SSI port as a "Stegmann Interface." It was formerly covered by the German patent DE 34 45 617 which expired in 1990. It is very suitable for applications demanding reliability and robustness in measurements under varying industrial environments.

It is different from the Serial Peripheral Interface Bus (SPI): An SSI is differential, simplex, non-multiplexed, and it relies on a time-out to frame the data. An SPI is single-ended, duplex, multiplex, and it uses a select-line to frame the data. However, SPI peripherals on microcontrollers can implement SSI with external differential driver-ICs and program-controlled timing.

This section describes the Synchronous Serial Interface (SSI) used by many position sensors and controllers. It is aimed at electrical or mechanical engineers who are designing a position sensing systems and want to understand how SSI works and gauge its merits, without getting too deeply in to the bits and bytes.

Background

Sensors for measuring linear or angular position are divided into two groups: incremental and absolute. It is possible to identify the group for which the sensor belongs by monitoring "What happens on power up?" If the sensor has to do a calibration step to find its position – it's incremental;

if it doesn't – it's absolute. Most engineers still specify incremental position sensors because they think absolute versions might be too costly but, nowadays, absolute sensors don't cost that much more. The proportion of absolute sensors is increasing because equipment and automation users are increasingly unwilling to wait for a protracted start-up routine.

In the 80s and 90s, many absolute position sensors gave a parallel output where the 1s and 0s of a binary value or Gray code were represented by a number of wires (often a ribbon cable) by high or low voltage. This parallel approach has died off because of its relatively high cost, size, complexity and reliability – especially in precision sensors. A 20-bit parallel output sensor would need at least 20 individual signal wires, which can be accepted if the application has a single sensor but it is completely unacceptable if the application, for example, robot, has lots of sensors, motors, and power lines.

Parallel interfaces have largely been replaced by Synchronous Serial Interface (SSI) where the 1s and 0s are fed in series (one after the other) along the signal wires. SSI is a well-established and widely used interface for industrial communications between a controller and sensor – especially for absolute position sensors. Importantly, the use of SSI is not restricted to a particular manufacturer or group of manufacturers and no membership or license fees are needed. It is – and, in the foreseeable future, is likely to be – the most widely used form of digital communications for absolute position sensors. Many types of transducers are available with SSI, including magnetostrictive displacement transducers (MDTs), absolute encoders, and laser measuring devices.

SSI has a number of advantages over other transducer interfaces, for example:

- High noise immunity
- Absolute position
- Supports a wide variety of transducers.
- Many SSI devices offer higher precision; for example, magnetostrictive transducers with SSI output are commonly available with resolutions to 1 μm, and some offer 0.1 μm.

6.11.1 SSI – General

SSI is based on the widely used RS-422 hardware standard. This standard specifies the electrical characteristics of the signaling circuits, but is limited to defining signal levels. In other words, it defines the basic hardware but

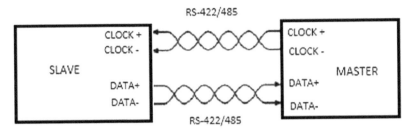

SSI BLOCK DIAGRAM

Figure 6.21 Simplified RS-422 DIAGRAM: SSI point-to-point communication.

other specifics such as electrical connectors, pins and wiring, are open to the designer's choice. The RS-422 circuit designs used by SSI allow data to be reliably and quickly transmitted over long distances in noisy environments without expensive or bulky electronics, cables or connectors. High data rates and long distances are achieved using balanced, or differential, signaling. Differential means when one line is high, the other is low and vice versa.

RS-422 uses a nominal 0 to 5 V signal levels and typically uses a cable made of two sets of twisted pair wires (one pair for data and one pair for clock signals) and a ground wire. While a double pair cable may be practical for many RS-422 applications, the RS-422 specification only defines one signal path and does not assign any function to it. Most SSI cable uses twisted pairs with a metal foil or mesh, as an electromagnetic shield, over each twisted pair and/or over the complete wire bundle underneath an overall cable sheath.

Generally, with SSI, data and clock transmissions over cable lengths of around 20 m need little or no special consideration. Distances of 20 m will cover most position sensor applications. At cable lengths of >20 m, it's a good idea to keep cable lengths between position sensor and controller as short as practical. High data rates of about 10 Mbits/second are readily achieved over cable lengths of 10 or 20 m (with 24 AWG cable) but as cable lengths increase above 20 m, sensor data rates should be reduced as per Figure 6.22:

The above data are based on 24 AWG wire and cable lengths can be increased if the cable is beefed up to say 22 or 20AWG. Maximum cable length is also influenced by the tolerable signal distortion, local electromagnetic noise levels, and differences in ground potential between the cable ends.

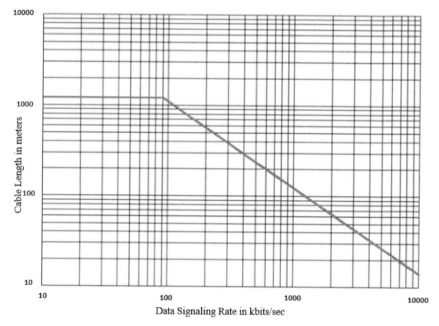

Figure 6.22 Noise in straight and twisted pair cable.

Relative to other interfaces, SSI's technical features are as follows:

- Cost effective data transfer: SSI is **inexpensive** due to
  - low count of electronic components
  - only 4 wires (5 wires if shield is needed)
  - slaves or sensors use master's clock and no need for precision oscillators
  - widely available connectors and cables
- **secure** data output with possibility for error detection and parity signaling
- Noise immunity: **resilient** to electromagnetic interference
- Transmission over long distances: **long cable lengths** of up to thousands of meters can be used
- **easy electrical isolation** between sensor and host
- **synchronized data transmission** to a clock signal
- **high baud rates** of up to 10 Mbits/s
- **flexible** – the number of bits in a message is not limited to a defined message size

- **multiple slaves** can be connected to a common clock.
- High solution. Down to 0.1 μm (approx. $0.000004''$) for linear SSI transducers.
- Transmission rate independent of data length and resolution

Important note: Engineers can sometimes get confused between SSI, RS-422 and RS-485. SSI is the serial communications method which uses RS-422 hardware standards. RS-485 is a more complex multi-node or bus system. RS-422 cannot implement a true multi-point communications network since there can be only one driver on each pair of wires; however, one clock signal can be used by more than one sensor.

6.11.2 SSI Design

The interface has a very simple design as illustrated in the above figure. It consists of two pairs of wires, one for transmitting the clock signals from the master and the other for transmitting the data from the slave. The clock sequences are triggered by the master when need arises. Different clock frequencies can be used ranging from 100 kHz to 2 MHz and the number of clock pulses depends on the number of data bits to be transmitted.

The simplest SSI slave interface uses a retriggerable monostable multivibrator (monoflop) to freeze the current value of the sensor. The current frozen values of the slave are stored in shift registers. These values are clocked out sequentially when initiated by the controller. The design is being revolutionized with the integration of microcontrollers, FPGAs and ASICs into the interface.

The data format is designed in such a way to ensure proper communication of data. The protocol for the data transmission is based on three different subsequent parts (Leading-"1" → Data-Bits → Trailing-"0"). The main significance of this type of format is to ensure the proper working of the interface and hence secure data transmission free from any hardware or software errors.

The SSI is initially in the idle mode, where both the CLOCK and DATA lines stay high and the slave keeps updating its data. Keeping the CLOCK and DATA outputs high while idle is useful in detecting broken wire contacts. This helps in observing the proper working condition of the interface.

As can be seen in Figure 6.23, the first falling edge after Tmu starts the Read Cycle and the transfer of data. Each rising edge of the CLOCK transmits the next data bit of the message, staring with Dn-1. After the last rising edge of the clock sequence, the data line is set by the Error Flag (if supported) for

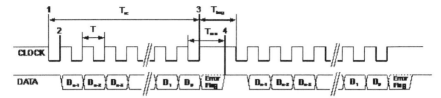

Figure 6.23   SSI timing diagram.

the period Tmu – 0.5Xt. After Tmu, the latest position data are now available for transmission in the next Read Cycle.

- T: Clock period (1/T = 100 kHz to 2 MHz)
- Trc: Read cycle time: This is defined as (n × T) + (0.5 × T)
- Tmu: Message update time. The time from last falling edge of clock to when new data are ready for transmission
- Timg: Intermessage gap time. Must be >Tmu otherwise position data will be indeterminate
- n: The number of bits in the message (not including the Error Flag).
- "tm" represents the transfer timeout (monoflop time). It is the minimum time required by the slave to realize that the data transmission is complete. After tm, the data line goes to idle and the slave starts updating its data in the shift register.
- "tp" represents the pause time. It is the time delay between two consecutive clock sequences from the master.
- "tw" represents the repetition time. It is the minimum time elapsed between retransmissions of the same data and is always less than tm.
- "T" represents the width of each clock cycle. It is the time taken between two falling or two rising edges in a continuous clock sequence.
- MSB: Most significant bit
- LSB: Least significant bit

Figure 6.23 illustrates the single data transmission using SSI protocol:

- The SSI is initially in the idle mode, where both the data and clock lines stay HIGH and the slave keeps updating its current data.
- The transmission mode is evoked when the master initiates a train of clock pulses. Once the slave receives the beginning of the clock signal (1), it automatically freezes its current data. With the first rising edge (2) of the clock sequence, the MSB of the sensor's value is transmitted and with consequent rising edges, the bits are sequentially transmitted to the output.

- After the transmission of complete data word (3) (i.e., LSB is transmitted), an additional rising edge of the clock sets the clock line HIGH. The data line is set to LOW and remains there for a period of time, tm, to recognize the transfer timeout. If a clock signal (data-output request) is received within that time, the same data will be transmitted again (multiple transmission).
- The slave starts updating its value and the data line is set to HIGH (idle mode) if there are no clock pulses within time, tm. This marks the end of single transmission of the data word. Once the slave receives a clock signal at a time, tp (=tm), the updated position value is frozen and the transmission of the value begins as described earlier.

After n-CLOCK pulses (rising edges) the data is completely transmitted. With the next CLOCK pulse (rising edge n+1) the sensor output goes to low level which can be used to detect a short circuit in the cable. If it is high even after n+1 rising edges then it means that the interface has a short circuit.

Readings from multiple slaves (up to three) can be enabled at the same time by connecting them to a common clock. However, to avoid ground loops and electrically isolate the slave, complete galvanic isolation by opto-couplers is needed.

6.11.3 SSI Clock and Data Transmission

The master (or controller) controls the clock signals and the slave (the position sensor) transmits the data/value. When invoked by the master, the data is clocked out from the slave's output – usually a shift register. The master and slave are synchronized by the clock. Data are transmitted using balanced or differential signals. Basically this means the CLOCK and DATA lines are twisted pair cables. The clock sequence is triggered by the master when data/value are required. Different clock frequencies can be used ranging from 100 kHz to 2 MHz and the number of clock pulses depends on how many data bits are to be transmitted.

The protocol for the data transmission is based on three different subsequent parts (Leading-"1" → Data-Bits → Trailing-"0"). This helps ensure reliable and secure data transmission, free from any hardware or software errors.

The SSI is initially in the idle mode, where both the CLOCK and DATA lines stay high and the slave keeps updating its data. Keeping the CLOCK and DATA outputs high while idle is useful in detecting broken wire contacts.

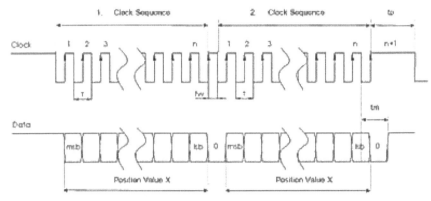

Figure 6.24   Multiple transmission.

Multiple Transmission

Multiple transmissions of the same data from the position sensor happen only if there is continuous clocking even after the transmission of the least significant bit. The initial sequences are the same as that of the single transmission. In the idle state, the CLOCK and DATA lines are high but with the arrival of the first falling edge the transmission mode is evoked and the similarly the data bits are transmitted sequentially starting with the most significant bit with every rising edge of the CLOCK. The transmission of the least significant bit means that the transmission of the data is complete. An additional rising edge pushes the data line to low, signifying the end of transmission of the data.

If there are continuous clock pulses even after the completion (i.e., the next clock pulses comes in time tw [<tm]) the value of the slave is not updated. This is because the monoflop output is still unsteady and the value in the shift register still contains the same value as before. So with the next rising edge, that is, after the (n+1) rising edge, the transmission of the same data continues, and the MSB of data transmitted earlier is re-transmitted. Then, it follows the same procedure as earlier transmissions, leading to multiple transmissions of the same data. The value of the slave is updated only when the timing between two clock pulses is more than the transfer timeout. Multiple transmission can be used to check the data integrity. The two consecutive received values are compared, and transmission failures are indicated by differences between the two values. The transmission of data is controlled by the master and the transmission can be interrupted at any time just by stopping the clock sequence, for a period longer than the time out

period. The slave automatically will recognize the transfer timeout and go into idle mode.

Some position sensor manufacturers have added additional information to the basic SSI protocol, in various efforts to ensure high integrity data transmission. For secure transmission and to indicate the end of data transmission CRC bits or parity bits can be added. They are used for identifying if the data from the position sensor has been correctly interpreted and received.

Multiple transmission is used to check the data integrity. The two consecutive received values are compared, and transmission failures are indicated by differences between the two values.

Interrupting transmission

The transmission of data is controlled by the master and the transmission can be interrupted at any time just by stopping the clock sequence, for a period longer than tm. The slave automatically will recognize the transfer timeout and go into idle mode.

6.11.4 Cabling – According to RS-422 Standards

Since SSI is based on RS-422 standards, it is necessary to select appropriate cables and to stay within the limits of cabling length and clock frequencies. The maximum allowable SSI cable length depends on the SSI clock rate. For SSI inputs on the UI/O module, wire delay compensation is available to allow longer lengths, as described in the Wire Delay Compensation section below.

The relation between the cable length and clock frequency is shown in Figure 6.25 and Table 6.4.

Figure 6.25 can be used as a conservative guide. This curve is based upon empirical data using a 24 AWG Standard, copper conductor, unshielded twisted-pair telephone cable with a shunt capacitance of 52.5 pF/meter (16 pF/foot) terminated in a 100 Ohm resistive load. The cable length restriction shown by the curve is based upon assumed load signal quality requirements of:

1. Signal rise and fall times equal to or less than one-half unit interval at the applicable data switching rate.
2. A maximum voltage loss between generator and load of 66%.

When high data rates are used, the application is limited to shorter cables. It is possible to use longer cables when low data rates are used. The DC resistance of the cable limits the length of the cable for low data rate applications by increasing the noise margin as the voltage drop in the cable

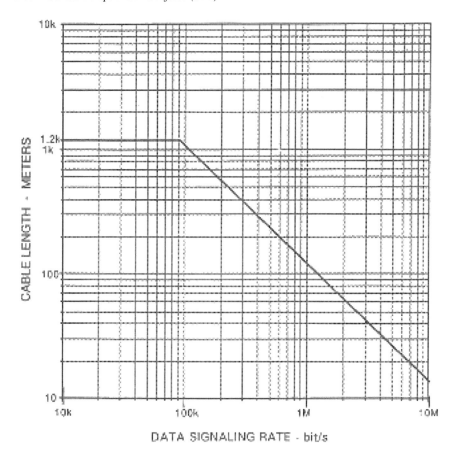

DATA SIGNALING RATE - bit/s

Cable length versus data signaling rate

Figure 6.25   Cable length versus signaling rate.

increases. The AC effects of the cable limit the quality of the signal and limit the cable length to short distances when high data rates are used. Examples of data rate and cable length combinations vary from 90 kbit/s at 1.2 km to 10 Mbit/s at 5 m for RS-422.

Cables having characteristics different from the twisted pair 24 AWG, 52.5 pF/m (16 pF/ft) can also be employed within bounds mentioned above. First, determine the absolute loop resistance and capacitance values of the typical 24 AWG cable provided by the cable length associated with the data signaling rate desired from the figure. Then convert those values to equivalent lengths of the cable actually used. For example, longer distances would be

Table 6.4 SSI cable length

Clock Rate	Maximum Cable Length*
100 kHz	2100 ft (640 m)
150 kHz	1360 ft (415 m)
230 kHz	850 ft (255 m)
250 kHz	770 ft (235 m)
375 kHz	475 ft (145 m)
400 kHz	450 ft (135 m)
500 kHz	325 ft (99 m)
625 kHz	225 ft (70 m)
921 kHz	120 ft (37 m)
971 kHz	110 ft (34 m)
1000 kHz	100 ft (30 m)
1500 kHz	25 ft (7.5 m)
2500 kHz	3 ft (1 m)

* The cable lengths are approximate and may be affected by the type of wire and transducer.

possible when using 19 AWG, while shorter distances would be necessary for 28 AWG.

The maximum permissible length of cable separating the master and slave is a function of data signaling rate and is influenced by the tolerable signal distortion, the amount of longitudinally coupled noise and ground potential differences introduced between the master and the slave circuit. Accordingly, users are advised to restrict cable length to a minimum. The type and length of the cable used must be capable of maintaining the necessary signal quality needed for the particular application. Furthermore, the cable balance must be such as to maintain acceptable crosstalk levels, both generated and received.

6.11.4.1 Wire delay compensation (RMC150 UI/O module only)

Wire delay compensation is available on some modules, for example, the RMC150 Universal I/O Module. The delay compensation is required for SSI wire runs that exceed the lengths given in the SSI Cable Length section above. If the wire to the SSI device is very long, there will be a significant delay between the clock signal and the returned data signal. As shown in Figure 6.27 if this delay exceeds one clock period, the module will not receive the correct data, unless the SSI Wire Delay parameter is used.

Minimal Delay

The timing diagram below shows an SSI system with very little delay. On the first rising edge of the Clock, the SSI device puts the first bit of data on the Data line. By the next rising edge of the Clock, when the RMC samples the data, the data is valid, and the read is successful.

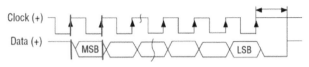

Figure 6.26   Effect of delay.

Excessive Delay

The timing diagram below shows an SSI system with a time delay of more than one clock period. On the first rising edge of the Clock, the SSI device puts the first bit of data on the Data line. By the next rising edge of the Clock, when the RMC samples the data, the data from the SSI device have not yet arrived, and the SSI input will not return the correct value.

To compensate for the delay, set SSI Wire Delay parameter. You can enter the wire length or enter the time delay directly. The SSI input will then use the delay value to correctly read the SSI input data.

Figure 6.27   Excessive delay effect.

6.11.5 Sensors using SSI

In this section, we use the term "sensor" rather than the more usual "encoder." This is deliberate because encoders are often, but incorrectly, thought of as optical devices, producing data in proportion to a measured position. In recent years, a new generation of non-contact encoders – especially absolute encoders – are not optical but rather inductive (sometimes referred to as "incoders"). Such devices use printed circuit board transformer constructions rather than the bulky and expensive transformer windings used in traditional inductive position sensors such as resolvers, LVDTs, RVDTs or synchros. These traditional devices have been the engineer's preferred choice in many harsh environments for many years due to their non-contact, reliable operation and excellent safety record. Incoders use the same basic physics of their traditional counterparts and so, unsurprisingly they are just as reliable and robust but are more accurate and easier to use. Their ease of use partly comes from the fact that they use SSI as a preferred communication method. They have gained a significant market share in through bore, bearing-less formats favored in high reliability, precision sensor applications in the defense, medical, aerospace and industrial sectors.

6.11.6 Derived protocols

Some manufacturers and organizations added additional information to the basic SSI protocol. It was done mainly to ensure proper data transmission. For secure transmission and to indicate the end of data transmission, CRC bits or parity bits can be added. In simple words, they were used for identifying if the byte has been correctly interpreted and received. In the original specification, multiple transmissions were used to ensure data integrity. In this case, two consecutive transmissions of the same data were initiated and compared for any error in transmission. But this however reduces the protocol efficiency by 50% compared to parallel data transmission.

6.11.7 Benefits

- Serial data transfer has reduced the wiring. This, in addition to the simplicity of SSI design due to use of minimal number of components, has considerably reduced the cost and created more transmission bandwidth for message bits.
- High Electromagnetic interference immunity due to RS-422 standards and higher reliability of data transmission due to differential signaling.
- Optimal galvanic isolation
- Complete protocol flexibility for the number of bits transferred. Not limited to a certain number of words and has an arbitrary choice of message size.
- Slaves use master's clock and hence don't need precision oscillators.
- The SSI allows to connect up to three slaves to a common clock. Therefore, it is possible to attain values from multiple sensors.

The limitations in this interface are negligible for applications in industrial automation. SSI can handle only short distance communication (up to 1.2 km) and supports only one master device. But, 1.2 km is rather a good distance for communication by automation industry standards. When compared to advanced communication systems based on field buses or Ethernet,

SSI is limited to a master–slave architecture and a simple point-to-point communication between a master and a slave. Another disadvantage is that there is no hardware slave acknowledgment, that is, detection of slave for communication.

References

Web Sources

[1] SPI Block Guide v3.06; Motorola/Freescale/NXP; 2003.
[2] "N5391B I^{2}C and SPI Protocol Triggering and Decode for Infiniium scopes".
[3] MICROWIRE Serial Interface National Semiconductor Application Note AN-452
[4] MICROWIRE/PLUS Serial Interface for COP800 Family National Semiconductor Application Note AN-579
[5] "QuadSPI flash: Quad SPI mode vs. QPI mode". *NXP community forums*. December 2014. Retrieved 2016-02-10.
[6] *Patterson, David (May 2012).* "Quad Serial Peripheral Interface (Quad-SPI) Module Updates" *(PDF) (Application note).* Freescale Semiconductor. Retrieved September 21, 2016.
[7] *Pell, Rich* (13 October 2011). "Improving performance using SPI-DDR NOR flash memory". EDN.
[8] Enhanced Serial Peripheral Interface (eSPI) Interface Base Specification (for Client and Server Platforms) *(PDF) (Report). Revision 1.0. Intel. January 2016. Document number 327432-004.* Retrieved 2017-02-05.
[9] Enhanced Serial Peripheral Interface (eSPI) Interface Specification (for Client Platforms) *(PDF) (Report). Revision 0.6. Intel. May 2012. Document Number 327432-001EN.* Retrieved 2017-02-05.

7

Universal Serial Bus (USB)

7.1 Introduction to USB

The Universal Serial Bus (USB)) is an industry standard that establishes specifications for cables, connectors, and protocols for connection, communication, and power supply between personal computers and their peripheral devices.

The USB specification contains all the information about the protocol such as the electrical signaling, the physical dimension of the connector, the protocol layer, and other important aspects.

The Universal Serial Bus is developed by Compaq, Intel, Microsoft and NEC, joined later by Hewlett–Packard, Lucent, and Philips. These seven companies formed the USB Implementers Forum (USB-IF) as a non-profit corporation to publish the specifications and organize further development in USB.

The team that worked on the standard at Intel succeeded to produce the first integrated circuits supporting USB in 1995. USB released in 1996 and the USB standard is currently maintained by the USB Implementers Forum (USB IF).

The goal of the USB-IF was to find a solution to the mixture of connection methods to the PC, in use at the time. Multitude of connectors were used at the back of PC to connect serial ports, parallel ports, keyboard, and mouse connections, joystick ports, midi ports and so on. None of these connectors satisfied the basic requirements of plug-and-play. Additionally many of these ports made use of a limited pool of PC resources, such as Hardware Interrupts, and DMA channels. So the USB was developed as a new means to connect a large number of devices to the PC, and eventually to replace the "legacy" ports. It was designed not to require specific Interrupt or DMA resources, and also to be "hot-pluggable." It standardized the connection of

245

peripherals like keyboards, pointing devices, digital still and video cameras, printers, mice, scanners, portable media players, disk drivers and network adapters to personal computers, both to communicate and to supply electric power. This means that while USB provides a sufficiently fast serial data transfer mechanism for data communications, it is also possible to obtain power through the connector making it possible to power small devices via the connector and this makes it even more convenient to use, especially 'on-the-go.'

It was important while developing USB to be sure that no special user-knowledge would be required to install a new device, and all devices would be distinguishable from all other devices, such that the correct driver software was always automatically used.

The USB specification allows for the connection of a maximum of 127 peripheral devices (including hubs) to the system, either on the same port or on different ports.

USB also supports Plug and Play installation and hot swapping. The **USB 1.1** standard supports both isochronous and asynchronous data transfers and has dual speed data transfer: 1.5 Mb/s (megabits per second) for **low-speed** USB devices and 12 Mb/s for **full-speed** USB devices (much faster than the original serial port). Cables connecting the device to the PC can be up to 5 m (16.4 ft) long. USB includes built-in power distribution for low power devices and can provide limited power (up to 500 mA of current) to devices attached on the bus.

The **USB 2.0** standard supports a signalling rate of 480 Mb/s, known as "**high speed**," which is 40 times faster than the USB 1.1 full-speed transfer rate.

USB 2.0 is fully forward- and backward-compatible with USB 1.1 and uses existing cables and connectors.USB 2.0 supports connections with PC peripherals that provide expanded functionality and require wider bandwidth. In addition, it can handle a larger number of peripherals simultaneously.

USB 2.0 enhances the user's experience of many applications, including interactive gaming, broadband Internet access, desktop, and Web publishing, Internet services and conferencing.

The Universal Serial Bus provides a very simple and effective means of providing connectivity, and as a result, it is very widely used. USB provides several benefits compared to other communication interfaces such as ease of use, low cost, low power consumption and, fast and reliable data transfer.

7.1.1 USB and RS-232

It is possible to add here that one of the main reasons behind the seven companies to push the development of USB was the huge amount of different RS-232 cable layouts for different purposes and the total lack of practical information about them. With USB the seven parties wanted to get rid of these problems: the large number of the different RS-232 cables layouts. In fact it was one of the three main motivations as described in the USB 1.1 specification.

USB has replaced the RS-232 and parallel communications in a lot of situations and has become commonplace on a wide range of devices. USB is now the most used interface to connect peripheral devices to personal computers.

USB connectors have been, also, increasingly replacing other types for battery chargers of portable devices.

Saying that USB has replaced the RS-232 does not mean that the two interfaces are completely compatible. Three differences may be identified between USB and RS-232. The differences are summarized here and discussed in detail in Section 7.4. The differences are:

1. USB is interface to connect and communicate with many kinds of devices and not only communicating with modems as it is the case of RS-232.
2. USB is not designed primarily for 1:1 communication like RS-232 or the parallel printer interface, but it is a bus architecture where more than two devices can be attached.
3. When saying that USB is serial communication as RS-232, it does not mean that they are similar or compatible. Serial is not a family of interchangeable communication interfaces. It just tells us that every bit of information is send in a specific time slot and that no two items of information can be send at one moment. This is not only the case for USB and RS-232. Ethernet networks also communicate in a serial way. Actually most communications – even high-speed – are performed in a serial way.

The above differences mean that, although RS-232 and USB (universal serial bus) are both serial communication standards to connect peripherals to computers, they are totally different in design. A simple cable is not enough to connect RS-232 devices to a computer with only USB ports.

There are however converter modules and cables that can be successfully used to connect RS-232 devices to computers via a USB port. Such convertors are discussed latter

7.2 USB Versions

USB has different versions, for example, USB 1.1 and USB 2.0, etc. Higher USB version numbers describe the USB interface with more features and higher speed. The USB interface is already under development for more than ten years. Version 0.7 of the USB interface definition was released in November 1994 and the first "real" definition of USB, USB 1.0, came out in January 1996 by Intel.

There have been three generations of USB specifications:

- USB 1.x
- USB 2.0, with multiple updates and addition.
- USB 3.x

The USB definition is in a lot of ways comparable with the RS-232 definition. It does not only specify things like communication speeds and low level interfacing, but also protocols, and the mechanical characteristics of the connectors to be used. This made USB different from other standards that had seen the light since RS-232, like the RS-422 and RS-485 that focused mainly on the low level interfacing and signal definition and less on the practical implementation. The necessity for a well-defined way of practical implementation has many times been overlooked by those developing standards.

7.2.1 USB 1.x

Released in January 1996. USB 1.0 specified data rates of 1.5 Mbit/s (*Low Bandwidth* or *Low Speed*) and 12 Mbit/s (*Full Speed*). It did not allow for extension cables or pass-through monitors, due to timing and power limitations. Few USB devices made it to the market until USB 1.1 was released in August 1998. USB 1.1 was the earliest revision that was widely adopted and led to what Microsoft designated the "Legacy-free PC."

Neither USB 1.0 nor 1.1 specified a design for any connector smaller than the standard type A or type B (See Connectors Type). Although many designs for a miniaturized type B connector appeared on many peripherals, conformity to the USB 1.x standard was hampered by treating peripherals that

Figure 7.1   A USB 2.0 PCI expansion card.

had miniature connectors as though they had a tethered connection (that is: no plug or receptacle at the peripheral end). There was no known miniature type A connector until USB 2.0 (revision 1.01) introduced one.

7.2.2 USB 2.0

USB 2.0 was released in April 2000, adding a higher maximum signaling rate of 480 Mbit/s *(High Speed* or *High Bandwidth)*, in addition to the USB 1.x *Full Speed* signaling rate of 12 Mbit/s. Due to bus access constraints, the effective throughput of the *High Speed* signaling rate is limited to 280 Mbit/s or 35 MB/s.

Modifications to the USB specification have been made via Engineering Change Notices (ECN). The most important of these ECNs are included into the USB 2.0 specification package available from USB.org:

- *Mini-A and Mini-B Connector;*
- *Micro-USB Cables and Connectors Specification 1.01;*
- *InterChip USB Supplement;*
- *On-The-Go Supplement 1.3* USB On-The-Go makes it possible for two USB devices to communicate with each other without requiring a separate USB host;
- *Battery Charging Specification 1.1* Added support for dedicated chargers, host chargers behavior for devices with dead batteries;
- *Battery Charging Specification 1.2.* with increased current of 1.5 A on charging ports for unconfigured devices, allowing High Speed communication while having a current up to 1.5 A and allowing a maximum current of 5 A;
- *Link Power Management Addendum ECN* which adds a *sleep* power state.

Note: The current specification is "Universal Serial Bus Specification, Revision 2". This can be obtained free of charge on the USB-IF website. USB 2.0 specification replaces the earlier 1.0 and 1.1 Specifications, which should no longer be used. The Revision 2.0 specification covers all three data speeds (see next) and maintains backwards compatibility. USB 2.0 does NOT mean High Speed.

7.2.3 USB 3.x

The USB 3.0 specification was released on 12 November 2008, with its management transferring from USB 3.0 Promoter Group to the USB Implementers Forum (USB-IF), and announced on 17 November 2008 at the SuperSpeed USB Developers Conference.

USB 3.0 adds a *SuperSpeed* transfer mode, with associated backward compatible plugs, receptacles, and cables. SuperSpeed plugs and receptacles are identified with a distinct logo and blue inserts in standard format receptacles.

The SuperSpeed bus provides for a transfer mode at a nominal rate of 5.0 Gbit/s, in addition to the three existing transfer modes. Its efficiency is dependent on a number of factors including physical symbol encoding and link level overhead. At a 5 Gbit/s (625 Mbyte/s) signaling rate with 8 b/10 b encoding, the raw throughput is 500 Mbyte/s. When flow control, packet framing and protocol overhead are considered, it is realistic for 400 Mbyte/s (3.2 Gbit/s) or more to be delivered to an application. Communication is full-duplex in SuperSpeed transfer mode; earlier modes are half-duplex, arbitrated by the host.

Low-power and high-power devices remain operational with this standard, but devices using SuperSpeed can take advantage of increased available current of between 150 mA and 900 mA, respectively.

USB 3.1, released in July 2013, preserves the existing *SuperSpeed* transfer rate under a new label *USB 3.1 Gen 1*, and introduces a new *SuperSpeed+* transfer mode, *USB 3.1 Gen 2* with the maximum data signaling rate to 10 Gbit/s (1250 MB/s, twice the rate of USB 3.0), which reduces line encoding overhead to just 3% by changing the encoding scheme to 128b/132b.

USB 3.2, released in September 2017, preserves existing USB 3.1 *SuperSpeed* and *SuperSpeed+* data modes but introduces two new *SuperSpeed+* transfer modes over the USB-C connector with data rates of 10 and 20 Gbit/s (1250 and 2500 MB/s). The increase in bandwidth is a result of multi-lane

Table 7.1   Performance figures for USB 3.0 and USB 3.1

Performance figures for USB 3.0 and USB 3.1			
USB Version	Duplex Status	Transfer Speed	Increase Over USB 2.0
USB 2.0	Half Duplex	480 Mbps	–
USB 3.0 – SuperSpeed	Full Duplex	5 Gbps	10 x
USB 3.1 – SuperSpeed+	Full Duplex	10 Gbps	20 x

operation over existing wires that were intended for flip-flop capabilities of the Type-C connector.

7.2.3.1 USB 3 capabilities

From the above, the USB 3.0, SuperSpeed and 3.1, SuperSpeed+ specifications enable much higher rates of data transfer. This is in keeping with requirements for downloading video and many other applications.

A summary of the basic data rates for USB 3.0 and USB 3.1 is given in the Table 7.1.

In addition to the basic speed improvements, USB 3.0 and 3.1 provide other significant benefits as well.

- *Transfer speed:* USB 3.0, SuperSpeed and USB 3.1 SuperSpeed+both provide significant increases in data transfer speed over previous versions. This enables them to maintain their performance for applications requiring higher levels of data to be transferred within a given time. A new transfer type called SuperSpeed or SS has been added and this enables data transfer rates of up to 5 Gbps on USB 3.0 and 10 Gbps on 3.1. Electrically, the techniques behind SuperSpeed are similar to those used for PCI Express 2.0 and SATA.
- *Duplex status:* USB 1 and USB 2 only had the capability for half duplex transmissions, that is, transmitting in one direction at any moment. USB 3.0 and USB 3.1 both have the capability for simultaneous transmission in both directions.
- *Power management:* Power management states U0 through U3 are supported.
- *Bus utilization improvements:* The operation of the bus has been improved by the addition of a new feature where packets NRDY and ERDY can be used to let a device asynchronously notify the host of its readiness without the need for polling.
- *Support to rotating media:* Bulk protocol is updated with a new feature called Stream Protocol that allows a large number of logical streams within an Endpoint.

7.2.3.2 USB 3 compatibility

One key issue with any advance in a standard is the issue of backward compatibility. For USB 3.0 and 3.1, a dual-bus architecture approach has been adopted. This enables both USB 2.0, including Full Speed, Low Speed, or High Speed, and USB 3.0 SuperSpeed operations to take place simultaneously.

The new USB 3.0 has also been developed to enable forward compatibility. As a result, it is possible to run USB 2.0 ports with USB 3.0 or 3.1 devices.

In terms of the topologies for links that can be used, these remain the same. This consists of a tiered star topology with a root hub at level 0. Then, hubs at lower levels can provide bus connectivity to devices.

7.2.3.3 USB 3.0/3.1 annotation

In order to attain the very high speeds, some differences are required in the interface, and this results in a different connector needing to be used. That said the connectors are compatible, although the additional connections required for the higher performance will not be used when interfacing to a USB 2.0 port, cable, or device.

In addition to the different connectors used on USB 3.0 cables, devices, etc., the cable are also distinguishable from their 2.0 counterparts by either the blue color of the ports or the SS initials on the plugs. Often devices with USB 3.0 capability are marked with the letters SS as well.

7.2.3.4 USB 3.0/3.1 connector pinouts

To accommodate the new SuperSpeed/SuperSpeed+ connections, USB 3.0 and 3.1 connectors have additional connections. The original D+ and D- connections remain, along with the power and ground, but new twisted pair balanced drivers and receivers are used for the SuperSpeed data transfer. These are labeled StdA_SSRX+ and StdA_SSRX- for the receive side and StdA_SSTX+ and StdA_SSTX- for the transmit side.

Table 7.2 gives USB3.0/USB 3.1 connector pinout.

7.2.3.5 Power management

The USB 3.0 specification incorporates new power management features that are far more sophisticated than before. It includes support for a variety of states including: idle, sleep, and suspend states. It also supports Link-, Device-, and Function-level power management.

Table 7.2 USB 3.0/ USB 3.1 connector pinout

USB 3.0/USB 3.1 Connector Pinout		
Pin Number	Connection Name	Wire Color
1	VBUS	Red
2	D-	White
3	D+	Green
4	GND	Black
5	StdA_SSRX-	Blue
6	StdA_SSRX+	Yellow
7	GND_DRAIN	GROUND
8	StdA_SSTX-	Purple
9	StdA_SSTX+	Orange
Shell	Shield	Connector shell

The bus power capability has also been increased to 900mA. This is an 80% increase over USB 2.0 which stood at 500mA.

USB 3, both as 3.0 and 3.1, is able to provide the data transfer speeds required for many applications for which USB connections are used. The speed increase is partially important in areas like data backups for PCs where the increased levels of data mean that the higher speeds of USB 3.0 and USB 3.1 are welcomed. The speed improvements are also felt in many other areas where large amounts of data need to be transferred rapidly over a USB interface.

7.2.4 USB-C or USB Type-C

USB-C or USB Type-C is a 24-pin USB connector system which was introduced in August 2014 to provide increased connectivity and smaller more robust connector.

The USB C or more correctly the USB Type-C connector provides many advantages over its predecessors. Providing improved connectivity, and robustness it is making a major introduction into the market.

The USB Type-C or USB-C connector itself can support new USB capabilities like USB 3.1 and also USB Power Delivery, USB PD. Indeed, its development has been closely linked with the introduction of these new USB standards. Typically, USB Type-C is used with USB3.0 and USB 3.1.

The standard USB connector used on items like flash memory cards and many computer connections of the Type-A, and there are other types as well that are used for camera connectors, smartphones and the like, but the type-A connector has remained the same for very many years.

7.2.4.1 USB-C connectors

USB Type-C connector is the new smaller connector standard for USB. It is about a third the size of the old type-A connectors.

The other advantage of the USB-C connector is that it is a single connector standard that can be used for all devices, so the proliferation of different connectors should reduce as USB-C becomes more widespread.

There has been a significant amount of development invested in the USB-C and it is considerably more flexible than previous iterations. It can be used for connections to low power devices like smartphones, cameras, and the like, or it can be used for computers and laptops.

The USB-C connector has some similarities to a micro-USB connector, although it is more oval-shaped and slightly thicker. Like the Lightning and MagSafe connectors, the USB-C connector has no up or down orientation. Line up the connector properly, and you don't have to flip it to plug it in.

The other difference is that the same USB-C connector is used at both ends of a cable, that is, it is reversible.

USB-C ports are considerably more flexible than their predecessors. The USB-C is able to support a variety of different protocols using what may be termed alternate modes. This allows the ability to have adapters that can output HDMI, VGA, DisplayPort, or other types of connections from that single USB port.

This provides a considerable level of convenience because the variety of USB, HDMI, DisplayPort, VGA, and power ports on typical laptops can be converged into the use of a single type of port.

7.2.4.2 USB-Type C and USB3.1

The new USB-C connector is also capable of carrying the very much higher data speeds that are now available with USB3.0/USB3.1.

With data speeds from a variety of sources increasing, and with the possibility of using USB-C for HDMI and other capabilities, the connector must be capable of supporting data rates of these speeds.

With USB 3.1 providing data transfer rates of 5Gbps, these speeds are commensurate with frequencies well into the microwave region.

7.2.4.3 USB-C and USB PD

The USB-Type-C connector has been developed to meet the needs of the USB PD specification. Many devices including smartphones, tablets, and other mobile devices use their USB connection for obtaining power for charging. However the old connectors were limited in the power they could pass. A USB 2.0 connection provides up to 2.5 watts of power. This is sufficient for most small devices, but not for larger ones like laptops.

The USB Power Delivery specification provides for the delivery of power levels up to 100 watts.

The USB-C is bi-directional, and so is the power delivery system. This means that using the same connector, a device can either send or receive power.

Power can also be delivered while data is being transmitted, meaning that, effectively, parallel operation of both capabilities is possible.

7.2.4.4 USB-C capabilities

The capabilities of the USB-C system are summarized in Table 7.3.

USB-C or USB Type-C is able to provide significantly improved connectivity for the USB system and is primarily intended to support the USB 3.0 and USB 3.1 versions of USB. In this way it helps to provide the SuperSpeed data transfers that can be achieved with these new standards.

7.2.4.5 USB version history

The next two tables, Tables 7.4 and 7.5, summarize the "Release versions" and "Power Related Specifications"

Table 7.3 USB Type-C connector specification summary

USB Type-C Connector Specification Summary	
Parameter	Specification
Receptacle opening	~8.3 × ~2.5 mm
Lifetime	10 000 cycles
Power delivery	3A for standard connectors5A for connectors
USB 2 compatibility	LS/FS/HS
USB 3 compatibility	Gen 1 (5 Gbps)Gen 2 (10 Gbps)
EMI	Improved over USB-A and USB-B
Enhanced power delivery	USB PD
Docking support	USB PD based interface configuration option

Table 7.4　Release versions

Release Name	Release Date	Maximum Transfer Rate	Note
USB 0.7	1994-11-11		Pre-release
USB 0.8	December 1994		Pre-release
USB 0.9	1995-04-13	Full Speed (12 Mbit/s)	Pre-release
USB 0.99	August 1995		Pre-release
USB 1.0-RC	November 1995		Release Candidate
USB 1.0	1996-01-15	Full Speed (12 Mbit/s), Low Speed (1.5 Mbit/s)	
USB 1.1	August 1998	Full Speed (12 Mbit/s)	
USB 2.0	April 2000	High Speed (480 Mbit/s)	
USB 3.0	November 2008	SuperSpeed (5 Gbit/s)	Also referred to as USB 3.1 Gen 1 and USB 3.2 Gen 1x1
USB 3.1	July 2013	SuperSpeed+ (10 Gbit/s)	Also referred to as USB 3.1 Gen 2 and USB 3.2 Gen 2x1
USB 3.2	September 2017	SuperSpeed+ (20 Gbit/s)	Includes new USB 3.2 Gen 1x2 and USB 3.2 Gen 2x2 multi-link modes

7.3　USB Data Speeds

There are four distinct data rates. Each new major USB specification introduced a new data rate. The four data rates are: Low speed, Full Speed, High Speed and Super Speed. Table 7.6 shows USB data rate types supported by the four USB specifications. Each new USB specification has been backward compatible.

Low-, full-, and high-speed devices are often advertised as 1.5 Mb/s, 12 Mb/s, and 480 Mb/s, respectively. However, these are bus rates and not data rates. The actual data rates are affected by bus loading, transfer type, overhead, OS, and so forth. The actual limits of the data transfer are listed in Table 7.6.

Table 7.7 shows maximum data rates for the four data rate types.

The USB specification defines four data speeds, shown in Table 7.7. These speeds are the fundamental clocking rates of the system and as such do not represent possible throughput, which will always be lower as the result of the protocol overheads.

Table 7.5 Power-related specifications

Release Name	Release Date	Nominal Voltage	Maximum Current	Maximum Power	Note
USB Battery Charging 1.0	2007-03-08	5 V	*1.5 A*	*7.5 W*	
USB Battery Charging 1.1	2009-04-15				
USB Battery Charging 1.2	2010-12-07	5 V	*5 A*	*25 W*	
USB Power Delivery revision 1.0 (version 1.0)	2012-07-05	20 V	*5 A*	*100W*	Using FSK protocol over bus power (V_{BUS})
USB Power Delivery revision 1.0 (version 1.3)	2014-03-11				
USB Type-C 1.0	2014-08-11	5 V	*3 A*	*15 W*	New connector and cable specification
USB Power Delivery revision 2.0 (version 1.0)	2014-08-11	20 V	*5 A*	*100 W*	Using BMC protocol over communication channel (CC) on type-C cables.
USB Type-C 1.1	2015-04-03	5 V	*3 A*	*15 W*	
USB Power Delivery revision 2.0 (version 1.1)	2015-05-07	20 V	*5 A*	*100 W*	
USB Power Delivery revision 2.0 (version 1.2)	2016-03-25	20 V	*5 A*	*100 W*	
USB Power Delivery revision 2.0 (version 1.3)	2017-01-12	20 V	*5 A*	*100 W*	
USB Power Delivery revision 3.0 (version 1.1)	2017-01-12	20 V	*5 A*	*100 W*	

Table 7.6   USB data rate

	Low-speed	Full Speed	High Speed	Super Speed
USB 1. 0	X	X		
USB 1.1	X	X		
USB 2.0	X	X	X	
USB 3.0	X	X	X	X

Table 7.7   Maximum data rate

	Low Speed	Full Speed	High Speed	Super Speed
Max data throughput	0.1875 Mb/s	1.5 Mb/s	60 Mb/s	625 Mb/s
Rating (1x = 0.15 Mb/s)	1.25x	10x	400x	4166.7x

Low Speed

This was intended for cheap, low data rate devices like mice. The low speed captive cable is thinner and more flexible than that required for full and high speed.

- Low-speed devices:
  - Examples: keyboards, mice, and game peripherals
  - Bus Rate: 1.5 Mb/s
  - Maximum Effective Data Rate: 800 B/s

Full Speed

This was originally specified for all other devices.

- Full-speed devices
  - Examples: phones, audio devices, and compressed video
  - Bus rate: 12 Mb/s
  - Maximum effective data rate: 1.2 Mb/s

High Speed

The high speed additions to the specification were introduced in USB 2.0 as a response to the higher speed of Firewire.

- Hi-speed devices
  - Examples: video, imaging, and storage devices
  - Bus Rate: 480 Mb/s
  - Maximum Effective Data Rate: 53 Mb/s

Table 7.8 USB data speed

Speed	Maximum Bandwidth	USB Specification	Note
Low speed	1.5 Mbit/s	1.0	Not typically used in contemporary designs
Full speed	12 Mbit/s	1.0	
High speed	480 Mbit/s	2.0	
Super speed	5 Gbit/s	3.0	Specification is also referred to as USB 3.1 Gen 1
Superspeed+	10 Gbit/s	3.1	USB 3.1 Gen 2

USB 3.1 specifications added another speeds:

Super Speed, and
Super Speed+

USB Speeds – Summary

USB has five supported speeds, each with a slightly different protocol and frame format. This is given in Table 7.8.

Speed Identification of USB

When a USB device is connected to a host, the speed of the device needs to be detected. This is done with pull-up resistors on the D+ or D- line. A 1.5-$k\Omega$ pull-up on the D+ line indicates that the attached device is a Full-speed device. A 1.5-$k\Omega$ pull-up resistor on the D- line indicates the attached device is a Low-speed device. Speed identification is discussed in Section 7.6.2.

Note: One common misconception about speed on a USB device is that a device listed as USB 2.0 indicates that the device is High-Speed. All Hi-Speed devices are USB 2.0, but this is because Hi-Speed support was added with USB 2.0. The USB 2.0 specification includes Full- and Low-Speed devices as well.

These speeds also have an effect on the bit timing for USB signaling, such as the End of Packet (EOP) signal. A low-speed and full-speed USB device will use a 48-MHz clock for the SIE and the other USB clocking purposes. This 48-MHz clock and the bus speed are what will determine USB bit times:

- **Full-speed:** 48 MHz/12 Mb/s = 4 clocks per bit time.
- **Low-speed:** 48 MHz/1.5 Mb/s = 32 clocks per bit time.

7.4 USB Architecture (or USB Topology)

The Universal Serial Bus is host controlled. There can only be one host per bus. The specification in itself does not support any form of multimaster arrangement. However the On-The-Go specification which is a tack on standard to USB 2.0 has introduced a Host Negotiation Protocol which allows two devices negotiate for the role of host. This is aimed at and limited to single point to point connections such as a mobile phone and personal organizer and not multiple hub, multiple device desktop configurations. The USB host is responsible for undertaking all transactions and scheduling bandwidth. Data can be sent by various transaction methods using a token-based protocol.

USB has a strict "tree" topology and "master-slave" protocol for addressing peripheral devices; single master which is a single host controller and up to 127 "slave" devices. Peripheral devices cannot interact with one another except via the host, and two hosts cannot communicate over their USB ports directly. The USB topology is called also **"tiered star topology." The topology is similar to that of 10Base T Ethernet**. The host learns about the device capabilities during enumeration, which allows the host operating system to load a specific driver for a particular USB device.

The host controller is connected to a hub, integrated within the PC, which allows a number of attachment points (often loosely referred to as ports). A further hub may be plugged into each of these attachment points, and so on (Figure 7.2). However there are limitations on this expansion.

Some extension to this limitation is possible through USB **On-The-Go**. A host cannot "broadcast" signals to all peripherals at once, each must be addressed individually. Some very high speed peripheral devices require sustained speeds not available in the USB standard. While converters exist between certain "legacy" interfaces and USB, they may not provide full implementation of the legacy hardware; for example, a USB to parallel port converter may work well with a printer, but not with a scanner that requires bi-directional use of the data pins.

The limitation of 127 devices (slaves), including hubs, that can be connected to the USB bus, is because the address field in a packet is 7 bits long (see latter), and the address 0 cannot be used as it has special significance. (In most systems the bus would be running out of bandwidth, or other resources, long before the 127 devices was reached.)

A device can be plugged into a hub, and that hub can be plugged into another hub and so on. However the maximum number of tiers permitted is six.

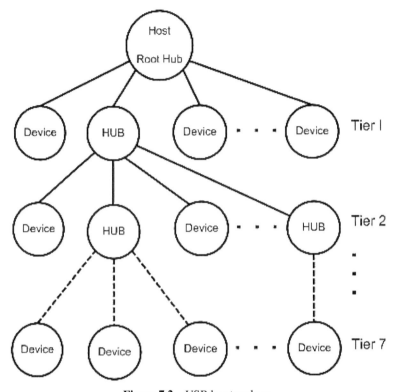

Figure 7.2 USB bus topology.

The length of any cable is limited to 5 m. This limitation is expressed in the specification in terms of cable delays etc., but 5 meters can be taken as the practical consequence of the specification. This means that a device cannot be further than 30 m from the PC, and even to achieve that will involve 5 external hubs, of which at least 2 will need to be self-powered.

So the USB is intended as a bus for devices near to the PC. For applications requiring distance from the PC, another form of connection is needed, such as Ethernet.

The tiered star topology, rather than simply daisy chaining devices together has some benefits. First, power to each device can be monitored and even switched off if an overcurrent condition occurs without disrupting other USB devices. High-, full-, and low-speed devices can be supported, with the hub filtering out high-speed and full-speed transactions so lower speed devices do not receive them.

Figure 7.3 Typical 4-port hub.

7.4.1 USB Components

The Universal Serial Bus (USB) consists of the following primary components (Figure 7.2):

- **USB Host**: The USB host platform is where the USB host controller is installed and where the client software/device driver runs. The *USB Host Controller* is the interface between the host and the USB peripherals. The host is responsible for detecting the insertion and removal of USB devices, managing the control and data flow between the host and the devices, providing power to attached devices and more.
- **USB Hub**: A USB device that allows multiple USB devices to attach to a single USB port on a USB host. Hubs on the back plane of the hosts are called *root hubs*. Other hubs are called *external hubs* (Figure 7.3).
- **USB Function (USB Device)**: A USB device that can transmit or receive data or control information over the bus and that provides a function. A function is typically implemented as a separate peripheral device that plugs into a port on a hub using a cable. However, it is also possible to create a *compound device*, which is a physical package that implements multiple functions and an embedded hub with a single USB cable. A compound device appears to the host as a hub with one or more non-removable USB devices, which may have ports to support the connection of external devices.

Next we discuss the three main components in some more details.

7.4.1.1 Host is master
USB Host

The USB host communicates with the devices using a USB host controller. The host is responsible for detecting and enumerating devices, managing bus access, performing error checking, providing and managing power, and exchanging data with the devices.

All communications on USB bus are initiated by the host: slaves (USB devices) cannot communicate directly.

A device cannot initiate a transfer, but must wait to be asked to transfer data by the host. The only exception to this is when a device has been put into "suspend" (a low power state) by the host then the device can signal a "remote wakeup."

7.4.1.2 Types of host controller

The USB host controllers have their own specifications. There are three commonly encountered types of USB host controller, each with its own history and characteristics.

With USB 1.1, there were two Host Controller Interface Specifications:

- UHCI (Universal Host Controller Interface)
- OHCI (Open Host Controller Interface)

UHCI (Universal Host Controller Interface)

Developed by Intel which puts more of the burden on software (Microsoft) and allowing for cheaper hardware. Requires a license from Intel. Low speed and full speed.

OHCI (Open Host Controller Interface)

Compaq, Microsoft and National Semiconductors cooperated to produce this standard host controller specification for USB 1.0 and USB 1.1. It is a more hardware oriented version than UHCI. Low speed and full speed.

EHCI (Extended Host Controller Interface)

When USB 2.0 appeared with its new high-speed functionality, the USB-IF insisted on there being a single host controller specification, to keep device development costs down. Significant contributors include Intel, Compaq, NEC, Lucent, and Microsoft so it would hopefully seem they have pooled together to provide us one interface standard and thus only one new driver to implement in our operating systems.

The EHCI handles high-speed transfers and hands off low- and full-speed transfers to either OHCI or UHCI companion controllers.

7.4.1.3 USB device

A USB device implements one or more USB functions where a function provides one specific capability to the system. Examples of USB functions are keyboards, webcam, speakers, or a mouse. The requirements of the USB functions are described in the USB class specification. For example, keyboards and mice are implemented using the Human Interface Device (HID) specification.

USB devices must also respond to requests from the host. For example, on power up, or when a device is connected to the host, the host queries the device capabilities during enumeration, using standard requests.

7.4.1.4 On-The-Go

An extension to the USB specification has been defined, to allow a device to also become a limited role host. This specification is known as On-The-Go. Section 7.6 covers this specification in detail.

7.5 USB Limitations

USB has some limitations. In Section 7.4, USB topology, we discussed the limitations due to the topology, for example, **the maximum number of tiers permitted is six**. Some other limitations are as follows:

- Maximum data rates
- Actual data throughput
- Cable length and total length
- Power

This section is dealing with cable length limitation.

7.5.1 USB Cable Length Limitations

USB cables are limited in length, as the standard was meant to connect to peripherals on the same table-top, not between rooms or between buildings. However, a USB port can be connected to a gateway that accesses distant devices, in other words to use a special type of USB cable known as an active or repeater cable. But before we get to active cables or hubs, how long can a USB cable be?

Maximum length of USB 2.0 cable: The 2.0 specification limits the length of a cable between USB 2.0 devices (Full speed or Hi-speed) to 5 m (or about 16 ft and 5 inches). In other words, the user cannot just connect a bunch of extension cables together (like taking a 6-ft cord and extending it with 4 other 6 foot extension cords) and run them 30 ft to another room. However, the user can connect a 6-ft cable with a 10-ft extension cable for a total of 16 ft, which is below the maximum cable length for USB 2.0.

Table 7.9   Maximum cable length

	USB 1.1	USB 2.0	USB 3.0
Maximum cable length	9.8 ft/3.0 m	16.4 ft/ 5.0 m	9.8 ft/ 3.0 m
Max total length (excluding device cable)	59.1 ft/18 m	98.4 ft/30 m	59.1 ft/18 m
Max total length (including device cable)	49.2 ft/15 m+ device cable length	82.0 ft/25 m+ device cable length	49.2 ft/ 15 m+ device cable length

Maximum length of USB 3.0/USB 3.1 cable: The 3.0/3.1 specification does not specify a maximum cable length between USB 3.0/3.1 devices (SuperSpeed or SuperSpeed+), but there is a recommended length of 3 m (or about 9 ft and 10 inches). However, the biggest limitation to the length of the cable is the quality of the cable. Results may vary, but with a high quality cable the user should be able to go beyond 3 m. However, to ensure achieving the best results possible, it is recommended to use an active cable when going more than 10 ft (3 m).

Table 7.9 shows the maximum cable and total lengths.

7.5.2 How to Break the USB Length Limitations

In spite of the fact that USB specs limit the length of cables, there is some ways to extend those limits. To go beyond these cable length limits (or recommended lengths), it is possible to use self-powered USB hubs or active (repeater) cables, both of which have their own limits as well. Other options such as USB over Ethernet or building your own USB bridge can extend the USB range further.

USB Hubs: It is possible to use extension cables and self-powered USB hubs connected together to extend the range of the USB device. However, it is important to remember when using 2.0 hubs and cables that the distance between each powered hub can be no more than 5 m (16 ft and 5 inches). When using 3.0/3.1 hubs and cables, do not exceed the recommended length of 3 m (9 ft and 10 inches) between hubs. Note: It is possible to use bus-powered USB hubs, but you must expect that you will quickly run out of power as you extend your setup.

Active (Repeater) Extension Cables: USB active extension cables contain electronics that regenerate the USB signal. Active cables are essentially 1 port

USB hubs. It is possible to use a regular USB cable in conjunction with an active cable as long as the regular cable is not more than 5 m (16 ft and 5 inches) long for 2.0 devices and not more than 3 m (9 ft and 10 inches) long for 3.0 devices. Note: Typically, active cables are bus-powered cables. To ensure you receive the full 500 mA power of a USB port, consider purchasing an active cable that includes a separate power adapter.

7.5.3 USB Hub Limits and Maximum Length of Active Cables

Just like there is a limit on a regular (passive) USB cable, there is also a limit on how long an active cable can be and how many USB hubs the user can use.

Maximum number of USB Hubs: The USB 2.0/3.0/3.1 specifications call for only 7 tiers of devices to be connected. When the user counts the devices on each end (the host and the peripheral device), that only leaves 5 tiers available and a USB hub is considered 1 tier. Thus, only a maximum of 5 USB hubs can be used for a total maximum length of 30 m (about 98 ft and 5 inches).

Maximum length of USB active (repeater) cable: This number depends on whether the user is using a regular cable with an active cable or not. If the user is not using a regular cable, then the maximum active cable length for USB 2.0 is 30 m (98 ft and 5 inches) and the maximum recommended length for USB 3.0/3.1 is 18 m (about 59 ft). If the user is using a regular cable (max length of 5 m for 2.0 and max length of 3 m for 3.0/3.1) with an active cable, then the maximum length for USB 2.0 is 25 m (about 82 ft) and the maximum recommended length for USB 3.0/3.1 is 15 m (about 49 ft).

Is there any way to go beyond the limit of active cables or hubs?

There are other ways that can be used to extend a USB signal beyond the 30 meter limit. One of them is to use USB over Ethernet to achieve distances up to 100 m (about 328 ft). Additionally, the user can build special (tailored) USB bridge to transmit data over different communication channels such as wireless methods.

Note: Case of USB 1.0/1.1

USB 1.0/1.1 has been superseded by USB 2.0 and USB 3.0/3.1. In addition, USB 2.0 cables are backward compatible so they will work just fine with any USB 1.0/1.1 devices. However, in case you are using a 1.0/1.1 USB host and device, there are limits to the maximum length of the cable.

The limit for USB 1.0/1.1 cable length is 3 m (about 9 ft and 10 inches) and the maximum total length should not exceed 18 m (about 59 ft).

Note: For more technical information see:

USB 2.0 Specification
USB 3.0/USB 3.1 Specifications

7.6 USB HARDWARE: Electrical Specifications, Cables, Connectors, Device Powering

7.6.1 Electrical: Cables

USB cables have been designed to ensure correct connections are always made. By having different connectors on host and device, it is impossible to connect two hosts or two devices together.

Unfortunately it is possible to buy non-approved cables and adapters with illegal combinations of connector. These may be useful in certain development situations, but can lead the unsuspecting user to make connections which can easily damage their equipment.

USB uses a differential transmission pair for data. This is encoded using NRZI and is bit stuffed to ensure adequate transitions in the data stream.

USB requires a shielded cable containing 4 wires (Figure 7.4):

1. D+ and D−: These two wires form a twisted pair responsible for carrying a differential data signal, as well as some single-ended signal states. (For low speed the data lines may not be twisted.)
2. GND wire: The signals on D+ and D- are referenced to the (third) GND wire.
3. VBUS: This is the fourth wire and it carries a nominal 5V supply, which may be used by a device for power.

On low and full speed devices, a differential '1' is transmitted by pulling D+ over 2.8 V with a 15 k ohm resistor pulled to ground and D− under 0.3 V with a 1.5 k ohm resistor pulled to 3.6 V. A differential '0' on the other

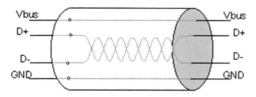

Figure 7.4   Makeup of USB cable.

hand is a D− greater than 2.8 V and a D+ less than 0.3 V with the same appropriate pull down/up resistors.

The receiver defines a differential '1' as D+ 200 mV greater than D- and a differential '0' as D+ 200 mV less than D−. The polarity of the signal is inverted depending on the speed of the bus. Therefore, the terms 'J' and 'K' states are used in signifying the logic levels. In low speed, a 'J' state is a differential 0. In high speed, a 'J' state is a differential 1.

USB transceivers will have both differential and single ended outputs. Certain bus states are indicated by single ended signals on D+, D- or both. For example, a single ended zero or SE0 can be used to signify a device reset if held for more than 10 mS. A SE0 is generated by holding both D− and D+ low (<0.3 V). Single ended and differential outputs are important to note if you are using a transceiver and FPGA as your USB device. You cannot get away with sampling just the differential output.

The low speed/full speed bus has a characteristic impedance of 90 ohms ± 15%. It is therefore important to observe the datasheet when selecting impedance matching series resistors for D+ and D−. Any good datasheet should specify these values and tolerances.

High speed (480 Mbits/s) mode uses a 17.78 mA constant current for signalling to reduce noise.

7.6.1.1 Cable types

The USB specification defines three forms of cable:

1. A high-/full-speed detachable cable with one end terminated with an *A* plug and the other end with a *B* or *mini-B* plug.
2. A captive high-/full-speed cable where one end is either hardwired to the vendor's equipment or connected via a vendor specific connector and the other end is terminated with an *A* plug.
3. A low speed version of 2.

The maximum length of a high-/full-speed cable is determined by the attenuation and propagation delay. But, for a low speed cable, it is the signal rise and fall times that determine the maximum length. This forces the maximum length for low-speed cable to be shorter than that for high/full speed.

7.6.2 Speed Identification

A USB device must indicate its speed by pulling either the D+ or D− line high to 3.3 V. A full speed device, shown in Figure 7.5, will use a pull up

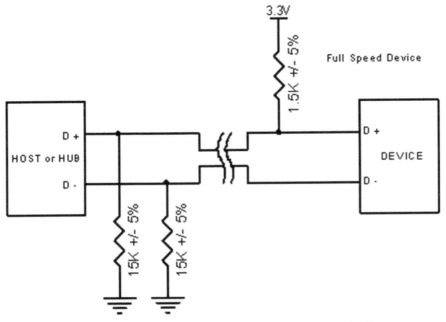

Figure 7.5 Full-speed device with pull-up resistor connected to D+.

resistor of 1.5 k ohm attached to D+ to specify itself as a full-speed device. A low-speed device (Figure 7.6) pulls D- to identify itself. These pull-up resistors at the device end will also be used by the host or hub to detect the presence of a device connected to its port. Without a pull-up resistor, USB assumes there is nothing connected to the bus. Some devices have this resistor built into its silicon, which can be turned on and off under firmware control, others require an external resistor.

For example, Philips Semiconductor has a SoftConnect[TM] technology. When first connected to the bus, this allows the microcontroller to initialize the USB function device before it enables the pull-up speed identification resistor, indicating a device is attached to the bus. If the pull-up resistor was connected to V_{bus}, then this would indicate a device has been connected to the bus as soon as the plug is inserted. The host may then attempt to reset the device and ask for a descriptor when the microprocessor hasn't even started to initialize the USB function device.

Other vendors such as Cypress Semiconductor also use a programmable resistor for Re-Numeration[TM] purposes in their EzUSB devices where one

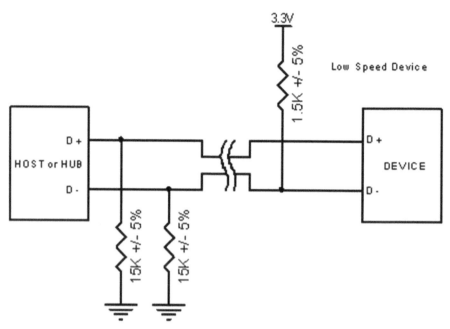

Figure 7.6   Low-speed device with pull-up resistor connected to D−.

device can be itemized for one function such as In field programming then be disconnected from the bus under firmware control, and itemize as another different device, all without the user lifting an eyelid. Many of the EzUSB devices do not have any Flash or OTP ROM to store code. They are bootstraped at connection.

Concerning high-speed identification, for high-speed mode, high-speed devices will start by connecting as a full-speed device (1.5 k to 3.3 V). Once it has been attached, it will do a high-speed chirp during reset and establish a high-speed connection if the hub supports it. If the device operates in high-speed mode, then the pull-up resistor is removed to balance the line.

A USB 2.0 compliant device is not required to support high-speed mode. This allows cheaper devices to be produced if the speed isn't critical. This is also the case for a low-speed USB 1.1 device which is not required to support full speed.

However, a high-speed device must not support low-speed mode. It should only support full-speed mode needed to connect first, then high-speed mode if successfully negotiated later. A USB 2.0 compliant downstream facing device (Hub or Host) must support all three modes, high speed, full speed, and low speed.

7.6.3 Power Distribution and Device Powering

Before discussing this matter, some terminology and facts will be given:

- A device (or hub) can only sink (consume) current from its upstream port.
- A 'self-powered' device is one which does not draw power from the bus. A device which draws its power from the bus is called a 'bus-powered' device. In normal operation, it may draw up to 100 mA, or 500 mA if permitted to do so by the host.
- A unit load is defined by the USB specification as 100 mA
- A device which has been 'Suspended', as a result of no bus activity, must reduce its current consumption to 0.5 mA or less.
  The USB 2.0 core specification specifies the value 0.5 mA for suspended standard load devices, but this value was superseded as a result of a later ECN (Engineering Change Note).
- Suspend mode is mandatory on all devices.

Device Powering

When it comes to USB power, there are two device categories: bus powered and self-powered. One of the benefits of USB is bus-powered devices, devices which obtain its power from the bus and require no external plug packs or additional cables. This benefit is due to the availability of the 5 V supply. However, before designing a bus-powered device it is well to consider the limitations of this approach. For example, the voltage supplied can fall to 4.35 V at the device. There can also be transients on this taking it 0.4 V lower, due to other devices being plugged in. Any device connected to the USB bus needs to cope with these voltage levels.

The standard unit load available, as mentioned before, is 100 mA. No device is permitted to take more than this before it has been *configured* by the host. It must also reduce its current consumption to 2.5 mA whenever it is 'suspended' by a lack of activity on the bus. However, it is not required to obey this rule for a period of 1 s from when it connects. (Note: *ECN removed the need to suspend during the first second after connect.*)

It should be remember that of this 2.5 mA, the required 1.5 k pull-up resistor is already drawing 0.3 mA. This leaves the designer a budget of

2.2 mA to power the rest of his device circuitry. If the device contains a microcontroller it will need a sleep mode which meets this requirement, but it is important to take into consideration the fact that a badly placed resistor can very easily draw current which the designer hadn't expected. It is important, to avoid this unexpected matter, to measure the suspend current with a meter.

The voltage supplied can fall to 4.35 V at the device. There can also be transients on this taking it 0.4 V lower, due to other devices being plugged in. Any device connected to the USB bus needs to cope with these voltage levels.

A USB device specifies its power consumption expressed in 2 mA units in the configuration descriptor which will discussed and examined in detail later. A device cannot increase its power consumption, greater than what it specifies during enumeration, even if it looses external power. There are three classes of USB functions:

- Low-power bus powered functions
- High-power bus powered functions
- Self-powered functions

Low-power bus powered functions draw all its power from the V_{BUS} and cannot draw any more than one unit load. Low power bus powered functions must also be designed to work down to a V_{BUS} voltage of 4.40 V and up to a maximum voltage of 5.25 V measured at the upsteam plug of the device. For many 3.3 V devices, LDO regulators are mandatory.

High-powered Devices: High-power bus powered device will draw all its power from the bus and cannot draw more than one unit load until it has been configured. After configuring the device as a high-power device, it may draw up to 500 mA, provided it asked for this in its descriptor. Being configured is dependent on the Hub being able to supply 500 mA, which implies a *self-powered* hub. So there is always a degree of uncertainty whether more than 100 mA will be available. It would be well to offer the option of external power via a socket on such a device.

High-power bus functions must be able to be detected and itemized at a minimum 4.40 V. When operating at a full unit load, a minimum V_{BUS} of 4.75 V is specified with a maximum of 5.25 V. Once again, these measurements are taken at the upstream plug.

Self-powered Devices: Devices requiring more than 500 mA are obliged to be self-powered. **Self-powered** devices may draw up to 1 unit load from the bus and derive the rest of its power from an external source. Should this external source fail, it must have provisions in place to draw no more than 1 unit load from the bus. Self-powered devices are easier to design to specification as there is not so much of an issue with power consumption. The 1 unit bus powered load allows the detection and enumeration of devices without mains/secondary power applied.

The practice of attempting to draw power from two adjacent USB ports, using a modified cable, is not permitted and can easily damage the ports.

When designing a self-powered device, the designer has to remember that he must not pull a D+ or D- line above the V_{bus} voltage supplied. This means that he must, at the very least, sense when V_{bus} is connected.

The D+ or D- resistor should, strictly speaking, be pulled up to a 3.3 V supply derived from Vbus, or controlled by Vbus in such a way that the resistor never sources current to the data line when Vbus is switched off.

If the designer pulls, say D+, high in the absence of V_{BUS} then he will risk faulty operation with On-The-Go hosts. (See later).

No USB device, whether bus powered or self- powered can drive the V_{BUS} on its upstream facing port. If V_{BUS} is lost, the device has a lengthy 10 s to remove power from the D+/D- pull-up resistors used for speed identification.

Hot – Pluggable: To achieve the goal of being able to plug a device into and out of a running system, some design rules must be followed to handle the "Inrush Current". Inrush current is the result of the capacitance added to the bus as a result of connecting a device between V_{BUS} and ground. The spec therefore specifies that the maximum decoupling capacitance the designer can have on the device is 10 uF. When the designer disconnects the device after current is flowing through the inductive USB cable, a large flyback voltage can occur on the open end of the cable. To prevent this, a 1 uF minimum V_{BUS} decoupling capacitance is specified.

Another factor to be considered is that when the user plugs a device in, any capacitance between Vbus and GND will cause a dip in voltage across the other ports of the hub to which the device is connecting. To limit the consequences of this (such as crashing other devices), the specification places a maximum on the value of capacitance across Vbus and GND of 10 uF. For the same reason, the hub port supply must be bypassed with at least 120 uF.

For the typical bus powered device, it cannot drain any more than 500 mA which is not unreasonable. If the current increases this limit the device will entre to the "Suspend Mode"

7.6.4 USB States and Relation with Power

Various USB states relate to USB power that a designer needs to know. These states are often seen in USB documentation and apply to the enumeration of a USB device.

- **Attached State:** Occurs when a device is attached to a host/hub, but does not give any power to the V_{BUS} line. This is commonly seen if the hub detects an over current event. The device is still attached, but the hub removes power to it.
- **Powered:** A device is attached to the USB and has been powered, but has not yet received a reset request.
- **Default:** A device is attached to the USB, is powered, and has been reset by the host. At this point, the device does not have a unique device address. The device responds to address 0.
- **Address:** A device is attached to the USB, is powered, has been reset, and has had a unique address assigned to it. The device however has not yet been configured.
- **Configured:** The device is attached to the USB, powered, has been reset, assigned a unique address, has been configured, and is not in a suspend state. At this point, bus-powered devices can draw more than 100 mA.
- **Suspend:** As mentioned earlier, occurs when the device is attached and configured, but has not seen activity on the bus for 3 ms.

The USB specification (Figure 7.1 in USB specification) has a diagram that illustrates how these power modes are related and transitioned to (see Figure 7.7).

USB power is communicated in 2 mA units for low-speed, full-speed, and high-speed USB devices. For example, a full-speed device that must have 100 mA of operational power communicates a value of 50 during enumeration.

When developing a USB design, consider how much power your device consumes from the bus. The root hub gets its power from the supply of the host PC. If the host is attached to AC power, then the USB specification requires the host provide 500 mA of power to each port on the hub. This is what causes the 500 mA limitation on bus-powered devices. If the host PC is battery powered, then it has the option of supplying either 100 mA or 500 mA to each port on the hub. When attaching a device to a hub that is bus-powered, the device must be low power and not consume more than 100 mA. A bus-powered hub has a total of 500 mA to distribute between all attached devices.

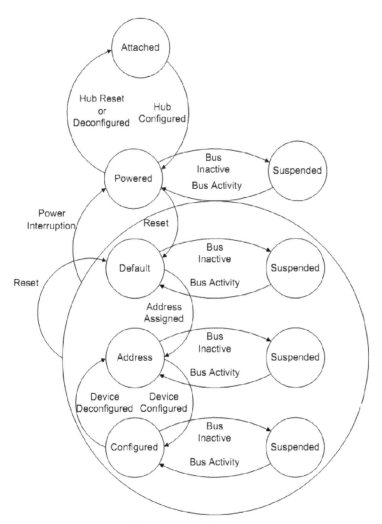

Figure 7.7 Device state.

7.6.5 Suspend Current

Suspend mode is mandatory on all devices. During suspend, additional constrains come into force. The maximum suspend current is proportional to the unit load. For a 1 unit load device (default), the maximum suspend current is 500 uA. This includes current from the pull-up resistors on the bus. At the hub, both D− and D+ have pull-down resistors of 15 k ohms. For the purposes of power consumption, the pull-down resistor at the device is in series with the 1.5 k ohm pull up, making a total load of 16.5 k ohms on a V_{TERM} of typically 3.3 V. Therefore this resistor sinks 200 uA before we even start.

Another consideration for many devices is the 3.3 V regulator. Many of the USB devices run on 3.3 V. The PDIUSBD11 is one such example. Linear regulators are typically quite inefficient with average quiescent currents in the order of 600 uA, and therefore, more efficient and thus expensive regulators are called for. In the majority of cases, the designer must also slow down or stop clocks on microcontrollers to fall within the 500 uA limit.

Exceeding the 500 uA limitation may cause some complications. The complications of exceeding the specified limits is coming from the fact that most hosts and hubs don't have the ability to detect such an overload of this magnitude, and thus if the device drain maybe 5 mA or even 10 mA, the system will not detect that and it should still be fine, in spite of the fact that the device violates the USB specification. However in normal operation, if you try to exceed the 100 mA or your designated permissible load, then expect the hub or host to detect this and disconnect your device, in the interest of the integrity of the bus.

Of course these design issues can be avoided if the user chooses to design a self- powered device. Suspend currents may not be a great concern for desk-top computers but with the introduction of the On-The-Go Specification we will start seeing USB hosts built into mobile phones and mobile organizers. The power consumption pulled from these devices will adversely affect the operating life of the battery.

7.6.6 Entering Suspend Mode

A USB device will enter suspend when there is no activity on the bus for greater than 3.0 ms. It then has a further 7 ms to shut down the device and draw no more than the designated suspend current and thus must be only drawing the rated suspend current from the bus 10 ms after bus activity stopped. In order to maintain connected to a suspended hub or host, the device must still provide power to its pull-up speed selection resistors during suspend.

USB has a start of frame packet or keep alive sent periodically on the bus. This prevents an idle bus from entering suspend mode in the absence of data.

- A high speed bus will have micro-frames sent every 125.0 Âţs ±62.5 ns.
- A full speed bus will have a frame sent down each 1.000 ms ±500 ns.

- A low speed bus will have a keep alive which is an EOP (End of Packet) every 1 ms only in the absence of any low speed data.

The term "Global Suspend" is used when the entire USB bus enters suspend mode collectively. However, selected devices can be suspended by sending a command to the hub that the device is connected too. This is referred to as a "Selective Suspend."

The device will resume operation when it receives any non-idle signalling. If a device has remote wakeup enabled then it may signal to the host to resume from suspend.

7.6.7 Data Signalling Rate

Another area which is often overlooked is the tolerance of the USB clocks. This is specified in the USB specification, Section 7.1.11.

- High-speed data are clocked at 480.00 Mb/s with a data signalling tolerance of $\pm$ 500 ppm.
- Full-speed data are clocked at 12.000 Mb/s with a data signalling tolerance of $\pm0.25\%$ or 2,500 ppm.
- Low-speed data are clocked at 1.50 Mb/s with a data signalling tolerance of $\pm1.5\%$ or 15,000 ppm.

This allows resonators to be used for low cost low-speed devices, but rules them out for full- or high-speed devices.

7.6.8 USB Connector

As mentioned before, the USB interface is one of the most used interfaces at this moment to connect peripheral equipment to computers. Although the USB interface itself is standard, and it is possible to connect every device to a USB-enabled computer if the appropriate driver exists, problems may arise to find the right cable. This is because especially for smaller equipment like cameras different models of USB connectors have been defined. This section contains information about all known USB connector types and will help the reader to find the right cable to connect equipment to his computer.

7.6.8.1 Basic design concepts of USB connectors

Much effort has been put in the design of the several USB connectors to make them useful for their purpose. The old Centronics connectors for parallel printers were bulky and needed clips to connect them securely to the devices.

The DB9 and DB25 connectors which are used for RS-232 ports and parallel ports on computers often have problems with connection bolts falling out of the computer case if someone has accidentally tightened the screws of the connector too far. People who have often disconnected and reconnected their VGA cable might have experienced that these densely populated connectors have very thin pins which bend easily.

Another problem with bad connector design is that you may accidentally connect them wrong. This is something which can happen, for example, with flat cable connectors and power connectors inside computers. As USB can power devices over the cable it is not only necessary that a USB connector cannot be connected in the wrong orientation, but the design must also not allow that two power providing USB devices are connected with each other as this may cause one or both power supplies to be damaged.

7.6.8.2 USB upstream and downstream
It is important to ensure that USB connections are made correctly and are able to follow the required protocols. To achieve this, all devices have an upstream connection to the host and all hosts have a downstream connection to the device. Upstream and downstream connectors are not mechanically interchangeable, thus eliminating illegal loopback connections at hubs such as a downstream port connected to a downstream port. There are commonly two types of connectors, called type A and type B, which are shown below.

7.6.8.3 Standard A and B USB connectors
Because of the reasons mentioned above, two USB connectors have been defined for basic use, the USB A connector which must be used on devices which provide power (mostly computers), and the USB B connector used on devices which receive power like most peripheral devices.

In the standard USB A and B connectors specified in the USB 1.1 and USB 2.0 specification, four pins are defined. Two pins are used for power and two pins are used for differential data transmission. If you look carefully at the connector you will see that the pins for the power connection (pin 1 and 4) are slightly longer. This is done on purpose to first connect the power supply when connecting a USB device, and only afterwards establish the data connection. With this sequence, the chance that the driver or receiver ports of the data connection receive awkward and possible dangerous voltages is lowered substantially.

Type A plugs always face upstream. Type A sockets will typically find themselves on hosts and hubs. For example, type A sockets are common

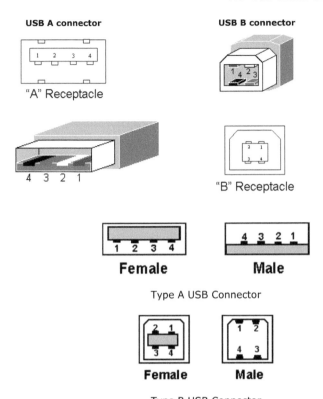

Figure 7.8 Types of USB connectors.

Table 7.10 Standard USB connectors

Standard USB A and B Connector Pin Names			
Pin	Signal Name	Color	Function
1	Vcc	Red	+5 V supply voltage
2	D−	White	Data- signal line
3	D+	Green	Data+ signal line
4	GND	Black	Supply ground
Shell	Shield		Drain Wire

on computer main boards and hubs. Type B plugs are always connected downstream, and consequently, type B sockets are found on devices.

It is interesting to find type A to type A cables wired straight through and an array of USB gender changers in some computer stores. This is in

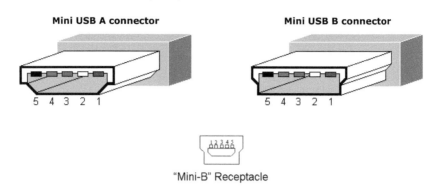

"Mini-B" Receptacle

Figure 7.9   Mini-USB connectors.

contradiction of the USB specification. The only type A plug to type A plug devices are bridges which are used to connect two computers together. Other prohibited cables are USB extensions which has a plug on one end (either type A or type B) and a socket on the other. These cables violate the cable length requirements of USB.

7.6.8.4 Mini USB A and B connectors

The good thing of a USB connector standard is that it is possible to design devices without need to think how that device should be connected to other devices. The USB A and B connectors proofed their usability with devices like printers, modems and scanners, but when the faster USB 2.0 was released and USB became not only a way to connect slow and bulky equipment but also faster and smaller devices like photo camera's and mobile telephones, especially the standard USB B connector was just too big to fit nicely on these smaller equipment. An update to the USB 2.0 specification was posted with the name Mini-B connector engineering change notice which defined a smaller version of the B connector. There has also existed a mini-USB A connector for some time, but as the USB A connector is used on the power sourcing side – mostly a larger piece of equipment like a computer – that connector was withdrawn from the standard and no new devices will receive certification any more if they contain such a connector. In practice, you won't find the mini-USB A connector any more.

Besides the size, the main difference between the standard USB A and B connectors and the mini-USB A and B versions is the extra pin which is called ID. In the mini-connector series this pin is normally not connected. It has been added for future enhancements of the USB standard.

Table 7.11 Mini-USB connectors

Mini USB A and B Connector Pin Names			
Pin	Signal Name	Cable Color	Function
1	Vcc	Red	+5 V supply voltage
2	D−	White	Data- signal line
3	D+	Green	Data+ signal line
4	ID	–	not connected
5	GND	Black	Supply ground
Shell	Shield		Drain Wire

The details on these connectors can be found in Mini-B Connector Engineering Change Notice.

7.6.8.5 Micro-USB AB and B connectors
In the modern world, small is never small enough and the mini-USB B connector soon was too large for new equipment like cell phones. Therefore in January 2007, the micro-USB connector was announced which could be easier integrated in thin devices than the mini-USB version. Although the micro-USB connector is much thinner than its mini-USB brother, it has been especially designed for rough use and the connector is specified for at least 10000 connect/disconnect cycles. One of the reasons is that with mobile devices like cell phones, PDA's and smartphones the number of mate cycles will be significantly higher than with static equipment like printers and mice. Furthermore the micro-USB connector is becoming the de facto standard to charge mobile devices and its use will therefore be even more widespread than of its mini-USB counterpart.

In the original USB specification, there was a strict separation between the host (mostly a computer) which acts as a master device, and the peripherals which have only slave functionality. As mobile devices get smart and often run their own operating system, the separation between the two types of devices has vanished. When connected to a PC a smartphone may be acting as a slave, but it could also be connected to a photo printer directly to print pictures made with the phone. In that case, the phone switches from its slave role to a master. To allow this, an extension to the USB 2.0 specification has been written which is called USB On-The-Go or more often USB OTG. This supplement provides means for easy switching between the master and slave role of a device. On-The-Go specification adds peer-to-peer functionality to USB.

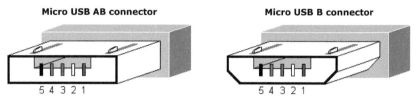

Figure 7.10 USB B and AB connectors.

Table 7.12 Micro-USB connector pin names

Micro-USB AB and B Connector Pin Names			
Pin	Name	Color	Function
1	Vcc	Red	+5 V supply voltage
2	D−	White	Data- signal line
3	D+	Green	Data+ signal line
4	ID	−	not connected: works as B connectorconnected to GND: works as A connector
5	GND	Black	Supply ground

Because most small devices which can both act as a master and a slave only have one USB connector, additions to the connector definition were necessary to allow a role change with only one type of cable. This is where the mini-USB AB and later the micro-USB AB connector are defined for. The mini-USB AB connector is now officially deprecated, but the micro-USB AB connector is replacing its place rapidly. Countries like China are even considering to make this micro-USB AB connector mandatory on all new cell phones sold. In this micro-USB AB connector, the ID pin is used to signal the master of slave function.

Micro-USB connectors. The only difference is that for the micro-USB AB connector the ID pin now has a function assigned to it.

8

USB Data Flow

8.1 Introduction

This chapter is continuation of our discussions on USB. The chapter concentrates on Data Flow in USB Devices. To give enough idea for the reader about this subject, the chapter introduces the following topics: USB Data Exchange, USB Data Transfer Types, USB Data Flow Model, USB On-the-Go (OTG), USB Class Devices, USB Enumeration, Device Drivers, USB Descriptors and the chapter ends by discussing RS-232 to USB converters.

8.1.1 USB is a Bus

Figure 7.2 of the previous chapter gives the topology of USB bus. Figure 8.1 is another way of drawing USB bus topology. When the host is transmitting a packet of data, it is sent to every device connected to an enabled port. It travels downwards via each hub in the chain which resynchronizes the data transitions as it relays it. Only one device, the addressed one, actually accepts the data. (The others all receive it but the address is wrong for them.)

One device at a time is able to transmit to the host, in response to a direct request from the host. Each hub repeats any data it receives from a lower device in an upward only direction.

Downstream direction ports are only enabled once the device connected to them is addressed, except that one other port at a time can reset a device to address 0 and then set its address to a unique value.

Transceivers

At each end of the data link between host and device is a transceiver circuit. The transceivers are similar, differing mainly in the associated resistors.

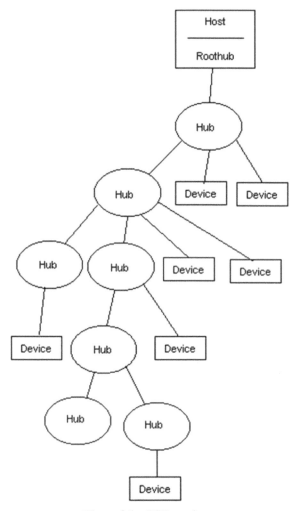

Figure 8.1   USB topology.

A typical upstream end transceiver is shown in Figure 8.2 with high-speed components omitted for clarity. By upstream, we mean the end nearer to the host. The upstream end has two 15 K pull-down resistors.

Each line can be driven low individually, or a differential data signal can be applied. The maximum 'high' level is 3.3 V.

The equivalent downstream end transceiver, as found in a device, is shown in Figure 8.3.

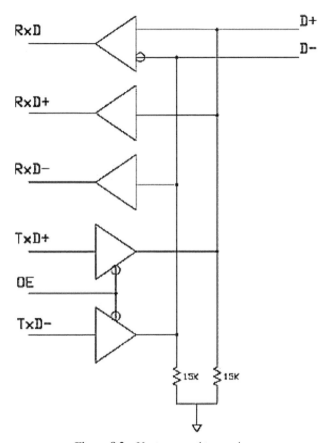

Figure 8.2 Upstream end transceiver.

When receiving, individual receivers on each line are able to detect single ended signals, so that the so-called Single Ended Zero (SE0) condition, where both lines are low, can be detected. There is also a differential receiver for reliable reception of data.

8.1.2 Speed Identification

As mentioned before, at the device end of the link a 1.5 k ohm resistor pulls one of the lines up to a 3.3 V supply derived from V_{BUS}.

This is on D− for a low speed device, and on D+ for a full speed device (Figure 8.4). (A high-speed device will initially present itself as a full speed device with the pull-up resistor on D+.)

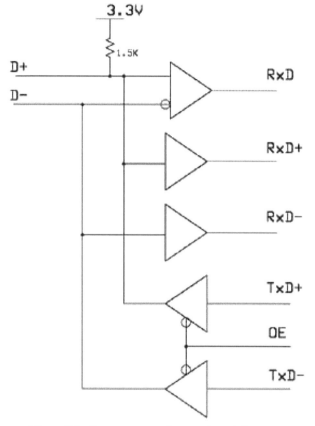

Figure 8.3   Downstream end transceiver (full speed).

The host can determine the required speed by observing which line is pulled high.

8.1.3 Line States

Given that there are just 2 data lines to use, it is surprising just how many different conditions are signaled using them, (see Figure 8.5):

Detached: When no device is plugged in, the host will see both data lines low, as its 15k ohm resistors are pulling each data line low.

Attached: When the device is plugged in to the host, the host will see either D+ or D− go to a '1' level, and will know that a device has been plugged

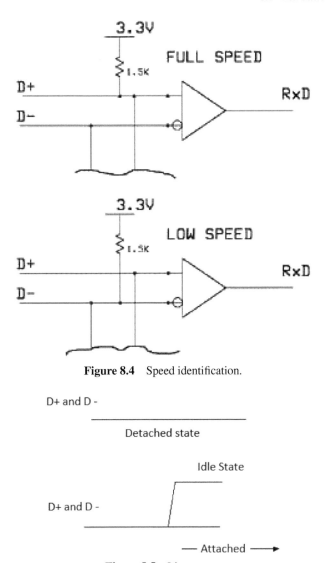

Figure 8.4 Speed identification.

D+ and D -

_____

Detached state

Idle State

D+ and D -

— Attached ⟶

Figure 8.5 Line states.

in. The '1' level will be on D− for a low speed device, and D+ for a full (or high) speed device.

Idle: The state of the data lines when the pulled up line is high, and the other line is low, is called the idle state. This is the state of the lines before and after a packet is sent.

Table 8.1   Bus states

Bus State	Levels
Differential '1'	D+ high, D− low
Differential '0'	D− high, D+ low
Single Ended Zero (SE0)	D+ and D− low
Single Ended One (SE1)	D+ and D− high
Data J State:	
Low-speed	Differential '0'
Full-speed	Differential '1'
Data K State:	
Low-speed	Differential '1'
Full-speed	Differential '0'
Idle State:	
Low-speed	D− high, D+- low
Full-speed	D+ high, D− low
Resume State	Data K state
Start of Packet (SOP)	Data lines switch from idle to K state
End of Packet (EOP)	SE0 for 2 bit times followed by J state for 1 bit time
Disconnect	SE0 for >= 2us
Connect	Idle for 2.5us
Reset	SE0 for >= 2.5 us

Note: This table has been simplified from the original in the USB specification. Please read the original table for complete information.

J, K and SEO States: To make it easier to talk about the states of the data lines, some special terminology is used. The 'J State' is the same polarity as the idle state (the line with the pull-up resistor is high, and the other line is low), but is being driven to that state by either host or device. The K state is just the opposite polarity to the J state. The Single Ended Zero (SE0) is when both lines are being pulled low. The J and K terms are used because for Full Speed and Low Speed links they are actually of opposite polarity.

Single Ended One (SE1): This is the illegal condition where both lines are high. It should never occur on a properly functioning link.

Reset: When the host wants to start communicating with a device it will start by applying a 'Reset' condition which sets the device to its default

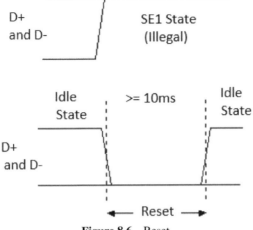

Figure 8.6 Reset.

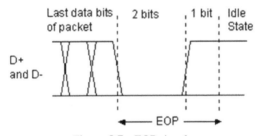

Figure 8.7 EOP signal.

unconfigured state. The Reset condition (Figure 8.6) involves the host pulling down both data lines to low levels (SE0) for at least 10 ms. The device may recognize the reset condition after 2.5 μs. This 'Reset' should not be confused with a micro-controller power-on type reset. It is a USB protocol reset to ensure that the device USB signaling starts from a known state.

EOP signal: The End of Packet (EOP) is an SE0 state for 2 bit times, followed by a J state for 1 bit time (Figure 8.7).

Suspend: One of the features of USB which is an essential part of today's emphasis of 'green' products is its ability to power down an unused device. It does this by suspending the device, which is achieved by not sending anything to the device for 3 ms. Normally a SOF packet (at full speed) or a Keep Alive signal (at low speed) is sent by the host every 1 ms, and this is what keeps the device awake. A suspended device may draw no more than 0.5 mA from

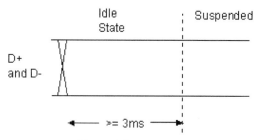

Figure 8.8   Suspend.

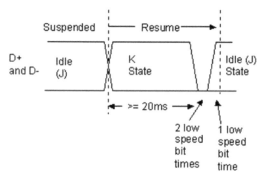

Figure 8.9   Resume state.

V_{bus}. A suspended device must recognize the resume signal, and also the reset signal.

Note: If a device is configured for high power (up to 500 mA), and has its remote wakeup feature enabled, it is allowed to draw up to 2.5 mA during suspend.

Resume: When the host wants to wake the device up after a suspend, it does so by reversing the polarity of the signal on the data lines for at least 20 ms. The signal is completed with a low speed end of packet signal (Figure 8.9). It is also possible for a device with its remote wakeup feature set, to initiate a resume itself. It must have been in the idle state for at least 5 ms, and must apply the wakeup K condition for between 1 and 15 ms. The host takes over the driving of the resume signal within 1 ms.

Keep Alive Signal: This is represented by a Low speed EOP. It is sent at least once every millisecond on a low speed link, in order to keep the device from suspending.

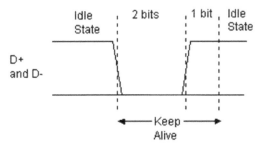

Figure 8.10 Keep alive signal.

Figure 8.11 Single packet.

Packets: The packet could be thought of as the smallest element of data transmission. Each packet conveys an integral number of bytes at the current transmission rate. Before and after the packet, the bus is in the idle state.

The designer does not need to concern with the detail of syncs, bit stuffing, and End of Packet conditions, unless he is designing at the silicon level, as the Serial Interface Engine (SIE) will deal with the details. The designer should just be aware that the SIE can recognize the start and end of a packet, and that the packet contains a whole number of bytes.

In spite of this packets often expect fields of data to cross byte boundaries. The important rule to remember is that all USB fields are transmitted **least significant bit first**. So if, for example, a field is defined by 2 successive bytes, the first byte will be the least significant, and the second byte transmitted will be the most significant.

A packet starts with a sync pattern to allow the receiver bit clock to synchronize with the data, Figure 8.11. It is followed, by the data bytes of the packet, and concluded with an End of Packet (EOP) signal. The data is actually NRZI encoded, and in order to ensure sufficiently frequent transitions, a zero is inserted after 6 successive 1's (this is known as bit stuffing).

Note 1: Serial Interface Engine (SIE): The complexities and speed of the USB protocol are such that it is not practical to expect a general purpose micro-controller to be able to implement the protocol using an instruction-driven basis. Dedicated hardware is required to deal with the time-critical portions

LSB							MSB
PID0	PID1	PID2	PID3	\PID0	\PID1	\PID2	\PID3

Figure 8.12 The PID is shown here in the order of transmission, LSB first.

of the specification, and the circuitry grouping which performs this function is referred to as the Serial Interface Engine (SIE).

8.1.4 Definitions: Endpoints, Pipe, and Transactions

Endpoints: Each USB device has a number of endpoints. Each endpoint is a source or sink of data. A device can have up to 16 OUT and 16 IN endpoints.

- OUT always means from host to device.
- IN always means from device to host.

Endpoint 0 is a special case which is a combination of endpoint 0 OUT and endpoint 0 IN, and is used for controlling the device.

Pipe: A logical data connection between the host and a particular endpoint, in which we ignore the lower level mechanisms for actually achieving the data transfers.

Transactions: Simple transfers of data called 'Transactions' are built up using packets.

Cyclic Redundancy Code (CRC): A CRC is a value calculated from a number of data bytes to form a unique value which is transmitted along with the data bytes, and then used to validate the correct reception of the data. The validation takes place by calculating, at the receiving end, the CRC for the received date and comparing this value with that received with the message. If the two are same, the received data are valid.

USB uses two different CRCs, one 5 bits long (CRC5) and one 16 bits long (CRC16). See the USB specification for details of the algorithms used.

8.1.5 Packet Formats

The first byte in every packet is a Packet Identifier (PID) byte. This byte needs to be recognized quickly by the SIE and so is not included in any CRC checks. It therefore has its own validity check. The PID itself is 4 bits long, and the 4 bits are repeated in a complemented form.

Table 8.2 PID values

/PID Type	PID Name	PID<3:0>*
Token	OUT	0001b
	IN	1001b
	SOF	0101b
	SETUP	1101b
Data	DATA0	0011b
	DATA1	1011b
	DATA2	0111b
	MDATA	1111b
Handshake	ACK	0010b
	NAK	1010b
	STALL	1110b
	NYET	0110b
Special	PRE	1100b
	ERR	1100b
	SPLIT	1000b
	PING	0100b
	Reserved	0000b

*Bits are transmitted LSB first.

There are 17 different PID values defined. This includes one reserved value, and one value which has been used twice with different meanings for two different situations.

Notice that the first 2 bits of a token which are transmitted, determine which of the 4 groups it falls into. This is why SOF is officially considered to be a token PID.

There are four different packet formats based on which PID the packet starts with.

Token Packet

Sync	PID	ADDR	ENDP	CRC5	EOP
	8 bits	7 bits	4 bits	5 bits	

Figure 8.13 Token packet format.

The Token Packet, Figure 8.13, is used for SETUP, OUT and IN packets. They are always the first packet in a transaction, identifying the targeted endpoint, and the purpose of the transaction. The token packet contains two addressing elements:

- Address (7 bits): This device address can address up to 127 devices. Address 0 is reserved for a device which has not yet had its address set.
- Endpoint number (4 bits): There can be up to 16 possible endpoints in a device in each direction. The direction is implicit in the PID. OUT and SETUP PIDs will refer to the OUT endpoint, and an IN PID will refer to the IN endpoint

The SOF packet is also defined as a Token packet, but has a slightly different format and purpose, which is described below.

Data Packet

Data packet format is given in Figure 8.14.

Sync	PID	DATA	CRC16	EOP
	8 bits	(0-1024) x 8 bits	16 bits	

Figure 8.14   Data packet format.

Used for DATA0, DATA1, DATA2 and MDATA packets. If a transaction has a data stage, this is the packet format used.

DATA0 and DATA1 PIDs are used in Low and Full speed links as part of an error-checking system. When used, all data packets on a particular endpoint use an alternating DATA0/DATA1 so that the endpoint knows if a received packet is the one it is expecting. If it is not it will still acknowledge (ACK) the packet as it is correctly received, but will then discard the data, assuming that it has been re-sent because the host missed seeing the ACK the first time it sent the data packet.

DATA0/DATA1 is used as a part of the error control protocol to (or from) a particular endpoint.

DATA2 and MDATA are only used for high speed links.

Handshake Packet

Sync	PID	EOP
	8 bits	

Figure 8.15   Handshake packet format.

Handshake packet, Figure 8.15, is used for *ACK, NAK, STALL* and *NYET* packets. This is the packet format used in the status stage of a transaction, when required.

ACK: Receiver acknowledges receiving error free packet.

NAK: Receiving device cannot accept data or transmitting device cannot send data.

STALL: Endpoint is halted, or control pipe request is not supported.

NYET: No response yet from receiver (high speed only)

SOF Packet

Sync	PID	Frame No.	CRC5	EOP
	8 bits	11 bits	5 bits	

Figure 8.16   SOF packet format.

The Start of Frame packet, Figure 8.16, is sent every 1 ms on full-speed links. The frame is used as a time frame in which to schedule the data transfers which are required. For example, an isochronous endpoint will be assigned one transfer per frame.

Frames: On a low-speed link, to preserve bandwidth, a Keep Alive signal is sent every millisecond, instead of a Start of Frame packet. In fact, Keep Alives may be sent by a hub on a low speed link whenever the hub sees a full speed token packet. At high speed, the 1 ms frame is divided into 8 microframes of 125 us. An SOF is sent at the start of each of these 8 microframes, each having the same frame number, which then increments every 1 ms frame.

8.1.6 Transactions

A successful transaction is a sequence of three packets which performs a simple but secure transfer of data.

For IN and OUT transactions used for isochronous transfers, there are only 2 packets; the handshake packet on the end is omitted. This is because error-checking is not required.

There are three types of transaction. In each of the illustrations below, the packets from the host are shaded, and the packets from the device are not, as shown below:

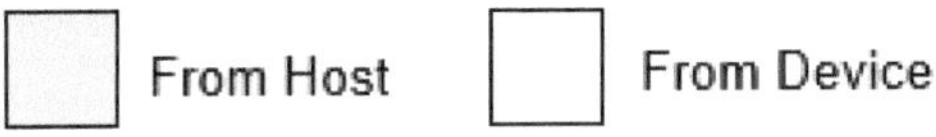

From Host   From Device

OUT Transaction

A successful OUT transaction, Figure 8.17, comprises two or three sequential packets. If it were being used in an *Isochronous Transfer* there would not be a handshake packet from the device.

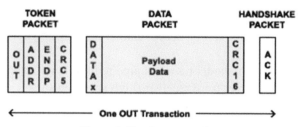

Figure 8.17   Out transaction.

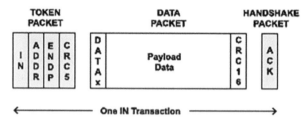

Figure 8.18   In transaction.

On a low or full speed link, the PID shown as DATAx will be either a DATA0 or a DATA1. An alternating.

IN Transaction

A successful IN transaction, Figure 8.18, comprises two or three sequential packets. If it were being used in an *Isochronous Transfer* there would not be a handshake packet from the host.

Here again, the DATAx is either a DATA0 or a DATA1.

SETUP Transaction

A successful SETUP transaction comprises three sequential packets. This is similar to an OUT transaction, but the data payload is exactly 8 bytes long, and the SETUP PID in the token packet informs the device that this is the first transaction in a *Control Transfer* (see Figure 8.19).

As will be seen below, the SETUP transaction always uses a DATA0 to start the data packet.

8.2 Data Flow in USB Devices: Pipes and Endpoints

During the operation of a USB device, the host can initiate a flow of data between the client software and the device.

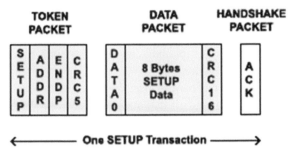

Figure 8.19 Setup transaction.

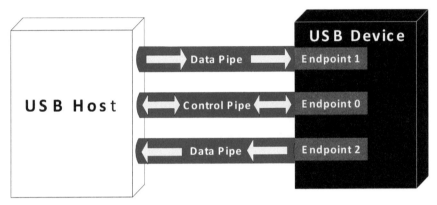

Figure 8.20 USB pipe model.

Data can be transferred between the host and only one device at a time (*peer to peer communication*). However, two hosts cannot communicate directly, nor can two USB devices (with the exception of On-The-Go (OTG) devices, where one device acts as the master (host) and the other as the slave.)

The data on the USB bus is transferred via *pipes* that run between software memory buffers on the host and *endpoints* on the device.

Data flow on the USB bus is half duplex, that is, data can be transmitted only in one direction at a given time.

USB Pipes

A **pipe** is a logical component that represents an association between an endpoint on the USB device and software on the host. Data is moved to and from a device through a pipe. A USB pipe is associated to a unique endpoint address, type of transfer, maximum packet size, and interval for transfer. The

USB specifications define two types of pipes based on the type of data transfer used in the pipe (communication mode):

- *Stream Pipes:* handle interrupt, bulk and isochronous transfers. Data carried over the pipe is unstructured.
- *Message Pipes* support the control transfer type. Data carried over the pipe has a defined structure.

USB Endpoints

An **endpoint** (or endpoint function), Figure 8.21, is a uniquely identifiable (addressable) entity on a USB device, which is the source or terminus of the data that flows from or to the device. Each USB device, logical or physical, has a collection of independent endpoints. The three USB speeds (low, full, and high) all support one bidirectional control endpoint (endpoint zero) and 15 unidirectional endpoints. Each unidirectional endpoint can be used for either inbound or outbound transfers, so theoretically there are 30 supported endpoints.

Each endpoint is identified using an endpoint address. The endpoint address of a device is fixed, and is assigned when the device is designed, as opposed to the device address, which is assigned by the host dynamically during enumeration. An endpoint address consists of an endpoint number field (0 to 15), and a direction bit that indicates if the endpoint sends data to the host (IN) or receives data from the host (OUT). The maximum number of endpoints allowed on a single device is 32.

Each endpoint has the following configurable attributes that define the behavior of a USB device:

- Bus access frequency requirements
- Bandwidth requirement
- Error handling mechanism
- Maximum packet size that the endpoint is able to send or receive
- Transfer type
- Direction in which data is sent and receive from the host

The USB Enumeration and Configuration section (Section 8.8) describes a step in which the device responds to the default address. This occurs before other descriptor information such as the endpoint descriptors are read by the host later in the enumeration process. During this enumeration sequence, a special set of endpoints are used for communication with the device. These special endpoints, collectively known as the Control Endpoint or Endpoint 0, are defined as Endpoint 0 IN and Endpoint 0 OUT. Even though Endpoint

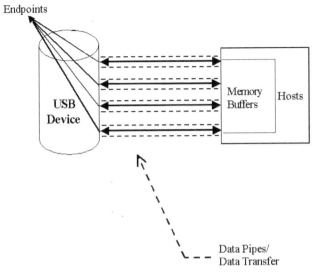

Endpoints

USB Device

Memory Buffers

Hosts

Data Pipes/ Data Transfer

Figure 8.21 USB endpoints.

0 IN and Endpoint 0 OUT are two endpoints, they look and act like one endpoint to the developer. Every USB device must support Endpoint 0. For this reason, Endpoint 0 does not require a separate descriptor.

Endpoint zero requirement

Endpoint zero (also known as Default Endpoint) is a bi-directional endpoint used by the USB host system to get information, and configure the device via standard requests. All devices must implement an endpoint zero configured for control transfers (see Section 8.4.1 for more information).

In addition to Endpoint 0, the number of endpoints supported in any particular device is based on its design requirements. A simple design such as a mouse may need only one IN endpoint. More complex designs may need several data endpoints. The USB specification sets a limit on the number of endpoints to 16 for each direction (16 IN/16 OUT – 32 Total) for High-Speed and Full-Speed devices, which does not include the control endpoints 0 IN and 0 OUT. Low-Speed devices are limited to two endpoints. USB Class devices may set a greater limit on the number of endpoints. For example, a Low-Speed human interface device (HID) design may have no more than two data endpoints – typically one IN endpoint and one OUT endpoint. Data endpoints are bidirectional by nature. It is not until they are configured that they take on a single direction (becoming unidirectional). Endpoint 1, for example, can be either an IN or OUT endpoint. It is in the device descriptors that it will officially make Endpoint 1 an IN endpoint.

Endpoints use cyclic redundancy checks (CRCs) to detect errors in transactions. The CRC is a calculated value used for error checking. The actual calculation equation is explained in the USB specification and the handling of these calculations is taken care of by the USB hardware so that the proper response can be issued. The recipient of a transaction checks the CRC against the data. If the two match, then the receiver issues an ACK. If the data and the CRC do not match, then no handshake is sent. This lack of a handshake tells the sender to try again.

The USB specification further defines four types of endpoints and sets the maximum packet size based on both the type and the supported device speed. Developers use the endpoint descriptor to identify the type of endpoint and maximum packet size based on their design requirements. The four types of endpoints and characteristics are:

Control Endpoint: These endpoints support control transfers, which all devices must support. Control transfers send and receive device information across the bus. The advantage of control transfers is guaranteed accuracy. Errors that occur are properly detected and the data is resent. Control transfers have a 10 percent reserved bandwidth on the bus in low and Full-Speed devices (20 percent at High-Speed) and give USB system level control.

Interrupt Endpoints: These endpoints support interrupt transfers. These transfers are used on devices that must use a high reliability method to communicate a small amount of data. This is commonly used in HID designs. The name of this transfer can be misleading. It is not truly an interrupt, but uses a polling rate. However, you get a guarantee that the host checks for data at a predictable interval. Interrupt transfers give guaranteed accuracy as errors are properly detected and transactions are retried at the next transaction. Interrupt transfers have a guaranteed bandwidth of 90% on Low- and Full-Speed devices and 80% on High-Speed devices. This bandwidth is shared with isochronous endpoints.

Interrupt endpoint maximum packet size is a function of device speed. High-speed capable devices support a maximum packet size of 1024 byes. Full-Speed capable devices support a maximum packet size of 64 bytes. Low-speed devices support a maximum packet size of 8 bytes.

Bulk Endpoints: These endpoints support bulk transfers, which are commonly used on devices that move relatively large amounts of data at highly variable times where the transfers can use any available bandwidth space. They are the most common transfer type for USB devices. Delivery time with a bulk transfer is variable because there is no set aside bandwidth for the transfer. The delivery time varies depending on how much bandwidth on the bus is available, which makes the actual delivery time unpredictable. Bulk transfers give guaranteed accuracy because errors are properly detected and transactions are resent. Bulk transfers are useful in moving large amounts of data that are not time sensitive.

In case of bulk endpoint, the maximum packet size is a function of device speed. High-speed capable devices support a maximum BULK packet size of 512 bytes. Full-speed capable devices support a maximum packet size of 64-bytes. Low-speed devices do not support bulk transfer types.

Isochronous Endpoints: These endpoints support isochronous transfers, which are continuous, real-time transfers that have a pre-negotiated bandwidth. Isochronous transfers must support streams of error tolerant data because they do not have an error recovery mechanism or handshaking. Errors are detected through the CRC field, but not corrected. With isochronous, arises tradeoff of guaranteed delivery versus guaranteed accuracy. Streaming music or video are examples of an application that uses isochronous endpoints because the occasional missed data is ignored by the human ears and eyes. Isochronous transfers have a guaranteed bandwidth of 90% on Low- and Full-speed devices (80% on High-speed devices) that is shared with interrupt endpoints.

High-speed capable devices support a maximum packet size of 1024 bytes. Full-speed devices support a maximum packet size of 1023 bytes. Low-speed devices do not support isochronous transfer types. There are special considerations with isochronous transfers. It is required to have 3x buffering to ensure data is ready. The three buffers are: one actively transmitting buffer, another buffer loaded and ready to transfer, and a third buffer being actively loaded.

8.3 USB Data Exchange

The USB standard supports two kinds of data exchange between a host and a device: functional data exchange and control exchange.

- **Functional Data Exchange** is used to move data to and from the device. There are three types of USB data transfers: Bulk, Interrupt, and Isochronous.

Table 8.3   Endpoint transfer type features

Transfer Type	Control	Interrupt	Bulk	Isochronous
Typical Use	Device Initialization and Management	Mouse and Keyboard	Printer and Mass Storage	Streaming Audio and Video
Low-Speed Support	Yes	Yes	No	No
Error Correction	Yes	Yes	Yes	No
Guaranteed Delivery Rate	No	No	No	Yes
Guaranteed Bandwidth	Yes (10%)	Yes (90%)[1]	No	Yes (90%)[1]
Guaranteed Latency	No	Yes	No	Yes
Maximum Transfer Size	64 bytes	64 bytes	64 bytes	1023 bytes (FS) 1024 bytes (HS)
Maximum Transfer Speed	832 KB/s	1.216 MB/s	1.216 MB/s	1.023 MB/s

[1] Shared bandwidth between isochronous and interrupt.

- **Control Exchange** is used to determine device identification and configuration requirements and to configure a device, and can also be used for other device-specific purposes, including control of other pipes on the device.Control exchange takes place via a control pipe, mainly the default *Pipe 0*, which always exists. The control transfer consists of a *setup stage* (in which a setup packet is sent from the host to the device), an optional *data stage* and a *status stage*.

8.4 USB Data Transfer Types

The USB device (function) communicates with the host by transferring data through a pipe between a memory buffer on the host and an endpoint on the device.

USB supports four transfer types that match the bandwidth and services requirements of the host and the device application using a specific pipe. Table 8.3 summarizes the endpoint transfer type features. A type is selected for a specific endpoint according to the requirements of the device and the software. The transfer type of a specific endpoint is determined in the endpoint descriptor. Each USB transfer encompasses one or more transactions that send data to and from the endpoint. The notion of transactions is related to the maximum payload size defined by each endpoint type. That is, when a transfer is greater than this maximum, it will be split into one or more transactions to fulfill the action.

There are four different ways to transfer data on a USB bus. Each has its own purposes and characteristics. Each one is built up using one or more transaction type.

Data Flow Type	Description
Control Transfer	Mandatory using Endpoint 0 OUT and Endpoint 0 IN.
Bulk Transfer	Error-free high volume throughput when bandwidth available.
Interrupt Transfer	Regular Opportunity for status updates, etc.Error-free Low throughput
Isochronous Transfer	Guaranteed fixed bandwidth.Not error-checked.

The USB specification provides for the following data transfer types.

8.4.1 Control Transfer

Control Transfer is mainly intended to support configuration, command and status operations between the software on the host and the device. This transfer type is used for low-, full- and high-speed devices. This is a bi-directional transfer which uses both an IN and an OUT endpoint. Each control transfer is made up of from 2 to several transactions.

Each USB device has at least one control pipe (default pipe), which provides access to the configuration, status and control information.

A default control pipe uses endpoint zero. The default control pipe is a bi-directional message pipe, that is, data can flow in both directions.

Control transfer is bursty, non-periodic communication.

Control transfer has a robust error detection, recovery and retransmission mechanism and retries are made without the involvement of the driver.

Control Transfer is divided into three stages, Table 8.4:

- The SETUP stage carries 8 bytes called the Setup packet. This defines the request, and specifies whether and how much data should be transferred in the DATA stage.

Table 8.4 Control transfer stages

Stage	Description
Setup	The Setup stage includes information about the request. This SETUP stage represents one transaction.
Data	The Data stage contains data associated with request. Some standard and class-specific request may not require a Data stage. This stage is an IN or OUT directional transfer and the complete Data stage represents one or more transactions. Figure 8.22 (taken from the USB specification) refers to a sequence of read and write transactions. Figure 8.23 is an example of control Read.
Status	The Status stage, representing one transaction, is used to report the success or failure of the transfer. The direction of the Status stage is opposite to the direction of the Data stage. If the control transfer has no Data stage, the Status stage always is from the device (IN).

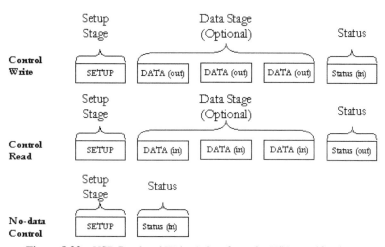

Figure 8.22 USB Read and Write (taken from the USB specification).

"(in)" indicates data flow from the device to the host.
"(out)" indicates data flow from the host to the device.

- The DATA stage is optional. If present, it always starts with a transaction containing a DATA1. The type of transaction then alternates between DATA0 and DATA1 until all the required data has been transferred.
- The STATUS stage is a transaction containing a zero-length DATA1 packet. If the DATA stage was IN then the STATUS stage is OUT, and vice versa.

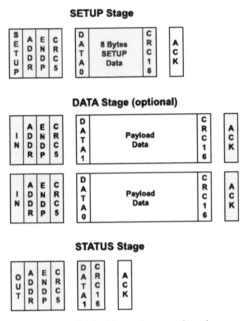

Figure 8.23 Example control read.

Control transfers are used for initial configuration of the device by the host, using Endpoint 0 OUT and Endpoint 0 IN, which are reserved for this purpose. They may be used (on the same endpoints) after configuration as part of the device-specific control protocol, if required.

The maximum packet size for control endpoints can be only 8 bytes for low-speed devices; 8, 16, 32, or 64 bytes for full-speed devices; and only 64 bytes for high-speed devices.

Error Control Flow
SETUP STAGE

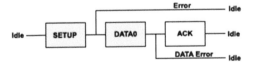

Figure 8.24 Setup stage.

Figure 8.24 shows the setup stage. Notice that it is not permitted for a device to respond to a SETUP with a NAK or a STALL.

DATA STAGE
(same as for bulk transfer)

STATUS STAGE
(same as for bulk transfer)

8.4.2 Isochronous Transfer

Isochronous Transfer is most commonly used for time-dependent information, such as multimedia streams and telephony. Isochronous transfers are, in general, used by devices that require data delivery at a constant rate with a certain degree of error-tolerance. It does not use any data checking, as there is not time to resend any data packets with errors - lost data can be accommodated better than the delays incurred by resending data. In other words isochronous transfers is used in applications where it is important to maintain the data flow, but not so important if some data gets missed or corrupted: Isochronous transfers have a guaranteed bandwidth, but error-free delivery is not guaranteed.

- This transfer type can be used by full-speed and high-speed devices, but not by low-speed devices.
- Isochronous transfer is periodic and continuous.
- The isochronous pipe is unidirectional, that is, a certain endpoint can either transmit or receive information. Bi-directional isochronous communication requires two isochronous pipes, one in each direction.
- USB guarantees the isochronous transfer access to the USB bandwidth (i.e., it reserves the required amount of bytes of the USB frame) with bounded latency, and guarantees the data transfer rate through the pipe, unless there is less data transmitted.
- Since timeliness is more important than correctness in this type of transfer, no retries are made in case of error in the data transfer. However, the data receiver can determine that an error occurred on the bus.
- An isochronous transfer uses either an IN transaction or an OUT transaction depending on the type of endpoint. The special feature of these transactions is that there is no handshake packet at the end.
- An isochronous packet may contain up to 1023 bytes at full speed, or up to 1024 at high speed. Isochronous transfers are not allowed at low speed.

Figure 8.25 is an example of isochronous transfer. Figure 8.26 shows the Error control flow.

Example Isochronous Transfer

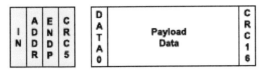

Figure 8.25 Isochronous transfer.

Error Control Flow

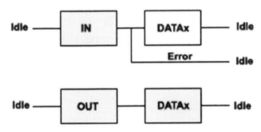

Figure 8.26 Error control flow.

8.4.3 Interrupt Transfer

- Interrupt transfers have nothing to do with interrupts. The name is chosen because they are used for the sort of purpose where an interrupt would have been used in earlier connection types.
- Interrupt Transfer is intended for devices that send and receive small amounts of data infrequently or in an asynchronous time frame, that is, devices using interrupt transfers can schedule data at any time, for example, mice or keyboard. Devices using interrupt transfers provide a polling interval which determines when the scheduled data is transferred over the bus.
- This transfer type can be used for low-, full-, and high-speed devices.
- Interrupt transfer type guarantees a maximum service period and that delivery will be re-attempted in the next period if there is an error on the bus.
- The interrupt pipe, like the isochronous pipe, is unidirectional and periodical.
- Interrupt transfers are regularly scheduled IN or OUT transactions, although the IN direction is the more common usage.

- Typically the host will only fetch one packet, at an interval specified in the *endpoint descriptor* (see below). The host guarantees to perform the IN transaction at least that often, but it may actually do it more frequently.
- The maximum packet size for interrupt endpoints can be 8 bytes or less for low-speed devices; 64 bytes or less for full-speed devices; and 1,024 bytes or less for high-speed devices.
- Interrupt transfer is normally used when it is needed to be regularly kept up to date of any changes of status in a device. Examples of their use are for a mouse or a keyboard.

Error control is very similar to that for bulk transfers.

Figure 8.27 is an example of Interrupt transfer. Figure 8.28 shows the Error control flow.

Example Interrupt Transfer

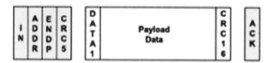

Figure 8.27   Interrupt transfer.

Error Control Flow

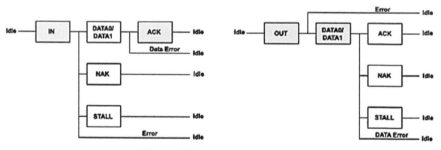

Figure 8.28   Error control flow.

8.4.4 Bulk Transfer

- Bulk Transfer, example is given in Figure 8.29, is typically used for devices that transfer large amounts of non-time sensitive data, and that can use any available bandwidth, such as printers and scanners. In other

Figure 8.29   Example bulk transfer.

words, it is designed to transfer large amounts of data with error-free delivery, but with no guarantee of bandwidth. The host will schedule bulk transfers after the other transfer types have been allocated.

- This transfer type can be used by full-speed and high-speed devices, but not by low-speed devices.
- Bulk transfer is non-periodic, large packet, bursty communication.
- Bulk transfer allows access to the bus on an "as-available" basis, guarantees the data transfer but not the latency, and provides an error check mechanism with retries attempts. If part of the USB bandwidth is not being used for other transfers, the system will use it for bulk transfer.
- Like the other stream pipes (isochronous and interrupt), the bulk pipe is also unidirectional, so bi-directional transfers require two endpoints.
- If an OUT endpoint is defined as using bulk transfers, then the host will transfer data to it using OUT transactions.
- If an IN endpoint is defined as using bulk transfers, then the host will transfer data from it using IN transactions.
- In this form of data transfer, variable length blocks of data are sent or requested by the Host. The max packet size is 8, 16, 32, or 64 at full speed and 512 for high speed. Bulk transfers are not allowed at low speed.
- The data integrity is verified using cyclic redundancy checking, CRC and an acknowledgement is sent. This USB data transfer mechanism is not used by time critical peripherals because it utilizes bandwidth not used by the other mechanisms.
- Bulk transfers are used when it is needed to shift a lot of data, as fast as possible, but where you would not have a large problem if there is a delay caused by insufficient bandwidth.

Figure 8.30 illustrate the possible flow of events in the face of errors.

Error Control – IN

- If the IN token packet is not recognized, the device will not respond at all. Otherwise, if it has data to send it will send it in a DATA0 or DATA1

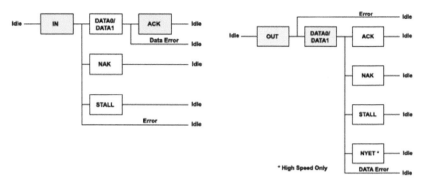

Figure 8.30   Bulk transfer error control flow.

packet, if it is not ready to send data it will send a NAK packet. If the endpoint is currently 'halted' then it will respond with a STALL packet.

- In the case of DATA0/1 being sent, the host will acknowledge with an ACK, unless the data is not validly received, in which case it does not send an ACK. (Note: the host never sends NAK!)

Error Control – OUT

- If the OUT token packet is not recognized, the device will not respond at all. It will then ignore the DATAx packet because it does not know that it has been addressed.
- If the OUT token is recognized but the DATAx packet is not recognized, then the device will not respond.
- If the data is received but the device can't accept it at this time, it will send a NAK, and if the endpoint is currently halted, it will send a STALL.

8.5 USB Data Flow Model

Figure 8.31 shows a graphical representation of the data flow model.

The flow of data as shown in the figure takes place in the following sequence:

1. The host software uses standard requests to query and configure the device using the default pipe. The default pipe uses endpoint zero (EP0).
2. USB pipes allow associations between the host application and the device's endpoints. Host applications send and receive data through USB pipes.

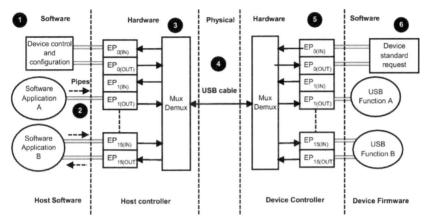

Figure 8.31 USB data flow.

3. The host controller is responsible for the transmission, reception, packing and unpacking of data over the bus.
4. Data is transmitted via the physical media.
5. The device controller is responsible for the transmission, reception, packing and unpacking of data over the bus. The USB controller informs the USB device software layer about several events such as bus events and transfer events.
6. The device software layer responds to the standard request, and implements one or more USB functions as specified in the USB class document.

Transfer Completion

The notion of transfer completion is only relevant for control, bulk and interrupt transfers as isochronous transfers occur continuously and periodically by nature. In general, control, bulk and interrupt endpoints must transmit data payload sizes that are less than or equal to the endpoint's maximum data payload size. When a transfer's data payload is greater than the maximum data payload size, the transfer is split into several transactions whose payload is maximum-sized except the last transaction which contains the remaining data. A transfer is deemed complete when:

• The endpoint transfers exactly the amount of data expected.
• The endpoint transfers a short packet that is a packet with a payload size less than the maximum.
• The endpoint transfers a zero-length packet.

8.6 USB On-the-Go (OTG): Uses and Support

((Reference: Dallas Semiconductor/Maxim and USB 2.0 specifications))

Introduction

The USB specification, as mentioned before, introduced a simple and inexpensive infrastructure for easily connecting multiple external peripherals to a PC. That was several years ago; today, more than 500 million peripheral devices are designed with USB ports, making USB the market's dominant I/O-connectivity standard. The widespread availability of USB peripherals is now driving non-PC applications, such as mobile, handheld, and embedded post-PC applications, to adopt USB for direct-I/O connections. In addition, many devices that have traditionally functioned as peripherals now require direct connections to other devices. USB's greatest limitation – its lack of support for point-to-point communication between devices – has deterred its use in consumer-electronic devices, such as mobile phones and PDAs. However, these devices are gaining popularity and intelligence, increasing the need for direct connections among them. The answer to this requirement is in a developing standard called **USB OTG (USB On-The-Go)**, a supplement to the USB specification that eliminates the requirement for a PC to act as host in exchanges of data among connected devices.

The OTG specification's goal is to enhance certain USB peripherals to enable them to also act as hosts for a selected set of peripherals. OTG introduces point-to-point communication between these enhanced USB peripherals. USB On-the-Go (OTG) allows two USB devices to talk to each other without requiring the services of a personal computer. Although OTG appears to add \"peer to peer\" connections to USB, it does not. Instead, USB OTG retains the standard USB host/peripheral model, where a single host talks to USB peripherals. OTG introduces the ***dual-role device*** (DRD), capable of functioning as either host or peripheral. Part of the magic of OTG is that a host and peripheral can exchange roles if necessary.

This tends to maintain the current USB host/peripheral architecture model. The OTG host always initiates communication with a normal bus-enumeration process (bus reset, acquisition of USB descriptors, and peripheral-device configuration (see Section 8.8). After these steps, the device serving as OTG host may transfer data to and from an OTG device performing as a peripheral. The OTG specification defines a mechanism for exchanging the roles of OTG host and peripheral. The initial role of each device is defined by which mini-plug a user inserts into its receptacle.

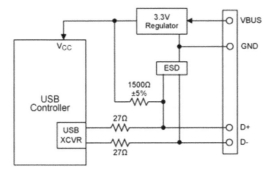

Figure 8.32 A USB peripheral controller and its associated circuitry.

Before OTG, the concept of an embedded host was already established in the USB world. Instead of duplicating the full UHCI/OHCI USB controllers and drivers built into personal computers, most embedded host chips provide limited hosting capabilities. This makes them better suited to the embedded environment than to the PC with its huge resources and infinite capacity for drivers and application software.

An OTG device may, or may not be capable of functioning as a host. It is likely, nonetheless, that most OTG devices will be dual-role.

8.6.1 USB Peripherals

Figure 8.32 illustrates the basic USB peripheral circuitry on which OTG builds. These example peripherals operate at low or full speed, and are commonly known as USB 1.1 devices. This nomenclature is used even though the USB 2.0 Specification includes the current USB 1.1 specification and introduces a third, higher speed.

The controller in Figure 8.32 might be a microprocessor plus USB SIE (Serial Interface Engine), an integrated microprocessor/USB chip, or an ASIC connected to a USB transceiver. A bus-powered peripheral requires a 3.3 V regulator, both to power the logic and to supply the proper voltage to a 1500Ω resistor connected to either the D+ or D− USB pins. This pull-up resistor signals the host that a device is connected, and indicates the device's operating speed. A pull-up to D+ indicates full speed; a pull-up to D− indicates low speed. The other end of the connection – host or hub – contains 15 kΩ pull-down resistors on D+ and D− so the pull-up resistor can be detected. Finally, an ESD protection circuit is advisable on D+, D−, and V_{BUS} pins because USB is designed to be hot-plugged.

How to be a Host

The Figure 8.32 circuit functions only as a USB peripheral device. To add OTG dual-role capability, the transceiver must be augmented to allow the OTG device to function as either host or peripheral. Adding the following to Figure 1 lets the system also function as a host:

- 15 kΩ pull-down resistors on D+ and D−
- A means to supply, rather than draw, power on V_{BUS}

The ASIC or controller must also contain logic to function as a USB host. Some of the host duties absent in a peripheral device are:

- Send SOF (Start of Frame) packets.
- Send SETUP, IN, and OUT packets.
- Schedule transfers within USB 1ms frames.
- Signal USB reset.
- Provide USB power management.

In addition to requiring a dual-role peripheral/host USB controller, OTG requires additional circuitry to support two new protocols, called HNP (Host Negotiation Protocol) and SRP (Session Request Protocol).

Host Negotiation Protocol

The OTG specification defines a negotiation protocol, the HNP (Host Negotiation Protocol). This protocol provides a means by which the A- and B-devices can exchange the OTG host and peripheral roles.

An OTG dual-role device can operate either as a host or peripheral. In OTG nomenclature, the *initial* host is called the A-Device, and the initial peripheral is called the B-Device. The word *initial* is important. Once connected, OTG dual-role devices can exchange roles – host and peripheral – by using the new Host Negotiation Protocol (HNP). HNP raises two obvious questions: (a) how are the initial roles determined; and (b) why is the role reversal necessary?

The cable orientation determines the initial roles (Figure 8.33). Dual-role devices use a new receptacle called the mini-AB. The mini-A plug, the mini-B plug and the mini-AB receptacle add a fifth pin (ID) to give different electrical identities to the cable ends. This fifth ID pin is connected to ground inside the mini-A plug and left floating in the mini-B plug. The OTG device receiving the grounded ID pin is the default A-Device (host); the device with the floating ID pin is the default B-Device (peripheral).

To understand the need for the HNP and host/peripheral role reversal, the example in Figure 8.34 shows two dual-role devices, a PDA and a printer.

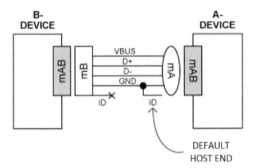

Figure 8.33 Fifth ID pin determines default host.

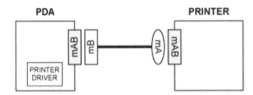

Figure 8.34 OTG cable is inserted backward.

The PDA has a printer driver inside. The two devices are connected with the new OTG cable as shown, making the printer the default host (A-Device) and the PDA the default peripheral (B-Device). But this setup is backward. The PDA, which has the printer driver, needs to act as USB host to the printer, which contains no driver. Rather than bothering the user to reverse the cable, HNP allows the devices' roles to reverse automatically and silently.

Session Request Protocol

The OTG Specification adds a second new protocol to USB, called Session Request Protocol (SRP). SRP allows a B-Device to request an A-Device to turn on V_{BUS} power and start a session.

An OTG session is defined as the time that the A-Device is furnishing V_{BUS} power. (Note: the A-Device always supplies V_{BUS} power, even if it is functioning as a peripheral due to HNP.) The A-Device can end a session by turning off V_{BUS} to conserver power, a very important requirement in a battery-powered device such as a cell phone.

Figure 8.35 shows a common OTG application: two cell phones connected together to exchange information. The right phone received the mini-A end of the cable, making it the A-Device and thus defaulting into the host role. The left phone is the B-Device, defaulting to peripheral. If there is no need

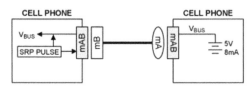

Figure 8.35   OTG session request protocol (SRP).

to communicate over USB, the A-Device can power down the V_{BUS} wire, which the B-Device can detect so that it too can enter a low-power state.

Now suppose that the user of the left phone presses a button to synchronize address books, or any other action that requires a USB session. The 'SRP Pulse' block in the left phone pulses first the D+ wire, and then the V_{BUS} wire to wake up the A-Device. (The A-Device can respond either to D+ or V_{BUS} pulsing.) The A-Device then detects the pulse, causing it to switch on V_{BUS} and start a session.

The SRP protocol is more complex than this simple illustration. The B-Device, for example, must first measure V_{BUS} to ensure that a session is not in progress. It must also be able to differentiate between a classic PC or an OTG device at the other end of the cable. It does this by delivering measured amounts of current to the V_{BUS} wire and noting the resulting voltage.

Once a session is underway, the devices may or may not use HNP.

OTG Transceiver

Next, we are going to examine the requirements for an OTG transceiver, illustrated in Figure 8.36.

The system given in Figure 8.36 builds on the Figure 8.31 example circuit. The ASIC block could also be a microprocessor or DSP with USB capability. Three additions make the transceiver OTG compatible:

1. Switchable pull-up and pull-down resistors on D+/D− to allow peripheral or host functionality.
2. Circuitry to monitor and supply 5V power on V_{BUS} as an A-Device, and to monitor and pulse V_{BUS} as a B-Device initiating SRP.
3. An ID input pin, which is made available as an output to the ASIC.

For this system to operate as a dual-role OTG device, the ASIC, DSP, or whatever is connected to the transceiver must be capable both of functioning as a peripheral or host, and of switching roles on-the-fly as a result of HNP.

Most of the added transceiver circuitry manages the V_{BUS} pin, which now must also supply 5V power at 8mA as a host, and perform V_{BUS} pulsing as

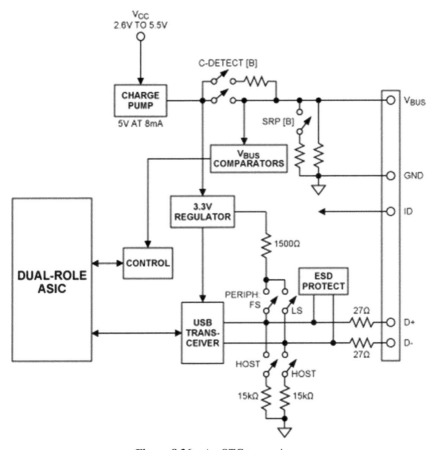

Figure 8.36 An OTG transceiver.

a peripheral. Analog switches configure the transceiver for the various roles that it must play.

8.6.2 Differences with USB 2.0

Because portable devices are typically battery-powered, forcing such devices to source a relatively large amount of current would result in very short battery life. The current-driving capability is often unnecessary anyway, because the device connected to the portable device is often self-powered. Instead of providing 100 or 500 mA at its port, a dual-role device must provide at least 8 mA. Dual-role devices are allowed to provide more current to accommodate additional peripherals but are not required to do so.

Another key difference is the V_{BUS} capacitance at the connector. OTG dual-role devices have a capacitance of 1 to 6.5 μF. Standard hosts have a capacitance of 96 μF or more. This difference is significant during the SRP. When an OTG device performs V_{BUS} pulsing, it drives a charge onto the V_{BUS} line through either a constant current source or a resistor. If the OTG device is connected to a standard host, the voltage will not be driven above 2V, thereby avoiding any possible damage to the host. However, because OTG devices have much smaller capacitance than standard hosts, another OTG device will have its V_{BUS} line driven above 2.1V.

Section 7.1.2.2 of the USB 2.0 specification defines four test points corresponding to the connectivity from host silicon (TP1) to host connector (TP2) to peripheral connector (TP3) to peripheral silicon (TP4). The maximum allowable delay is 30 ns. Section 7.1.16 of the USB 2.0 specification limits the cable delay (from TP2 to TP3) to 26 ns. Because of the possible use of adapters, which are allowed a 1-ns delay, any cable with a mini-A plug is limited to a cable delay of 25 ns. Using a 5 m standard-B-to-standard-A cable with a standard-A-receptacle-to-mini-A-plug adapter violates, by 1 ns, the 26-ns limitation. However, when operating as peripherals, OTG devices have an only 1-ns delay between TP3 and TP4, and, when operating as OTG hosts, a 1-ns delay between TP1 and TP2. Therefore, the 30-ns delay between host and peripheral silicon is preserved.

Notes: The following two comments are completing the discussion of this section: OTG.

[1] Cables and connectors

Concerning Connectors: The USB 2.0 specification defines three types of connector pairs (plugs and receptacles): the standard-A, the standard-B, and the mini-B. The mini-B connector was developed for smaller peripherals, such as mobile phones. The OTG specification introduces a fourth plug, the mini-A, and two receptacles, mini-A and mini-AB. Mini-A and mini-B connectors are designed such that one of them cannot fit into the receptacle of the other, however the mini-AB receptacle accepts any of them. Also the design lets the mini-A plug's ID pin is shorted to ground.

Concerning Cables: The USB 2.0 specification defines two cables: the standard-A to standard-B and the standard-A to mini-B. OTG defines two more cables: the mini-A to standard-B and the mini-A to mini-B. The end-to-end delay of the mini-A-to-mini-B cables has been reduced to allow the use of adapters at the A end of the cable.

Adapters are required to allow for connecting different combinations of devices.

[2] Dual-role devices

Because each OTG dual-role device incorporates a mini-AB receptacle, a mini-A-to-mini-B cable can directly interconnect the devices. Users will perceive no difference in the devices based on the cable connection. That is, users will not know which device is the initial OTG host and which is initially the peripheral.

A dual-role device's initial role is defined by which end of the cable a user inserts into the device's mini-AB receptacle. The dual-role device with the mini-A plug is the initial OTG host, also known as the A-device. Conversely, the dual-role device with the mini-B plug is the initial peripheral, also called the B-device. The dual-role device determines which end of the cable to insert by whether or not the ID pin is shorted to the GND pin.

The A-device must supply the voltage on V_{BUS} when a communication session is in progress. This key difference between the A- and the B-devices causes the two connected devices to be unequal and prevents a peer-to-peer connection. Because the A-device supplies the voltage on V_{BUS} and, therefore, controls when a communication session occurs, the B-device requires a mechanism for requesting a communication session. This mechanism is the SRP (Session Request Protocol).

8.7 USB Class Devices

The USB Implementers Forum has a list of recognized and approved USB device classes. The most common device classes are:

- Human Interface Device (HID)
- Mass Storage Device (MSD)
- Communication Device Class (CDC)
- Vendor (Vendor Specific)

There are several considerations to think about when developing an application for a certain class. First, each class has a fixed maximum bandwidth. Second, each class has limitations on the supported transfer types and certain commands that must be supported. However, the biggest advantage to using a predefined USB device class is the cross platform support across various operating systems. All major operating systems include a driver in the OS for most of the predefined USB classes that eliminates the need to create a

Table 8.5 Drivers used with Cypress products

Feature	HID	CDC	WinUSB	LibUSB	CYUSB
Driver Support in Windows	Yes	Need inf.	Need inf.[1]	No	No
Support for 64-Bit	Yes	Yes	Yes	Yes	Yes
Support for Control Transfers	Yes	No	Yes	Yes	Yes
Support for Interrupt Transfer	Yes	No	Yes	Yes	Yes
Support for Bulk Transfers	No	Yes	Yes	Yes	Yes
Support for Isochronous Transfers	No	Yes	No	Yes	Yes
Maximum Speed (Full-Speed)	~64 KB/s	~80 KB/s	~1 MB/s	~1 MB/s	~1 MB/s

[1]. WinUSB is not native to Windows XP; it must be installed with the WinUSB co-installer.

custom driver. Table 8.5 shows some of the more common drivers that are used with Cypress products and some of the capabilities of those drivers.

Devices that do not meet the definition of a specific USB device class are called vendor-specific devices. These devices allow developers to create applications with their own creativity and customization options, which are not bound by a specific USB class, but still conform to the USB specification. Devices that fall under a vendor-specific device use WinUSB, CYUSB, LibUSB, or another type of vendor-specific driver. The advantage to using WinUSB is that it is Windows own vendor-specific driver and does not need to undergo Windows Hardware Quality Labs (WHQL) testing for driver signing. WHQL testing is discussed later in this application note. LibUSB is an open source driver project with support for Windows, Mac, and Linux operating systems. CyUSB is Cypress' own vendor-specific driver. The advantage to using this driver in an application is the broad range of example applications, supporting documentation, and direct support from Cypress.

In the USB Descriptors section (see USB specifications and also Section 8.10 of this chapter), notice that the fourth byte in the device descriptor and the sixth byte in the interface descriptor are where the class of the USB device is defined. The USB specification defines many different USB classes and the device class codes that go along with them. Table 8.6 shows some USB class codes that can be used in these bytes to give an idea of the various USB classes that are available.

Table 8.6 USB class codes

Class	Usage	Description	Examples
00h	Device	Unspecified	Device class is unspecified, interface descriptors are used to determine needed drivers
01h	Interface	Audio	Speaker, microphone, sound card, MIDI
02h	Both	Communications and CDC Control	Modem, Ethernet adapter, Wi-Fi adapter
03h	Interface	Human Interface Device (HID)	Keyboard, mouse, joystick
05h	Interface	Physical Interface Device (PID)	Force feedback joystick
06h	Interface	Image	Camera, scanner
07h	Interface	Printer	Printers, CNC machine
08h	Interface	Mass Storage	External hard drives, flash drives, memory cards
09h	Device	USB Hub	USB hubs
0Ah	Interface	CDC-Data	Used in conjunction with class 02h.
0Bh	Interface	Smart Card	USB smart card reader
0Dh	Interface	Content Security	Fingerprint reader
0Eh	Interface	Video	Webcam
0Fh	Interface	Personal Healthcare	Heart rate monitor, glucose meter
DCh	Both	Diagnostic Device	USB compliance testing device
E0h	Interface	Wireless Controller	Bluetooth adapter
EFh	Both	Miscellaneous	ActiveSync device
FEh	Interface	Application Specific	IrDA Bridge, Test & Measurement Class (USBTMC), USB DFU (direct firmware update)
FFh	Both	Vendor Specific	Indicates a device needs vendor specific drivers

8.8 USB Enumeration

How the Host Learns about Devices

The Universal Serial Bus (USB), as mentioned before, has seen enormous success in PC systems and is replacing the older parallel and serial ports.

For a standard serial port, the communication is directly performed by the application running on the computer. In order to be Plug-and-Play and Hot-Plug, the USB bus introduces a process that uniquely identifies a device to the Host computer in order for it to learn the capabilities of the device and to load the appropriate driver. This identification process is called the Enumeration process and uses a standard set of commands (setup requests- standard requests) described in the Chapter 9 of the USB specification, "USB Device Framework". This involves a mixture of hardware techniques for detecting something is present and software to identify what has been connected.

From the user's perspective, enumeration is invisible and automatic but may display a message that announces the new device and whether the attempt to configure it succeeded. Sometimes on first use, the user needs to assist in selecting a driver or telling the host where to look for driver files. Under Windows, when enumeration is complete, the new device appears in the Device Manager. (Right-click Computer, click Manage, and in the Computer Management pane, select Device Manager.) On detaching, the device disappears from Device Manager. In a typical device, firmware decodes and responds to received requests for information. Some controllers can manage enumeration entirely in hardware except possibly for vendor-provided values in EEPROM or other memory. On the host side, the operating system handles enumeration.

Getting to the Configured state

The USB 2.0 specification defines six device states. During enumeration, a device moves through the Powered, Default, Address, and Configured states. (The other defined states are Attached and Suspend.) In each state, the device has defined capabilities and behavior.

Typical USB 2.0 sequence

The steps below are a typical sequence of events that occurs during enumeration of a USB 2.0 device under Windows. Device firmware shouldn't assume that enumeration requests and events will occur in a particular order. Different OSes and different OS editions might use a different sequence. To function successfully, a device must detect and respond to any control request or other bus event at any time as required by the USB specifications. Figure 8.37 shows received requests and other events during a device enumeration.

1. The system has a new device. A user attaches a device to a USB port, or the system powers up with a device attached. The port may be on the root hub at the host or on a hub that connects downstream from the host. The hub

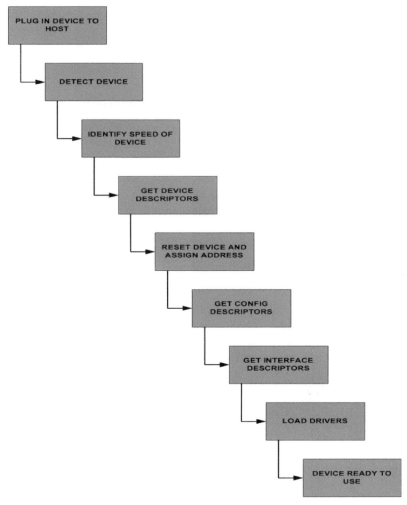

Figure 8.37 Events during a device enumeration.

provides power to the port, and the device is in the Powered state. The device is in the Attached state and can draw up to 100 mA from the bus.

2. The hub detects the device. The hub monitors the voltages on the signal lines (D+ and D−) at each of its ports. The hub has a pull-down resistor of 14.25k–24.8kW on each line. A device has a pull-up resistor of 900–1575W on D+ for a full-speed device or on D− for a low-speed device. High-speed-capable devices attach at full speed. On attaching to a port, the device connects to the bus by bringing the appropriate pull-up line high so the hub

can detect that a device is attached. Except for some devices with weak or dead batteries, the device must connect within 1 s after detecting that VBUS is at least 0.8 V. A device can continue to draw 100 mA of bus current for 1 s after connecting regardless of whether the upstream bus segment is suspended. On detecting a device, the hub continues to provide power but doesn't yet transmit USB traffic to the device.

3. The host learns of the new device. Each hub uses its interrupt endpoint to report events at the hub. The report indicates only whether the hub or a port (and if so, which port) has experienced an event. On learning of an event, the host sends the hub a Get Port Status request to find out more. Get Port Status and the other hub-class requests used during enumeration are standard requests that all hubs support. The information returned tells the host when a device is newly attached.

4. The hub detects whether a device is low or full speed. Just before resetting the device, the hub determines whether the device is low or full speed by detecting which signal line has a higher voltage when idle. The hub sends the information to the host in response to the next Get Port Status request. A USB 1.1 hub may instead detect the device's speed just after a bus reset. USB 2.0 requires speed detection before the reset so the hub knows whether to check for a high-speed-capable device during reset as described below.

5. The hub resets the device. When a host learns of a new device, the host sends the hub a Set Port Feature request that asks the hub to reset the port. The hub places the device's USB data lines in the Reset condition for at least 10 ms. Reset is a special condition where both D+ and D− are logic low. (Normally, the lines have opposite logic states.) The hub sends the reset only to the new device. Other hubs and devices on the bus don't see the reset.

6. The host learns if a full-speed device supports high speed. Detecting whether a device supports high speed uses two special signal states. In the Chirp J state, only the D+ line is driven, and in the Chirp K state, only the D− line is driven.

During the reset, a device that supports high speed sends a Chirp K. A high-speed-capable hub detects the Chirp K and responds with a series of alternating Chirp K and Chirp J. On detecting the pattern KJKJKJ, the device removes its full-speed pull-up and performs all further communications at high speed. If the hub doesn't respond to the device's Chirp K, the device knows it must continue to communicate at full speed. All high-speed devices must be capable of responding to control-transfer requests at full speed.

7. The hub establishes a signal path between the device and the bus. The host verifies that the device has exited the reset state by sending a Get Port Status request. A bit in the returned data indicates whether the device is still in the reset state. If necessary, the host repeats the request until the device has exited the reset state.

When the hub removes the reset, the device is in the Default state. The device's USB registers are in their reset states, and the device is ready to respond to control transfers at endpoint zero. The device communicates with the host using the default address of 0x00.

8. The host sends a Get Descriptor request to learn the maximum packet size of the default pipe. The host sends the request to device address 0x00, endpoint zero. Because the host itemizes only one device at a time, only one device will respond to communications addressed to device address 0x00 even if several devices attach at once.

The eighth byte of the device descriptor contains the maximum packet size supported by endpoint zero. The host may request 64 bytes but after receiving just one packet (whether or not it has 64 bytes), may begin the Status stage of the transfer.

On completing the Status stage, Windows may request the hub to reset the device as in step 5 above. The USB 2.0 specification does not require a reset here. The reset is a precaution that ensures that the device will be in a known state when the reset ends. Windows 8 and later skip the second reset for high-speed devices because these devices typically don't require a second reset. If enumeration fails without the second reset, Windows includes the reset on the next enumeration attempt.

9. The host assigns an address. When the reset is complete, the host controller assigns a unique address to the device by sending a Set Address request. The device completes the Status stage of the request using the default address and then implements the new address. The device is now in the Address state. All communications from this point on use the new address. The address is valid until the device is detached, a hub resets the port, or the system reboots. On the next enumeration, the host may assign a different address to the device.

10. The host learns about the device's abilities. The host sends a Get Descriptor request to the new address to read the device descriptor. This time the host retrieves the entire descriptor. The descriptor contains the maximum packet size for endpoint zero, the number of configurations the device supports, and other information about the device.

The host continues to learn about the device by requesting the configuration descriptor(s) specified in the device descriptor. A request for a configuration descriptor is actually a request for the configuration descriptor followed by all of its subordinate descriptors up to the number of bytes requested.

If the host requests 255 bytes, the device responds by sending the configuration descriptor followed by all of the configuration's subordinate descriptors, including interface descriptor(s), with each interface descriptor followed by any endpoint descriptors for the interface. Some configurations also have class- or vendor-specific descriptors.

One of the configuration descriptor's fields is the total length of the configuration descriptor and its subordinate descriptors. If the value is greater than 255, the device returns 255 bytes. Windows then requests the configuration descriptor again, this time requesting the number of bytes in the total length specified in the configuration descriptor.

Earlier Windows editions began by requesting just the configuration descriptor's nine bytes to retrieve the total length value, then requesting the complete descriptor set.

11. The host requests additional information from the device. The host then may request additional descriptors from the device. In every case, a device that doesn't support a requested descriptor should return STALL.

When the device descriptor reports that the device is USB 2.1 or higher, the host requests a BOS descriptor. If the device returns the BOS descriptor, the host uses the descriptor's total length value to request the BOS descriptor followed by its subordinate descriptor(s).

The host requests string descriptor zero, which contains one or more codes indicating what languages additional strings use.

If the device descriptor reports that the device contains a serial number string descriptor, the host requests that descriptor.

If the device descriptor indicates that the device contains a Product string descriptor, the host requests that descriptor.

For USB 2.0 and higher devices, if Windows doesn't have a record of previously retrieving a Microsoft-specific MS OS string descriptor, the OS may request that descriptor.

If a BOS descriptor or a Microsoft OS string descriptor indicates support for additional Microsoft-defined descriptors, the host may request these descriptors.

For USB 2.0 or higher devices operating at full speed with an upstream USB 1.1 hub, the host requests a device qualifier descriptor. A device that returns this descriptor is capable of operating at high speed if all upstream ports are USB 2.0 or higher.

12. The host assigns and loads a device driver (except for composite devices). After learning about a device from its descriptors, the host looks for the best match in a driver to manage communications with the device. Windows hosts use INF files to identify the best match. The INF file may be a system file for a USB class or a vendor-provided file that contains the device's Vendor ID and Product ID. Chapter 9 has more about INF files and selecting a driver.

For devices that have been itemized previously, Windows may use stored information instead of searching the INF files. After the operating system assigns and loads a driver, the driver may request the device to resend descriptors or send other class-specific descriptors.

An exception to this sequence is composite devices, which can have different drivers assigned to multiple interfaces in a configuration. The host can assign these drivers only after enabling the interfaces, so the host must first configure the device as described below.

13. The host's device driver selects a configuration. After learning about a device from the descriptors, the device driver requests a configuration by sending a Set Configuration request with the desired configuration number. Many devices support only one configuration. When a device supports multiple configurations, many drivers just select the first configuration, but a driver can decide based on information the driver has about how the device will be used, or the driver can ask the user what to do. On receiving the request, the device implements the requested configuration. The device is now in the Configured state and the device's interface(s) are enabled.

For composite devices, the host can now assign drivers. As with other devices, the host uses the information retrieved from the device to find a driver for each active interface in the configuration. The device is then ready for use.

Hubs are also USB devices, and the host itemizes a newly attached hub in the same way as other devices. If the hub has devices attached, the host itemizes these after the hub informs the host of their presence.

Attached state. If the hub isn't providing power to a device's VBUS line, the device is in the Attached state. The absence of power may occur if the hub has detected an over-current condition or if the host requests the hub to remove power from the port. With no power on VBUS, the host and device can't communicate, so from their perspective, the situation is the same as when the device isn't attached.

Suspend State. A device enters the Suspend state after detecting no bus activity, including SOF markers, for at least 3 ms. In the Suspend state, the device should limit its use of bus power. Both configured and unconfigured devices must support this state. Chapter 17 has more about the Suspend state.

8.8.1 Enhanced SuperSpeed Differences

[Note: **Enhanced SuperSpeed is the USB 3.1 spec's term for SuperSpeed and SuperSpeedPlus**.]

Enumerating Enhanced SuperSpeed devices has some differences compared to USB 2.0:

- On detecting a downstream Enhanced SuperSpeed termination at a port, a hub initializes and trains the port's link. Enumeration then proceeds at SuperSpeed or SuperSpeedPlus with no need for further speed detecting.
- The host isn't required to reset the port after learning of a new device.
- The bus-current limits are 150 mA before configuration and 900 mA after configuration.
- The host sends a Set Isochronous Delay request to inform the device of the bus delay for isochronous packets.
- The host sends a Set SEL request to inform the device of the system exit latency (the amount of time required to transition out of a low-power state).
- Protocols for entering and exiting the Suspend state differ.
- For hubs, the host sends a Set Hub Depth request to set the hub-depth value.

Device removal

When a user removes a device from the bus, the hub disables the device's port. The host knows that the removal occurred after the hub notifies the host that an event has occurred, and the host sends a Get Port Status request to learn what the event was. The device disappears from Device Manager and the device's address becomes available to another newly attached device.

8.9 Device Drivers

This section provides the reader with a general introduction to device drivers and an idea about the structural elements of a device driver.

8.9.1 Device Driver Overview

Device drivers are the software segments that provides an interface between the operating system and the specific hardware devices such as terminals, disks, tape drives, video cards and network media. The device driver brings the device into and out of service, sets hardware parameters in the device, transmits data from the kernel to the device, receives data from the device and passes it back to the kernel, and handles device errors.

A driver acts like a translator between the device and programs that use the device. Each device has its own set of specialized commands that only its driver knows. In contrast, most programs access devices by using generic commands. The driver, therefore, accepts generic commands from a program and then translates them into specialized commands for the device.

8.9.2 Classification of Drivers According to Functionality

There are numerous driver types, differing in their functionality. This subsection briefly describes three of the most common driver types.

8.9.2.1 Monolithic drivers

Monolithic drivers, Figure 8.38, are device drivers that embody all the functionality needed to support a hardware device. A monolithic driver is accessed by one or more user applications, and directly drives a hardware device. The driver communicates with the application through I/O control commands (IOCTLs) and drives the hardware using calls to the different WDK, ETK, DDI/DKI functions.

Monolithic drivers are supported in all operating systems including all Windows platforms and all Unix platforms.

8.9.2.2 Layered drivers

Layered drivers, Figure 8.39, are device drivers that are part of a stack of device drivers that together process an I/O request. An example of a layered driver is a driver that intercepts calls to the disk and encrypts/decrypts all data being transferred to/from the disk. In this example, a driver would be hooked on to the top of the existing driver and would only do the encryption/decryption.

Layered drivers are sometimes also known as filter drivers, and are supported in all operating systems including all Windows platforms and all Unix platforms.

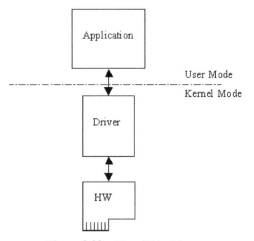

Figure 8.38   Monolithic drivers.

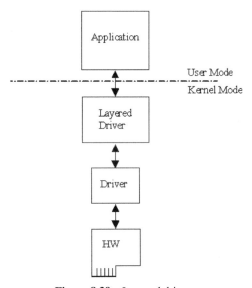

Figure 8.39   Layered drivers.

8.9.2.3 Miniport drivers

A Miniport driver, Figure 8.40, is an add-on to a class driver that supports miniport drivers. It is used so the miniport driver does not have to implement all of the functions required of a driver for that class. The class driver provides the basic class functionality for the miniport driver.

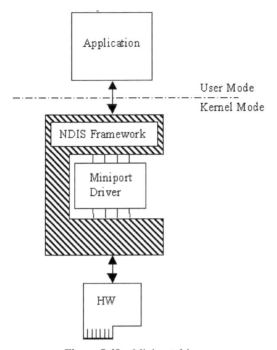

Figure 8.40   Miniport drivers.

A class driver is a driver that supports a group of devices of common functionality, such as all HID devices or all network devices.

Miniport drivers are also called miniclass drivers or minidrivers, and are supported in the Windows NT (2000) family, namely Windows 7/Vista/Server 2008/Server 2003/XP/2000/NT 4.0.

Windows 7/Vista/Server 2008/Server 2003/XP/2000/NT 4.0 provide several driver classes (called ports) that handle the common functionality of their class. It is then up to the user to add only the functionality that has to do with the inner workings of the specific hardware. The NDIS miniport driver is one example of such a driver. The NDIS miniport framework is used to create network drivers that hook up to NT's communication stacks and are therefore accessible to common communication calls used by applications. The Windows NT kernel provides drivers for the various communication stacks and other code that is common to communication cards. Due to the NDIS framework, the network card developer does not have to write all of this code, only the code that is specific to the network card he is developing.

8.9.3 Classification of Drivers According to Operating Systems

8.9.3.1 WDM drivers

Windows Driver Model (WDM) drivers are kernel-mode drivers within the Windows NT and Windows 98 operating system families. The Windows NT family includes Windows 7/Vista/Server 2008/Server 2003/XP/2000/NT 4.0, and the Windows 98 family includes Windows 98 and Windows Me.

WDM works by channeling some of the work of the device driver into portions of the code that are integrated into the operating system. These portions of code handle all of the low-level buffer management, including DMA and Plug- and -Play (Pnp) device enumeration.

WDM drivers are PnP drivers that support power management protocols, and include monolithic drivers, layered drivers and miniport drivers.

8.9.3.2 VxD drivers

VxD drivers are Windows 95/98/Me Virtual Device Drivers, often called VxDs because the file names end with the .vxd extension. VxD drivers are typically monolithic in nature. They provide direct access to hardware and privileged operating system functions. VxD drivers can be stacked or layered in any fashion, but the driver structure itself does not impose any layering.

8.9.3.3 Unix device drivers

In the classic Unix driver model, devices belong to one of three categories: character (char) devices, block devices and network devices. Drivers that implement these devices are correspondingly known as char drivers, block drivers or network drivers. Under Unix, drivers are code units linked into the kernel that run in privileged kernel mode. Generally, driver code runs on behalf of a user-mode application. Access to Unix drivers from user-mode applications is provided via the file system. In other words, devices appear to the applications as special device files that can be opened.

Unix device drivers are either layered or monolithic drivers. A monolithic driver can be perceived as a one-layer layered driver.

8.9.3.4 Linux device drivers

Linux device drivers are based on the classic Unix device driver model. In addition, Linux introduces some new characteristics.

Under Linux, a block device can be accessed like a character device, as in Unix, but also has a block-oriented interface that is invisible to the user or application.

Traditionally, under Unix, device drivers are linked with the kernel, and the system is brought down and restarted after installing a new driver. Linux introduces the concept of a dynamically loadable driver called a module. Linux modules can be loaded or removed dynamically without requiring the system to be shut down. A Linux driver can be written so that it is statically linked or written in a modular form that allows it to be dynamically loaded. This makes Linux memory usage very efficient because modules can be written to probe for their own hardware and unload themselves if they cannot find the hardware they are looking for.

Like Unix device drivers, Linux device drivers are either layered or monolithic drivers.

8.9.4 The Entry Point of the Driver

Every device driver must have one main entry point, like the main() function in a C console application. This entry point is called DriverEntry() in Windows and init_module() in Linux. When the operating system loads the device driver, this driver entry procedure is called.

There is some global initialization that every driver needs to perform only once when it is loaded for the first time. This global initialization is the responsibility of the DriverEntry()/init_module() routine. The entry function also registers which driver callbacks will be called by the operating system. These driver callbacks are operating system requests for services from the driver. In Windows, these callbacks are called *dispatch routines*, and in Linux they are called *file operations*. Each registered callback is called by the operating system as a result of some criteria, such as disconnection of hardware, for example.

8.9.5 Associating the Hardware with the Driver

Operating systems differ in the ways they associate a device with a specific driver.

In Windows, the hardware–driver association is performed via an INF file, which registers the device to work with the driver. This association is performed before the DriverEntry() routine is called. The operating system recognizes the device, checks its database to identify which INF file is associated with the device, and according to the INF file, calls the driver's entry point.

In Linux, the hardware–driver association is defined in the driver's init_module() routine. This routine includes a callback that indicates which hardware the driver is designated to handle. The operating system calls the driver's entry point, based on the definition in the code.

8.9.6 Communicating with Drivers

Communication between a user-mode application and the driver that drives the hardware, is implemented differently for each operating system, using the custom OS Application Programming Interfaces (APIs).

On Windows, Windows CE, and Linux, the application can use the OS file-access API to open a handle to the driver (e.g., using the Windows CreateFile() function or using the Linux open() function), and then read and write from/to the device by passing the handle to the relevant OS file-access functions (e.g., the Windows ReadFile() and WriteFile() functions, or the Linux read() and write() functions).

The application sends requests to the driver via I/O control (IOCTL) calls, using the custom OS APIs provided for this purpose (e.g., the Windows DeviceIoControl() function, or the Linux ioctl() function). The data passed between the driver and the application via the IOCTL calls is encapsulated using custom OS mechanisms. For example, on Windows, the data are passed via an I/O Request Packet (IRP) structure and are encapsulated by the I/O Manager.

8.10 USB Descriptors

USB devices report their attributes using **descriptors**, which are data structures with a defined structure and format in which the data is transferred. A complete description of the USB descriptors can be found in Chapter 9 of the USB Specification (see http://www.usb.org for the full specification). Each descriptor begins with a byte-wide field containing the total number of bytes in the descriptor followed by a byte-wide field identifying the descriptor type.

It is best to view the USB descriptors as a hierarchical structure with four levels:

- The *Device* level
- The *Configuration* level
- The *Interface* level (this level may include an optionalsub-level called *Alternate Setting*)
- The *Endpoint* level

There is only one device descriptor for each USB device. Each device has one or more configurations, each configuration has one or more

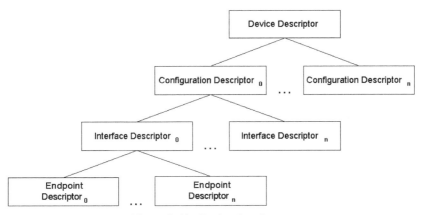

Figure 8.41 Device descriptors.

interfaces, and each interface has zero or more endpoints, as demonstrated in Figure 8.41.

- **Device Level**: The device descriptor includes general information about the USB device, that is, global information for all of the device configurations. The device descriptor identifies, among other things, the device class (HID device, hub, locator device, etc.), subclass, protocol code, vendor ID, device ID and more. Each USB device has one device descriptor.
- **Configuration Level**: A USB device has one or more configuration descriptors. Each descriptor identifies the number of interfaces grouped in the configuration and the power attributes of the configuration (such as self-powered, remote wakeup, maximum power consumption and more). Only one configuration can be loaded at a given time. For example, an ISDN adapter might have two different configurations, one that presents it with a single interface of 128 Kb/s and a second that presents it with two interfaces of 64 Kb/s each.
- **Interface Level**: The interface is a related set of endpoints that present a specific functionality or feature of the device. Each interface may operate independently. The interface descriptor describes the number of the interface, the number of endpoints used by this interface and the interface-specific class, subclass and protocol values when the interface operates independently.
  In addition, an interface may have **alternate settings**. The alternate settings allow the endpoints or their characteristics to be varied after the device is configured.

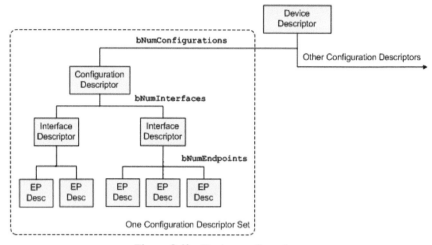

Figure 8.42   Device configuration.

- **Endpoint Level**: The lowest level is the endpoint descriptor, which provides the host with information regarding the endpoint's data transfer type and maximum packet size. For isochronous endpoints, the maximum packet size is used to reserve the required bus time for the data transfer – that is, the bandwidth. Other endpoint attributes are its bus access frequency, endpoint number, error handling mechanism and direction. The same endpoint can have different properties (and consequently different uses) in different alternate settings.

When a USB device is attached to the USB bus, the host uses **bus enumeration** to identify and configure the device. The USB Host sends setup requests as soon as the device has joined the USB network. The device will be instructed to select a configuration and an interface to match the needs of the application running on the USB Host. Once a configuration and an interface have been selected, the device must service the active endpoints to exchange data with the USB Host.

The usual number of descriptors used is given as follows:

- One Device Descriptor
- One Configuration Descriptor
- One Interface Descriptor
- One or more Endpoint_Descriptors
- String Descriptors describe the above-mentioned descriptors in human readable format.

Alternative information that is needed when the device can operate in different speed modes can be defined in a Device Qualifier Descriptor.

Complex devices have multiple interfaces. Each interface can have a number of endpoints representing a functional unit. For example, a voice-over-IP phone might have:

- One audio class interface with two isochronous endpoints for transferring audio data in each direction.
- One HID interface with a single IN interrupt endpoint for a built-in keypad.

Provisions have been made in the USB component to give the user the option to **override the USB descriptors** if necessary. This can be the case when the device class needs to be changed at runtime or other reports need to be created.

In the following, some of the descriptors are given.

8.10.1 Device Descriptor

The **Device Descriptor** (USB_DEVICE_DESCRIPTOR) is the root of the descriptor tree and contains basic device information. The unique numbers, *idVendor* and *idProduct*, identify the connected device. The Windows operating system uses these numbers to determine which device driver must be loaded.

idVendor is the number assigned to each company producing USB-based devices. The USB Implementers Forum is responsible for administering the assignment of Vendor IDs.

The *idProduct* is another 16-bit field containing a number assigned by the manufacturer to identify a specific product.

Offset	Field	Type	Size	Value	Description
0	bLength	uint8_t	1	Number	Size of this descriptor in bytes.
1	bDescriptorType	uint8_t	1	Constant	Device Descriptor Type = 1.
2	bcdUSB	uint16_t	2	BCD	USB Specification Release Number in Binary-Coded Decimal (i.e., 2.10 is 210h). This field identifies the release of the USB Specification with which the device and its descriptors are compliant.

Offset	Field	Type	Size	Value	Description
4	bDeviceClass	uint8_t	1	Class	Class code (assigned by the USB-IF). If this field is • reset to zero, each interface within a configuration specifies its own class information and the various interfaces operate independently. • set to a value between 1 and FEh, the device supports different class specifications on different interfaces and the interfaces may not operate independently. This value identifies the class definition used for the aggregate interfaces. • set to FFh, the device class is vendor specific.
5	bDeviceSubClass	uint8_t	1	SubClass	Subclass code (assigned by the USB-IF). These codes are qualified by the value of the *bDeviceClass* field. If *bDeviceClass* is • reset to zero, this field must also be reset to zero. • not set to FFh, all values are reserved for assignment by the USB-IF.
6	bDeviceProtocol	uint8_t	1	Protocol	Protocol code (assigned by the USB-IF). These codes are qualified by the value of the *bDeviceClass* and *bDeviceSubClass* fields. If a device supports class-specific protocols on a device basis as opposed to an interface basis, this code identifies the protocols that the device uses as defined by the specification of the device class.

Offset	Field	Type	Size	Value	Description
					If this field is
					• reset to zero, the device does not use class specific protocols on a device basis. However, it may use class specific protocols on an interface basis. • set to FFh, the device uses a vendor specific protocol on a device basis.
7	bMaxPacketSize0	uint8_t	1	Number	Maximum packet size for Endpoint zero (only 8, 16, 32, or 64 are valid).
8	idVendor	uint16_t	2	ID	Vendor ID (assigned by the USB-IF).
10	idProduct	uint16_t	2	ID	Product ID (assigned by the manufacturer).
12	bcdDevice	uint16_t	2	BCD	Device release number in binary-coded decimal.
14	iManufacturer	uint8_t	1	Index	Index of string descriptor describing manufacturer.
15	iProduct	uint8_t	1	Index	Index of string descriptor describing product.
16	iSerialNumber	uint8_t	1	Index	Index of string descriptor describing the device's serial number.
17	bNumConfigurations	uint8_t	1	Number	Number of possible configurations.

8.10.2 Configuration Descriptor

The **Configuration Descriptor** (USB_CONFIGURATION_DESCRIPTOR) contains information about the device power requirements and the number of interfaces it can support. A device can have multiple configurations. The host can select the configuration that best matches the requirements of the application software.

Offset	Field	Type	Size	Value	Description
0	bLength	uint8_t	1	Number	Size of this descriptor in bytes.
1	bDescriptorType	uint8_t	1	Constant	Configuration Descriptor Type = 2.
2	wTotalLength	uint16_t	2	Number	Total length of data returned for this configuration. Includes the combined length of all descriptors (configuration, interface, endpoint, and class or vendor specific) returned for this configuration.
4	bNumInterfaces	uint8_t	1	Number	Number of interfaces supported by this configuration.
5	bConfigurationValue	uint8_t	1	Number	Value to select this configuration with *SetConfiguration()*.
6	iConfiguration	uint8_t	1	Index	Index of string descriptor describing this configuration.
7	bmAttributes	uint8_t	1	Bitmap	Configuration characteristics • D7: Reserved (must be set to **one** for historical reasons) • D6: Self-powered • D5: Remote Wakeup • D4…0: Reserved (reset to zero) A device configuration that uses power from the bus and a local source reports a non-zero value in *bMaxPower* to indicate the amount of bus power required and sets D6. The actual power source at runtime can be determined using the GetStatus(DEVICE) request. If a device configuration supports remote wakeup, D5 is set to 1.
8	bMaxPower	uint8_t	1	mA	Maximum power consumption of the USB device from the bus in this specific configuration when the device is fully operational. Expressed in 2 mA units (i.e., 50 = 100 mA).

8.10.3 Interface Descriptor

The **Interface Descriptor** (USB_INTERFACE_DESCRIPTOR) defines the collection of endpoints. This interface supports a group of pipes that are suitable for a particular task. Each configuration can have multiple interfaces. The interface can be selected dynamically by the USB Host. The **Interface Descriptor** can associate its collection of pipes with a device class, which in turn has an associated class device driver within the host operating system. Typically, the device class is a functional type such as a printer class or mass storage class.

An interface descriptor never includes Endpoint 0 in the numbering of endpoints. If an interface uses only Endpoint 0, then the field *bNumEndpoints* must be set to zero.

If no class type has been selected for the device, then none of the standard USB drivers is loaded, and the developer has to provide its own device driver.

Offset	Field	Type	Size	Value	Description
0	bLength	uint8_t	1	Number	Size of this descriptor in bytes.
1	bDescriptorType	uint8_t	1	Constant	Interface Descriptor Type = 4.
2	bInterfaceNumber	uint8_t	1	Number	The number of this interface. Zero-based value identifying the index in the array of concurrent interfaces supported by this configuration.
3	bAlternateSetting	uint8_t	1	Number	Value used to select an alternate setting for the interface identified in the prior field. Allows an interface to change the settings on the fly.
4	bNumEndpoints	uint8_t	1	Number	Number of endpoints used by this interface (excluding endpoint zero).
					• If this value is zero, this interface uses the Default Control Pipe only.

Offset	Field	Type	Size	Value	Description
5	bInterfaceClass	uint8_t	1	Class	Class code (assigned by the USB-IF). • A value of zero is reserved for future standardization. • If this field is set to FFh, the interface class is vendor specific. • All other values are reserved for assignment by the USB-IF.
6	bInterfaceSubClass	uint8_t	1	SubClass	Subclass code (assigned by the USB-IF). If *bInterfaceClass* • is reset to zero, this field must also be reset to zero. • is not set to FFh, all values are reserved for assignment by the USB-IF.
7	bInterfaceProtocol	uint8_t	1	Protocol	Protocol code (assigned by the USB). If an interface supports class-specific requests, this code identifies the protocols that the device uses as defined in the device class. If this field • is reset to zero, the device does not use a class-specific protocol on this interface. • is set to FFh, the device uses a vendor specific protocol for this interface.
8	iInterface	uint8_t	1	Index	Index of string descriptor describing this interface.

For example, two devices with different interfaces are needed.

The first interface, *Interface0*, has the field *bInterfaceNumber* set to 0. The next interface, *Interface1*, has the field *bInterfaceNumber* set to 1 and the field *bAlternativeSetting* also set to 0 (default). It is possible to define an alternative setting for this device, by leaving the field *bInterfaceNumber* set to 1 and with the field *bAlternativeSetting* set to 1 instead of 0.

The first two interface descriptors with *bAlternativeSettings* equal to 0 are used. However, the host can send a *SetInterface()* request to enable the alternative setting.

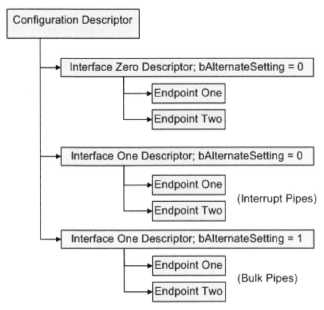

Figure 8.43 Alternative interface.

8.10.4 Endpoint Descriptor

The **Endpoint Descriptor** (USB_ENDPOINT_DESCRIPTOR) specifies the transfer type, direction, polling interval, and maximum packet size for each endpoint. Endpoint 0 (zero), the default endpoint, is always assumed to be a control endpoint and never has a descriptor.

Offset	Field	Type	Size	Value	Description
0	bLength	uint8_t	1	Number	Size of this descriptor in bytes.
1	bDescriptorType	uint8_t	1	Constant	Endpoint descriptor type = 5.

2	bEndpointAddress	uint8_t	1	Endpoint	The address of the endpoint on the USB device described by this descriptor. The address is encoded as follows:
					
					• Bit 3...0: The endpoint number
					• Bit 6...4: Reserved, reset to zero
					• Bit 7: Direction, ignored for control endpoints.
					
					  – 0 = OUT endpoint
					  – 1 = IN endpoint

Offset	Field	Type	Size	Value	Description
3	bmAttributes	uint8_t	1	Bitmap	The endpoint attribute when configured through *bConfigurationValue*.
					• Bits 1…0: Transfer type
					– 00 = Control – 01 = Isochronous – 10 = Bulk – 11 = Interrupt
					For non-isochronous endpoints, bits 5…2 must be set to zero. For isochronous endpoints, they are defined as:
					• Bits 3..2: Synchronization type
					– 00 = No synchronization – 01 = Asynchronous – 10 = Adaptive – 11 = Synchronous
					• Bits 5..4: Usage type
					– 00 = Data – 01 = Feedback – 10 = Implicit feedback – 11 = Reserved
					All other bits are reserved and must be reset to zero.
4	wMax PacketSize	uint16_t	2	Number	Is the maximum packet size of this endpoint. For isochronous endpoints, this value is used to reserve the time on the bus required for the per-(micro)frame data payloads.
					• Bits 10…0 = max. packet size (in bytes).
					For high-speed isochronous and interrupt endpoints:
					• Bits 12…11 = number of additional transaction opportunities per micro-frame:
					• 00 = None (1 transaction per micro-frame) • 01 = 1 additional (2 per micro-frame) • 10 = 2 additional (3 per micro-frame) • 11 = Reserved
					• Bits 15…13 are reserved and must be set to zero.

Offset	Field	Type	Size	Value	Description
6	bInterval	uint8_t	1	Number	Interval for polling endpoint for data transfers. Expressed in frames or micro-frames depending on the operating speed (1 ms or 125 μs units).

- For full-/high-speed isochronous endpoints, this value must be in the range from 1 to 16. The *bInterval* value is used as the exponent for a $2^{\text{bInterval}-1}$ value, for example, a *bInterval* of 4 means a period of 8 (2^{4-1}).
- For full-/low-speed interrupt endpoints, the value of this field may be from 1 to 255.
- For high-speed interrupt endpoints, the *bInterval* value is used as the exponent for a $2^{\text{bInterval}-1}$ value, for example, a *bInterval* of 4 means a period of 8 (2^{4-1}). This value must be from 1 to 16.
- For high-speed bulk/control OUT endpoints, the *bInterval* must specify the maximum NAK rate of the endpoint. A value of 0 indicates the endpoint never NAKs. Other values indicate at most 1 NAK each *bInterval* number of microframes. This value must be in the range from 0 to 255.

8.10.5 String Descriptor

String descriptors (USB_STRING_DESCRIPTOR) are optional and add human readable information to the other descriptors. If a device does not support string descriptors, all references to string descriptors within device, configuration, and interface descriptors must be set to zero.

String descriptors are encoded in Unicode so that multiple languages can be supported with a single product. When requesting a string descriptor, the requester specifies the desired language using a 16-bit language ID (LANGID) defined by the USB-IF (refer to Language Identifiers (LANGIDs)). String index zero is used for all languages and returns a string descriptor that contains an array of two-byte LANGID codes supported by the device.

The array of LANGID codes is not NULL-terminated. The size of the array (in byte) is computed by subtracting two from the value of the first byte to the descriptor.

Offset	Field	Type	Size	Value	Description
0	bLength	uint8_t	N + 2	Number	Size of this descriptor in bytes.
1	bDescriptorType	uint8_t	1	Constant	String descriptor type
2	wLANGID[0]	uint8_t	2	Number	LANGID code zero (for example, 0x0407 German [Standard]).
...	...	...	...	...	...
N	wLANGID[x]	uint8_t	2	Number	LANGID code zero x (for example, 0x0409 English [United States]).

The UNICODE string descriptor is not NULL-terminated. The string length is computed by subtracting two from the value of the first byte of the descriptor.

Offset	Field	Type	Size	Value	Description
0	bLength	uint8_t	1	Number	Size of this descriptor in bytes.
1	bDescriptorType	uint8_t	1	Constant	String Descriptor Type
2	bString	uint8_t	N	Number	UNICODE encoded string.

8.10.6 Device Qualifier Descriptor

A high-speed capable device that has different device information for full speed and high speed must have a **Device Qualifier Descriptor** (USB_DEVICE_QUALIFIER_DESCRIPTOR). For example, if the device is currently operating at full speed, the **Device Qualifier** returns information about how it would operate at high speed and vice versa.

The fields for the vendor, product, device, manufacturer, and serial number are not included. This information is constant for a device regardless of the supported speeds.

If a full-speed only device receives a *GetDescriptor()* request for a *device_qualifier*, it must respond with a request error. Then, the host must not make a request for an *other_speed_configuration descriptor*.

Offset	Field	Type	Size	Value	Description
0	bLength	uint8_t	1	Number	Size of this descriptor in bytes.
1	bDescriptorType	uint8_t	1	Constant	Device Qualifier Descriptor Type = 6.
2	bcdUSB	uint16_t	2	BCD	USB Specification Release Number in Binary-Coded Decimal (i.e., 2.10 is 210h). This field identifies the release of the USB Specification with which the device and its descriptors are compliant. At least V2.00 is required to use this descriptor.
4	bDeviceClass	uint8_t	1	Class	Class code (assigned by the USB-IF). If this field is • reset to zero, each interface within a configuration specifies its own class information and the various interfaces operate independently. • set to a value between 1 and FEh, the device supports different class specifications on different interfaces and the interfaces may not operate independently. This value identifies the class definition used for the aggregate interfaces. If this field is set to FFh, the device class is vendor specific.
5	bDeviceSubClass	uint8_t	1	SubClass	Subclass code (assigned by the USB-IF). These codes are qualified by the value of the *bDeviceClass* field. If *bDeviceClass* is • reset to zero, this field must also be reset to zero. • not set to FFh, all values are reserved for assignment by the USB-IF.

Offset	Field	Type	Size	Value	Description
6	bDeviceProtocol	uint8_t	1	Protocol	Protocol code (assigned by the USB-IF). These codes are qualified by the values of the *bDeviceClass* and *bDeviceSubClass* fields. If a device supports class-specific protocols on a device basis as opposed to an interface basis, this code identifies the protocols that the device uses as defined by the specification of the device class. If this field is • reset to zero, the device does not use class-specific protocols on a device basis. However, it may use class-specific protocols on an interface basis. • set to FFh, the device uses a vendor specific protocol on a device basis.
7	bMaxPacketSize0	uint8_t	1	Number	Maximum packet size for other speed.
8	bNumConfigurations	uint8_t	1	Number	Number of other-speed configurations.
9	bReserved	uint8_t	1	Zero	Reserved for future use, must be zero.

8.11 RS-232 to USB Converters

8.11.1 USB to RS-232 Conversion

This section discusses the question of how an RS-232 connector can be soldered to a USB cable. Unfortunately, although RS-232 and USB (universal serial bus) are both serial communication standards to connect peripherals to computers, they are totally different in design. A simple cable is not enough to connect RS-232 devices to a computer with only USB ports. There are however converter modules and cables that can be successfully used to connect RS-232 devices to computers via a USB port. These adapters and cables contain electronics, and the success rate depends on the capabilities of this electronics and the device driver software that is shipped with the converter to communicate with these electronics over the USB bus. This section introduces to the reader information needed to help him to get the proper USB to RS-232 converter.

8.11.2 Differences Between USB and RS-232 from the Application Point of View

RS-232 is a definition for serial communication on a 1:1 base. RS-232 defines the interface layer, but not the application layer. To use RS-232 in a specific situation, application specific software must be written on devices on both ends of the connecting RS-232 cable. The developer is free to define the protocol used to communicate. RS-232 ports can be either accessed directly by an application, or via a device driver in the operating system.

USB on the other hand is a bus system which allows more than one peripheral to be connected to a host computer via one USB port. Hubs can be used in the USB chain to extend the cable length and allow for even more devices to connect to the same USB port. The standard not only describes the physical properties of the interface, but also the protocols to be used. Because of the complex USB protocol requirements, communication with USB ports on a computer is always performed via a device driver.

It is easy to see where the problems arise. Developers have lots of freedom where it comes to defining RS-232 communications and ports are often directly, or almost directly accessed in the application program. Settings like baud-rate, data bits, hardware software flow control can often be changed within the application. The USB interface does not give this flexibility. When however an RS-232 port is used via a USB to RS-232 converter, this flexibility should be present in some way. Therefore to use an RS-232 port via a USB port, a second device driver is necessary which emulates a RS-232 UART, but communicates via USB.

Many applications expect a certain timing with RS-232 communications. With ports directly fitted in a computer this is most of the time no problem. The application communicates directly, or via a thin device driver layer with the UART, and everything happens within a well-defined time frame. The USB bus is however shared by several devices. Communication congestion may be the result of this, and the timeframe in which specific RS-232 actions are performed might not be so well defined as in the direct port approach. Also, the double device driver layer with an RS-232 driver working on top of the complex USB driver might add extra overhead to the communications, resulting in delays.

8.11.3 Hardware Specific Problems

RS-232 ports which are physically mounted in a computer are often powered by three power sources: +5 V for the UART logic, and −12 and +12 V for the output drivers. USB however only provides a +5 V power source. Some USB to RS-232 converters use integrated DC/DC converters to create the appropriate voltage levels for the RS-232 signals, but in very cheap implementations, the +5 V is directly used to drive the output. This may sound strange, but many RS-232 ports recognize a voltage above 2 V as a space signal, where a voltage of 0 V or less is recognized as a mark signal. This is not according to the original standard, because in the original RS-232 standard, all voltages between -3 V and +3 V result in an undefined signal state. The well-known Maxim MAX232 series of RS-232 driver chips have this non-standard behavior, for example. Although the outputs of these drivers swings between −10 V and +10 V, the inputs recognize all signals swinging below 0 V and above 2 V as valid signals.

This non-standard behavior of RS-232 inputs makes it even more difficult to select the right RS-232 to USB converter. If you connect and test an RS-232 to USB converter over a serial line with another device, it might work with some devices, but not with others. This can particularly become a problem with industrial applications. Low-cost computers are often equipped with cheap RS-232 drivers and when you test the RS-232 to USB converter with such a computer, it might work. But the same converter may fail if you try it in an industrial environment. The chances that RS-232 ports from low-cost computers accept signals in the 0.5 V range are higher than with industrial equipment which is often specifically designed to be immune for noise.

Another hardware specific problem arises from handshaking to prevent buffer overflows at the receiver's side. RS-232 applications can use, as mentioned before, two types of handshaking, either with control commands in the data stream, called software flow control, or with physical lines, called hardware flow control. Not all USB to RS-232 converters provide these hardware flow control lines. It is not always easily identified if an application needs them. Some applications do not use hardware flow control at all, and those cheap USB to RS-232 converters will work without problems. Other applications use hardware flow control, but infrequently. Only with large data bursts, or in situations where the CPU is busy performing other tasks, hardware flow control might kick in to prevent data loss. In those situations, communications may seem error free, but with sometimes bytes lost, or unspecified errors in the communications.

8.11.4 USB to RS-232 Converter Selection Criteria

Resuming, when choosing the right USB to RS-232 converter, look at the following potential problems:

- Does your application have very tight timing requirements? In that case it might be better to use an internal RS-232 port, instead of a USB to RS-232 converter. The extra layer at the device driver level and bus congestion might make the communications less reliable.
- What are the RS-232 output voltages of the converter. Do they meet the requirements for the equipment you want to connect?
- What are the handshaking requirements for your application? If hardware flow control is required, make sure that these inputs and outputs on the converter are present.

Resources Used in Writing This Chapter

- USB Implementers Forum
- Universal Serial Bus Revision 2.0 Specification
- Universal Serial Bus Revision 3.0 Specification
- Wikipedia – Universal Serial Bus
- Everything USB
- Windows USB FAQ: Introductory Level
- Beyond Logic – USB in a Nutshell – Making sense of the USB standard

Sources/References

[1] Executive Comment: The future of mobile connectivity lies in USB On-The-Go/David Murray and Terry Remple/EBN/November, 13 2001
[2] USB OTG Spec Signals Developers to Proceed with a New Generation of Mobile Products Capable of Point to Point Data Exchange/USB-IF Press Release/December 18, 2001
[3] USB On-The-Go: P-to-P Communications in Mobile Devices/Kosta Koeman and David Murray/Electronic News/August 27, 2001
[4] OTG Supplement to the USB 2.0 Specification, Rev 1.0a/June 24, 2003

9

Wi-Fi Technology

9.1 Introduction to Wireless Communication

Wireless means transmitting signals using radio waves as the medium instead of wires. Wireless technologies are used for tasks as simple as switching off the television or as complex as supplying the sales force with information from an automated enterprise application while in the field. Now cordless keyboards and mice, PDAs, pagers, and digital and cellular phones have become part of our daily life.

Some of the inherent characteristics of wireless communications systems which make it attractive for users are given below:

- **Mobility:** A wireless communications system allows users to access information beyond their desk and conduct business from anywhere without having a wire connectivity.
- **Reachability:** Wireless communication systems enable people to be stay connected and be reachable, regardless of the location they are operating from.
- **Simplicity:** Wireless communication systems are easy and fast to deploy in comparison with cabled network. Initial setup cost could be a bit high but other advantages overcome that high cost.
- **Maintainability:** In a wireless system, you do not have to spend too much cost and time to maintain the network setup.
- **Roaming services:** Using a wireless network system, you can provide service anywhere any time including train, buses, aero planes, etc.
- **New services:** Wireless communication systems provide various smart services like SMS and MMS.

9.1.1 Wireless Network Topologies

There are basically three ways to set up a wireless network.

Point-to-point Bridge: A bridge is used to connect two networks. A point-to-point bridge interconnects two buildings having different networks. For example, a wireless LAN bridge can interface with an Ethernet network directly to a particular access point (as shown in Figure 9.1).

Point-to-multipoint Bridge: This topology is used to connect three or more LANs that may be located on different floors in a building or across buildings (Figure 9.2).

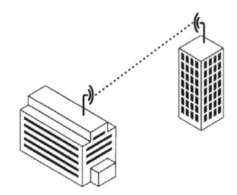

Figure 9.1　Point-to-point bridge.

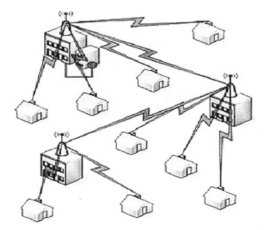

Figure 9.2　Point-to-multipoint bridge.

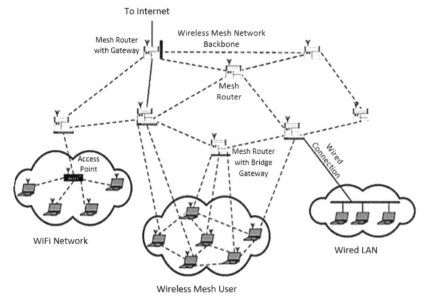

Figure 9.3   Mesh network.

Mesh or ad hoc network: This network is an independent local area network that is not connected to a wired infrastructure and in which all stations are connected directly to one another (Figure 9.3).

Wireless Technologies: Wireless technologies can be classified into different ways depending on their range. Each wireless technology is designed to serve a specific usage segment. The requirements for each usage segment are based on a variety of variables, including bandwidth needs, distance needs, and power.

Wireless Wide Area Network (WWAN): This network enables you to access the Internet via a wireless wide area network (WWAN) access card and a PDA or laptop. These networks provide a very fast data speed compared with the data rates of mobile telecommunications technology, and their range is also extensive. Cellular and mobile networks based on CDMA and GSM are good examples of WWAN.

Wireless Personal Area Network (WPAN): These networks are very similar to WWAN except their range is very limited.

Wireless Local Area Network (WLAN): This network enables you to access the Internet in localized hotspots via a wireless local area network

(WLAN) access card and a PDA or laptop. It is a type of local area network that uses high-frequency radio waves rather than wires to communicate between nodes. These networks provide a very fast data speed compared with the data rates of mobile telecommunications technology, and their range is very limited. Wi-Fi is the most widespread and popular example of WLAN technology.

Wireless Metropolitan Area Network (WMAN): This network enables you to access the Internet and multimedia streaming services via a wireless region area network (WRAN). These networks provide a very fast data speed compared with the data rates of mobile telecommunication technology as well as other wireless network, and their range is also extensive.

9.1.2 Issues with Wireless Networks

There are following three major issues with wireless networks.

- **Quality of Service (QoS):** One of the primary concerns about wireless data delivery is that, unlike the Internet through wired services, QoS is inadequate. Lost packets and atmospheric interference are recurring problems of the wireless protocols.
- **Security Risk:** This is another major issue with a data transfer over a wireless network. Basic network security mechanisms like the *service set identifier (SSID) and Wireless Equivalency Privacy (WEP)* may be adequate for residences and small businesses, but they are inadequate for the entities that require stronger security.
- **Reachable Range:** Normally, wireless network offers a range of about 100 m or less. Range is a function of antenna design and power. Nowadays the range of wireless is extended to tens of miles so this should not be an issue any more.

9.1.3 Wireless Broadband Access (WBA)

Broadband wireless is a technology that promises high-speed connection over the air. It uses radio waves to transmit and receive data directly to and from the potential users whenever they want it. Technologies, such as 3G, Wi-Fi, WiMAX, and UWB, work together to meet unique customer needs.

WBA is a point-to-multipoint system which is made up of base station and subscriber equipment. Instead of using the physical connection between the base station and the subscriber, the base station uses an outdoor antenna to send and receive high-speed data and voice-to-subscriber equipment.

WBA offers an effective, complementary solution to wireline broadband, which has become globally recognized by a high percentage of the population.

In this chapter and next chapter, we are introducing two of the wireless technologies: Wi-Fi and WiMAX.

9.2 Introduction to Wireless Network Protocols

It is common that people are referring to wireless networking as "Wi-Fi" even when the network uses a totally unrelated kind of wireless technology. For users, it would be better if all devices worldwide are using one common protocol such as Wi-Fi, but in reality today's networks are supporting a wide variety of different wireless protocols. The main reason is the fact that no one protocol provides an optimal solution for all the different application in use by the costumers. Some are better optimized to conserve battery on mobile devices, while others offer higher speeds or more reliable and longer-distance connections.

Next is given a summary for some of the wireless network protocols that have proven popular and useful in costumer devices and business environments.

LTE (Long-term Evolution): Before newer smartphones adopted so-called fourth-generation ("4G") wireless networking, phones employed a dizzying variety of older generation cellular communication protocols with names such as HSDPA, GPRS, and EV-DO. Phone carriers and the industry have invested large sums of money to upgrade cell towers and other network equipment to support 4G, standardizing on a communication protocol called Long-term Evolution (LTE) that emerged as a popular service starting in 2010.

LTE technology was designed to significantly improve the low data rates and roaming problems with older phone protocols. The protocol can carry more than 100 Mbps of data, although the network bandwidth is normally regulated to levels below 10 Mbps for individual users. Due to the significant cost of equipment, plus some government regulatory challenges, phone carriers have not yet deployed LTE in many locations. LTE is also not suitable for home and other local area networking, being designed to support a larger number of customers across much longer distances (and corresponding higher cost).

Long Term Evolution (LTE) is a wireless broadband technology designed to support roaming internet access by cellphones and other handheld devices.

Because LTE offers significant improvements over older cellular communication standards, some refer to it as a 4G technology, along with WiMAX.

Concerning the current use of LET, the first devices that supported LTE technology appeared in 2010. Most high-end smartphones and many tablets are equipped with the right interfaces for LTE connections. Older mobile phones usually don't offer LTE service. Check with your service provider.

LTE is based on IP technology to support internet connections, not voice calls. Some voice-over IP technologies work with LTE service, but some cellular providers configure their phones to switch seamlessly to a different protocol for phone calls.

Concerning the advantages of LTE connections, LTE service offers an improved online experience on mobile devices. LTE offers the following:

- Much faster upload and download speeds
- Enhanced support for mobile devices
- Low data transfer latency

Wi-Fi: Wi-Fi is widely associated with wireless networking as it has become the de facto standard for home networks and public hotspot networks. Wi-Fi became popular starting in the late 1990s as the networking hardware required to enable PCs, printers, and other consumer devices became widely affordable and the supported data rates were improved to acceptable levels (from 11 Mbps to 54 Mbps and above).

Although Wi-Fi can be made to run over longer distances in carefully controlled environments, the protocol is practically limited to work within single residential or commercial buildings and outdoor areas within short walking distances. Wi-Fi speeds are also lower than for some other wireless protocols. Mobile devices increasingly support both Wi-Fi and LTE (plus some older cellular protocols) to give users more flexibility in the kinds of networks they can use.

Wi-Fi Protected Access security protocols add network authentication and data encryption capabilities to Wi-Fi networks. Specifically, WPA2 is recommended for use on home networks to prevent unauthorized parties from logging into the network or intercepting personal data sent over the air.

Wi-Fi is the subject of this chapter.

Bluetooth: One of the oldest wireless protocols still broadly available, bluetooth was created in the 1990s to synchronize data between phones and other battery-powered devices. Bluetooth requires a lower amount of power to operate than Wi-Fi and most other wireless protocols. In return, bluetooth connections only function over relatively short distances, often 30 ft (10 m) or less and support relatively low data rates, usually 1–2 Mbps. Wi-Fi has replaced bluetooth on some newer equipment, but many phones today still support both of these protocols.

Bluetooth is a radio communication technology that enables low-power, short distance wireless networking between phones, computers, and other network devices. The name Bluetooth is borrowed from King Harald Gormsson of Denmark who lived more than 1,000 years ago. The king's nickname meant "Bluetooth," supposedly because he had a dead tooth that looked blue. The bluetooth logo is a combination of the two Scandinavian runes for the King's initials.

Bluetooth technology was designed primarily to support networking of portable consumer devices and peripherals that run on batteries, but bluetooth support can be found in a wide range of devices including:

- Cell phones
- Wireless headsets (including hands-free car kits)
- Wireless keyboards
- Printers
- Wireless speakers
- Computers

Compared with Wi-Fi, bluetooth networking is slower, is more limited in range, and supports fewer peer devices. Bluetooth is the subject of Chapter 8 of our book "Microcontroller Networks."

60 GHz Protocols – WirelessHD and WiGig: One of the most popular activities on computer networks is streaming of video data, and several wireless protocols that run on 60 Gigahertz (GHz) frequencies have been built to better support this and other usages which require large amounts of network bandwidth. Two different industry standards called WirelessHD and WiGig were created in the 2000s both using 60 GHz technology to support high-bandwidth wireless connections: WiGig offers between 1 and 7 Gbps of bandwidth while WirelessHD supports between 10 and 28 Gbps.

Although basic video streaming can be done over Wi-Fi networks, best-quality high-definition video streams demand the higher data rates these protocols offer. The very high signaling frequencies of WirelessHD and WiGig compared to Wi-Fi (60 GHz vs.2.4 or 5 GHz) greatly limit connection range, generally shorter than bluetooth, and typically to within a single room (as 60 GHz signals do not penetrate walls effectively).

The high frequency band used by *WirelessHD* and *WiGig* lets the two systems having some technical advantages compared to other network protocols like Wi-Fi but also some limitations. It increases the amount of network bandwidth and effective data rate they can support. In return for increased speed, 60 Gbps protocols sacrifice network range. A typical 60 Gbps wireless protocol connection can only function at distances of 30 ft (about 10 m) or less. Extremely high-frequency radio signals are not able to pass through most physical obstructions and so indoor connections are also generally limited to a single room. On the other hand, the greatly reduced range of these radios also means that they are much less likely to interfere with other nearby 60 GHz networks, and makes remote eavesdropping and network security break-ins much more difficult for outsiders.

Wireless Home Automation Protocols – Z-Wave and ZigBee: Various network protocols have been created to support home automation systems that allow remote control of lights, home appliances, and consumer gadgets. Two prominent wireless protocols for home automation are Z-Wave and ZigBee. To achieve the extremely low energy consumption required in home automation environments, these protocols and their associated hardware support only low data rates – 0.25 Mbps for ZigBee and only about 0.01 Mbps for Z-Wave. While such data rates are obviously unsuitable for general-purpose networking, these technologies work well as interfaces to consumer gadgets which have simple and limited communication requirements.

In the past, home automation was confronted with distance barriers in large homes and commercial buildings because the network was limited in how far the signals could travel. Differences in electrical wiring, called phases, required you to use phase couplers to bridge the signals from one electrical circuit to another. Large homes with longer wiring distances experienced weak signals and sporadic performance.

Bluetooth and ZigBee are handled in detail in five chapters of our book "Smart Home and Microcontroller Networks". Wi-Fi is the subject of this chapter.

9.3 What is Wi-Fi: IEEE 802.11

9.3.1 Overview

Let us start with the origin of the name: Wi-Fi. Many people are thinking that the name stands for **Wi**reless **Fi**delity. Even though the term "Wireless Fidelity" often appears in many documents, the truth is that this is an incorrect explanation of the term. The term Wi-Fi was coined as a brand name by the Wi-Fi Alliance when they were formed and took on board the promotion of the standard.

The technical name of Wi-Fi is IEEE 802.11. It is one of the family of IEEE 802.11 standards and is primarily a local area networking (LAN) technology designed to provide in-building broadband coverage. As a matter of fact, local area networks of all forms use Wi-Fi as one of the main forms of communication along with Ethernet. 802.11 is a family of different variant, for example, 802.11c and 802.11n. All the different variants are different standards within the overall IEEE 802.11 family. By releasing updated variants, the overall technology has been able to keep pace with the ever growing requirements for more data and higher speeds, etc. Technologies including gigabit Wi-Fi are now widely used.

Wi-Fi is a wireless based technology that allows devices like laptops, smart phones, TVs, and gaming devices, to connect at high speed to the internet without the need for a physical wired connection. In addition to this, computers, laptops, tablets, cameras, and very many other devices use Wi-Fi.

Wi-Fi access is available in many places via Wi-Fi access points or small DSL/Ethernet routers. Homes, offices, shopping centers, airports, coffee shops, and many more places offer Wi-Fi access.

Wi-Fi is now one of the major forms of communication for many devices and with home automation increasing, even more devices are using it. Home Wi-Fi is a big area of usage of the technology with most homes that use broadband connections to the Internet using Wi-Fi access as a key means of communication. For the home, office, and many other areas, Wi-Fi is a major carrier of data.

Wi-Fi has become the *de facto* standard for *last mile* broadband connectivity in homes, offices, and public hotspot locations. Systems can typically provide a coverage range of only about 1,000 ft from the access point.

Current Wi-Fi systems support a peak physical-layer data rate of 54 Mbps and typically provide indoor coverage over a distance of 100 ft.

The technology uses license free allocations so that it is free for all to use without the need for a wireless transmitting license. Typically Wi-Fi uses the 2.4 and 5 GHz Industrial, Scientific and Medical (ISM) bands as these do not require a license, but it also means they are open to other users as well and this can mean that interference exists.

Power levels are also low. Typically they are around 100 or 200 mW, although the maximum levels depend upon the country in which the equipment is located. Some allow maximum powers of a watt or more on some channels.

Figure 9.4   Typical modern Wi-Fi router.

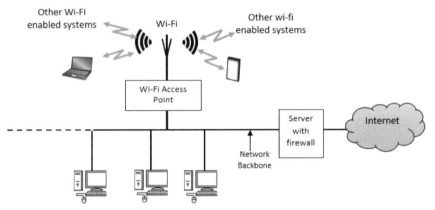

Figure 9.5   How a Wi-Fi access point can be connected on an office local area network.

9.3.2 Wi-Fi Access Point

The core of any Wi-Fi system is known as the Access Point (AP). The Wi-Fi access point is essentially the base station that communicates with the Wi-Fi enabled devices – data can then be routed onto a local area network, normally via Ethernet and typically links onto the Internet (Figures 9.5 and 9.6).

Public Wi-Fi access points are typically used to provide local Internet access often on items like smartphones or other devices without the need for having to use more costly mobile phone data. They are also often located within buildings where the mobile phone signals are not sufficiently strong.

Home Wi-Fi systems often use an Ethernet router: this provides the Wi-Fi access point as well as Ethernet communications for desktop computers, printers, and the like as well as the all-important link to the Internet via a firewall. Being an Ethernet router, it transcribes the IP addresses to provide a firewall capability.

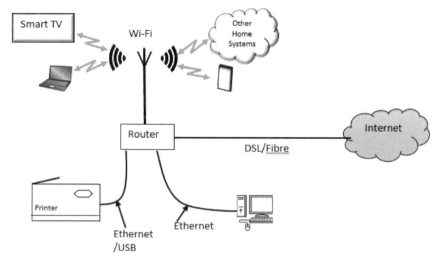

Figure 9.6 Typical home Wi-Fi with local area network.

Although Wi-Fi links are established on either of the two main bands, 2.4 GHz and 5 GHz, many Ethernet routers and Wi-Fi access points provide dual band Wi-Fi connectivity and they will provide 2.4 GHz and 5 GHz Wi-Fi. This enables the best Wi-Fi links to be made regardless of usage levels and interference on the bands.

There will typically be a variety of different Wi-Fi channel that can be used. The Wi-Fi access point or Wi-Fi router will generally select the optimum channel to be used. If the access point or router provides dual band Wi-Fi capability, a selection of the band will also be made. These days, this selection is normally undertaken by the Wi-Fi access point or router, without user intervention so there is no need to select 2.4 GHz or 5 GHz Wi-Fi as on older systems.

In order to ensure that the local area network to which the Wi-Fi access point is connected remains secure, a password is normally required to be able to log on to the access point. Even home Wi-Fi networks use a password to ensure that unwanted users do not access the network.

Many types of device can connect to Wi-Fi networks. Today devices like smartphones, laptops, and the like expect to use Wi-Fi, and therefore, it is incorporated as part of the product – no need to do anything apart from connect. A lot of other devices also have Wi-Fi embedded in them: smart TVs, cameras, and many more. Their set up is also very easy.

Occasionally some devices may need a little more attention. These days, most desktop PCs will come ready to use with Ethernet, and often they have Wi-Fi capability included. Some may not have Wi-Fi incorporated and therefore that may need additional hardware if they are required to use Wi-Fi links. An additional card in the PC or an external dongle should suffice for this.

In general, most devices that need to communicate data electronically will have a Wi-Fi capability.

9.3.3 Wi-Fi Network Types

Although most people are familiar with the basic way that a home Wi-Fi network might work, it is not the only format for a Wi-Fi network.

Essentially there are two basic types of Wi-Fi network:

- *Local area network-based network:* This type of network may be loosely termed a LAN based network. Here a Wi-Fi access point (AP) is linked onto a local area network to provide wireless as well as wired connectivity, often with more than one Wi-Fi hotspot.

  The infrastructure application is aimed at office areas or to provide a "hotspot." The office may even work wirelessly only and just have a Wireless Local Area Network (WLAN). A backbone-wired network is still required and is connected to a server. The wireless network is then split up into a number of cells, each serviced by a base station or access point (AP) which acts as a controller for the cell. Each access point may have a range of between 30 and 300 m dependent upon the environment and the location of the access point.

  Normally a LAN-based network will provide both wired and wireless access. This is the type of network that is used in most homes, where a router which has its own firewall is connected to the Internet, and wireless access is provided by a Wi-Fi access point within the router. Ethernet and often USB connections are also provided for wired access.

- *Ad hoc network:* The other type of Wi-Fi network that may be used is termed an Ad-Hoc network. These are formed when a number of computers and peripherals are brought together. They may be needed when several people come together and need to share data or if they need to access a printer without the need for having to use wire connections. In this situation the users only communicate with each other and not with a larger wired network.

As a result, there is no Wi-Fi access point and special algorithms within the protocols are used to enable one of the peripherals to take over the role of master to control the Wi-Fi network with the others acting as slaves. This type of network is often used for items like games controllers / consoles to communicate.

9.4 Wi-Fi History

Although it is possible to trace the history of Wi-Fi back to many developments in radio or wireless technology, the first release of IEEE 802.11 occurred in 1997. This was a time when the Internet was in its infancy, and most personal computers were desktop computers. The first release of IEEE 802.11 was for a system that provided 1 or 2 Mbps transfer rates using frequency hopping or direct sequence spread spectrum. The standard was only referred to as IEEE 802.11 and there were no suffix letters as we see today.

Then in 1999, the 802.11b specification was released. This provided raw data rates of 11 Mbps and used the 2.4 GHz ISM band; the first products were released in 2000.

The release of 802.11b was followed by 802.11a and this used an OFDM waveform and could transfer data at rates of between 1.5 and 54 Mbps, and it uses RF channels in the 5 GHz ISM band where there was far more available space.

Further releases took place as time progressed, each one providing improved performance or different capabilities, the major ones being: 802.11g (2003); 802.11n (2009), 802.11ac (2013), and 802.11ax (2019).

Another major milestone in the development of Wi-Fi 802.11 was the formation of the Wi-Fi Alliance in 1999. This is an industry body that works towards greater levels of adoption of Wi-Fi as well as ensuring that all devices can inter-operate successfully. It is separate from the IEEE which develops the standards, but naturally it works with them.

9.5 Wi-Fi Standards: IEEE 802.11 Variants

As Wi-Fi is used for so many different purposes and Wi-Fi capabilities are incorporated into a huge number of devices made by different manufacturers, it is of great importance that it has internationally agreed standards and specifications.

By having standards that define the exact operation of the technology, it is possible to ensure that equipment made by different manufacturers will communicate satisfactorily. Using common standards enables reliable interoperation, and this enables the technology to be more widely accepted and used.

Wi-Fi, IEEE 802.11 is a prime example of how an accessible standard has enabled multiple manufacturers to make equipment for it and together ensure that the whole Wi-Fi technology is considerably more widely used.

As Wi-Fi has developed, many new variants or standards have been developed to accommodate the increasing speeds and performance. The various standards under the IEEE 802.11 umbrella cover everything from the bearers to elements of the system required for interworking, for example, security, hotspots, quality of service, roaming, and the like.

The main IEEE 802.11 standards are listed next:

- ***802.11a:*** This was the first Wi-Fi standard in the 802.11 series. Released in 1999, it defined a wireless network bearer operating in the 5 GHz ISM band using orthogonal frequency division multiplexing with data rate up to 54 Mbps. This was done with the hope of encountering less interference since many devices (like most wireless phones) also use the 2.4 GHz band.

  Although 802.11a was used, it was as widely used as the 802.11b version. Although the 5 GHz band was much wider and accommodated many more channels, the technology was more expensive at the time and this reduced its use considerably.

- ***802.11b:*** The 802.11b standard was far more widely used than 11a. Although the maximum raw data rates were much lower at 11 Mbps, the standard used the 2.4 GHz ISM band and technology for this at the time was much cheaper. Also the usage of Wi-Fi was much less and interference was not the issue it would be today.

- ***802.11e:*** One of the key areas of sending data over any medium is what is termed the quality of service or QoS and the prioritization of data. IEEE 802.11e addresses this topic so that a defined approach can be taken.

- ***802.11f:*** IEEE 802.11f is a recommendation that describes an optional extension to IEEE 802.11 to enable wireless access point communications among multivendor systems. IEEE 802.11F was issues for a trialing its use, but it was not taken up across the industry and therefore it was withdrawn in 2006.

- ***802.11g:*** The 802.11b standard came as a result of the demand for faster Wi-Fi using the 2.4 GHz band. 802.11g utilizes OFDM technology and enabled 54 Mbps raw data transfer rates. It was also backward compatible allowing communication with DSSS but at the lower rate of 802.11b. Backwards compatibility was a requirement in view of the number of older access points and computers that might only have the older standard available, a requirement that is always of importance.
- ***802.11h:*** The IEEE 802.11h-2003 specification defines the power control required for Wi-Fi. It governs Spectrum and Transmit Power Management Extensions and addresses issues including the possible interference with satellites and radar that also use the 5 GHz ISM band. The standard originally provided for Dynamic Frequency Selection (DFS) and Transmit Power Control (TPC) to the 802.11a PHY, but it has also been integrated into the full IEEE 802.11-2007 standard.
- ***802.11i:*** Security is a major issue for Wi-FI as many Wi-Fi hotspots are in public areas and open to the possibility of hackings gaining unwanted access to the devices of people using the hotspot. The IEEE 802.11i standard is used to facilitate secure end-to-end communication for wireless local area networks. The IEEE 80211i standard improves mechanisms for wireless authentication, encryption, key management, and detailed security.
- ***802.11j:*** IEEE 802.11j-2004 is an amendment to the basic standard that extends wireless communication and signaling for 4.9 and 5 GHz band operations in Japan.
- ***802.11k:*** The IEEE 802.11 standard extends Radio Resource Measurement (RRM) mechanisms for wireless local area networks. It provides some recommendations about optimizing the WLAN performance.
- ***802.11n:*** 802.11n or, more fully, IEEE 802.11n-2009 is a Wi-Fi standard that operates in the 2.4 and 5 GHz ISM bands with data rates up to 600 Mbps. It uses MIMO technology along with frame aggregation, and it also provides security improvements over previous wireless bearer standards. Wi-Fi Alliance have also labeled the technology for the standard as Wi-Fi 4.
- ***802.11s:*** This IEEE 802.11 standard amendment addresses the topic of mesh networking. It details how Wi-Fi devices can interconnect to create a WLAN mesh network, which may be used for relatively fixed non-mobile topologies and wireless ad hoc networks.
- ***802.11u:*** IEEE 802.11u-2011 is an amendment to the IEEE 802.11-2007 standard. It adds features that are used for interworking with external

networks. It is used for roaming and it is also used for the Hotspot2.0 initiative.

- *802.11ac:* IEEE 802.11ac gave a major leap in terms of performance when it was introduced. The standard was released in 2013, but even though many companies had sight of the standard as it was released it took a short while after its release before products were seen and it became widely used. The standard defines a Wi-Fi "wireless network bearer" that operates below 6 GHz and provides data rates of at least 1 Gbps per second for multi-station operation and 500 Mbps on a single link. The standard has been labeled as Wi-Fi 5 by Wi-Fi Alliance in view of its features and performance.

- *802.11ad:* 802.11ad also known as WiGi or Gigabit Wi-Fi and it is designed to provide extremely high throughput data and uses millimeter wave bands where there are large amounts of bandwidth to achieve this. It is defined as a Multiple Gigabit Wireless System (MGWS) standard, and it operates at frequencies up to 60 GHz frequency – it is a networking standard for WiGig networks.
  In view of the very high frequencies used, ranges are very limited – often just a few meters and it is severely attenuated by objects like walls, etc., that would allow signals from lower frequencies through.

- *802.11af:* There is often a lot of what is termed White Space in the regions where television transmitters require guard regions so that transmitters using the same frequency do not interfere. In these regions where there is the white space, low power signals can be used for a variety of other services as their power level means they will not travel to far and cause interference to the primary users. One use for this white space is Wi-Fi, and IEEE 802.11af has been defined to operate in these regions. In view of its application and method of frequency use, it is often called White-Fi.

- *802.11ah:* Although the 2.4 and 5 GHz bands are most widely used for Wi-Fi, there are also some ISM allocations below 1 GHz. IEEE 802.11ah seeks to use the unlicensed spectrum below 1 GHz. One advantage is that it will be able to provide long range communications and hence give support for the Internet of Everything. The drawback of these bands is that they are relatively narrow and this can limit the data speed.

- *802.11ax:* 802.11ax is seen as the future successor to 802.11ac. Using technologies including OFDMA, MU-MIMO and others its aim is to increase spectral efficiency and hence the overall usability.

In addition to the standards mentioned above, the IEEE and its working groups are working towards developing new Wi-Fi standards. These will ensure that the technology moves forwards in line with the requirements of the industry and IEEE 802.11 Wi-Fi is able to meet the needs of the future.

Although the network bearer standards like IEEE 802.11g, 802.11n, IEEE 802.11ac, etc. are possibly the most widely known, they are all linked by the common basic technology behind 802.11. As can be seen by the list above, there are many 802.11 standards that address topic common to all Wi-Fi systems. Security, quality of service, authentication and the like are all important and are required to build a strong environment for the development and use of Wi-Fi technology.

Brief history of the wireless standards is given in Table 9.1, and technical comparison between the three major standards is given in Table 9.2.

Table 9.1 A brief history of wireless standards

IEEE standard	802.11a	802.11b	802.11g	802.11n	802.11ac	802.11ax
Year released	1999	1999	2003	2009	2014	2019
Frequency	5 GHz	2.4 GHz	2.4 GHz	2.4 GHz and 5 GHz	2.4 GHz & 5 GHz	2.4 GHz and 5 GHz
Maximum data rate	54 Mbps	11 Mbps	54 Mbps		1.3 Gbps	10–12 Gbps

Table 9.2 Technical comparison between the three major Wi-Fi standards

Feature	Wi-Fi (802.11b)	Wi-Fi (802.11a/g)
Primary application	Wireless LAN	Wireless LAN
Frequency Band	2.4 GHz ISM	2.4 GHz ISM (g)5 GHz U-NII (a)
Channel bandwidth	25 MHz	20 MHz
Half-/Full-duplex	Half	Half
Radio technology	Direct sequenceSpread spectrum	OFDM(64-channels)
Bandwidth Efficiency	<=0.44 bps/Hz	==2.7 bps/Hz
Modulation	QPSK	BPSK, QPSK, 16-, 64-QAM
FEC	None	Convolutional Code
Encryption	Optional-RC4m (AES in 802.11i)	Optional-RC4(AES in 802.11i)
Mobility	In development	In development
Mesh	Vendor proprietary	Vendor proprietary
Access protocol	CSMA/CA	CSMA/CA

9.5.1 Wi-Fi Alliance's Wireless Standard Naming System

The Wi-Fi Alliance argue that the 802.11 terminology is confusing for consumers. They are right, updating one or two letters doesn't give users much information to work with. Wi-Fi Alliance, to solve this conflict, introduced Wi-Fi Alliance's naming system and named it Wi-Fi 6.

The Wi-Fi Alliance naming system runs concurrently with the IEEE 802.11 convention. Here is how the naming standards correlate:

- **Wi-Fi 6:** 802.11ax (coming in 2019)
- **Wi-Fi 5:** 802.11ac (2014)
- **Wi-Fi 4:** 802.11n (2009)
- **Wi-Fi 3:** 802.11g (2003)
- **Wi-Fi 2:** 802.11a (1999)
- **Wi-Fi 1:** 802.11b (1999)
- **Legacy:** 802.11 (1997)

9.6 Wi-Fi Channels, Frequencies, Bands, Bandwidths, and Modulation

There are several unlicensed frequencies and bands within the radio spectrum that are used for the Wi-Fi, and within these, there are many channels that have been designated with numbers so they can be identified. These Wi-Fi bands and their channels are described in this section starting by introducing the ISM bands.

Although many Wi-Fi channels are selected automatically, it sometimes helps to have an understanding of the Wi-Fi spectrum, bands, frequencies, and the channels with their channel numbers to enable the best performance to be gained through establishing a successful Wi-Fi links. Also when office Wi-Fi access points are installed, it helps to understand the bands, their characteristics, and the channels available (Figure 9.7).

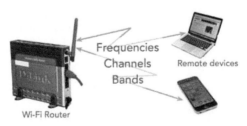

Figure 9.7 Wi-Fi router.

Table 9.3 Summary of major ISM bands

Lower Frequency (MHz)	Upper Frequency (MHz)	Comments
2400	2500	Often referred to as the 2.4 GHz band, this spectrum is the most widely used of the bands available for Wi-Fi, used by 802.11b, g, and n. It can carry a maximum of three non-overlapping channels. This band is widely used by many other non-licensed items including microwave ovens, bluetooth, etc.
5725	5875	This 5 GHz Wi-Fi band or to be more precise the 5.8 GHz band provides additional bandwidth, and being at a higher frequency, equipment costs are slightly higher, although usage, and hence interference is less. It can be used by 802.11a and n. It can carry up to 23 non-overlapping channels, but gives a shorter range than 2.4 GHz. 5 GHz Wi-Fi is preferred by many because of the number of channels and the bandwidth available. There are also fewer other users of this band.

ISM bands: Wi-Fi is using the unlicensed Industrial, Scientific and Medical (ISM) bands. These bands have been internationally agreed, and unlike most other bands, they can be used without the need for a transmitting license. This gives access to everyone to use them freely.

The ISM bands are not only used by Wi-Fi, but everything from microwave ovens to many other forms of wireless connectivity and many industrial, scientific and medical uses.

While the ISM bands are available globally, there are some differences and restrictions that can occur in some countries.

The main bands used for carrying Wi-Fi are those in Table 9.3.

802.11 systems and frequency bands: The different 802.11 variants in use are mentioned above. Different 802.11 variants use different bands. A summary of the bands used by the 802.11 systems is given in Table 9.4.

9.6.1 2.4 GHz 802.11 Channels

There are a total of 14 channels defined for use by Wi-Fi 802.11 for the 2.4 GHz ISM band. Not all of the Wi-Fi channels are allowed in all countries: 11 are allowed by the FCC and used in what is often termed the North American domain, and 13 are allowed in Europe where channels have been defined by ETSI. The WLAN / Wi-Fi channels are spaced 5 MHz apart (with the exception of a 12-MHz spacing between the last two channels).

Table 9.4   802.11 types and frequency bands

IEEE 802.11 Variant	Frequency Bands Used
802.11a	5 GHz
802.11b	2.4 GHz
802.11g	2.4 GHz
802.11n	2.4 and 5 GHz
802.11ac	Below 6 GHz
802.11ad	Up to 60 GHz
802.11af	TV white space (below 1 GHz)
802.11ah	700 MHz, 860 MHz, 902 MHz, etc. ISM bands dependent upon country and allocations
802.11ax	

The 802.11 Wi-Fi standards specify a bandwidth of 22 MHz and channels are on a 5 MHz incremental step. Often nominal figures of 20 MHz are given for the Wi-Fi channels. The 20/22 MHz bandwidth and channel separation of 5 MHz means that adjacent channels overlap and signals on adjacent channels will interfere with each other.

The 22 MHz Wi-Fi channel bandwidth holds for all standards even though 802.11b WLAN standard can run at variety of speeds: 1, 2, 5.5, or 11 Mbps and the newer 802.11g standard can run at speeds up to 54 Mbps. The differences occur in the RF modulation scheme used, but the WLAN channels are identical across all of the applicable 802.11 standards.

When using 802.11 to provide Wi-Fi networks and connectivity for offices, installing Wi-Fi access points, or for any WLAN applications, it is necessary to ensure that parameters such as the channels are correctly set to ensure the required performance is achieved. On most Wi-Fi routers these days, this is set automatically.

Wi-Fi routers often use two bands to provide dual band Wi-Fi, the 2.4 GHz band is one of the primary bands, and it is most commonly used with the 5 GHz Wi-Fi band.

2.4 GHz Wi-Fi channel frequencies: Table 9.5 provides the frequencies for the total of fourteen 802.11 Wi-Fi channels that are available around the globe. Not all of these channels are available for use in all countries.

2.4 GHz Wi-Fi channel overlap and selection: The channels used for Wi-Fi are separated by 5 MHz in most cases but have a bandwidth of 22 MHz. As a

Table 9.5 2.4 GHz band channel numbers and frequencies

Channel Number	Lower Frequency (MHz)	Center Frequency (MHz)	Upper Frequency (MHz)
1	2401	2412	2423
2	2406	2417	2428
3	2411	2422	2433
4	2416	2427	2438
5	2421	2432	2443
6	2426	2437	2448
7	2431	2442	2453
8	2436	2447	2458
9	2441	2452	2463
10	2446	2457	2468
11	2451	2462	2473
12	2456	2467	2478
13	2461	2472	2483
14	2473	2484	2495

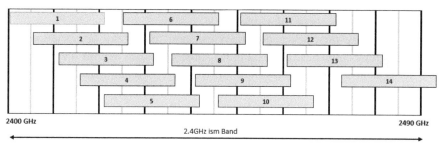

Figure 9.8 2.4 GHz Wi-Fi channels, frequencies, etc., showing overlap and which ones can be used as sets.

result, the Wi-Fi channels overlap and it can be seen that it is possible to find a maximum of three non-overlapping ones. Therefore, if there are adjacent pieces of WLAN equipment, for example, in a Wi-Fi network consisting of multiple access points that need to work on non-interfering channels, there is only a possibility of three. There are five combinations of available non-overlapping channels given in Figure 9.8.

From the diagram above, it can be seen that Wi-Fi channels 1, 6, 11, or 2, 7, 12, or 3, 8, 13 or 4, 9, 14 (if allowed) or 5, 10 (and possibly 14 if allowed) can be used together as sets. Often, Wi-Fi routers are set to channel 6 as the default, and therefore, the set of channels 1, 6, and 11 is possibly the most widely used.

As some energy spreads out further outside the nominal bandwidth, if only two channels are used, then the further away from each other the better the performance.

It is found that when interference exists, the throughput of the system is reduced. It therefore pays to reduce the levels of interference to improve the overall performance of the WLAN equipment.

With the use of IEEE 802.11n, there is the possibility of using signal bandwidths of either 20 MHz or 40 MHz. When 40 MHz bandwidth is used to gain the higher data throughput, this obviously reduces the number of channels that can be used.

Figure 9.9 shows the 802.11n 40 MHz signals. These signals are designated with their equivalent center channel numbers.

2.4 GHz Wi-Fi channel availability: In view of the differences in spectrum allocations around the globe and different requirements for the regulatory authorities, not all the WLAN channels are available in every country. Table 9.6 provides a broad indication of the availability of the different Wi-Fi channels in different parts of the world.

Figure 9.9 only provides a general view, and there may be variations between different countries. For example, some countries within the European zone and Spain have restrictions on the Wi-Fi channels that may be used (France: channels 10–13 and Spain channels 10 and 11) and do not allow many of the channels that might be thought to be available, although the position is likely to change.

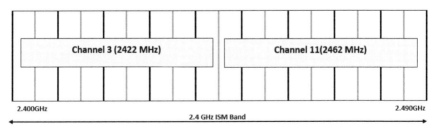

Figure 9.9 IEEE 802.11n 2.4 GHz Wi-Fi 40 MHz channels, frequencies, and channel numbers.

Table 9.6   2.4 GHz Wi-Fi channel availability

Channel Number	Europe (ETSI)	North America (FCC)	Japan
1	✓	✓	✓
2	✓	✓	✓
3	✓	✓	✓
4	✓	✓	✓
5	✓	✓	✓
6	✓	✓	✓
7	✓	✓	✓
8	✓	✓	✓
9	✓	✓	✓
10	✓	✓	✓
11	✓	✓	✓
12	✓	No	✓
13	✓	No	✓
14	No	No	802.11b only

9.6.2 3.6 GHz Wi-Fi Band

This band of frequencies is only allowed for use within the USA under a scheme known as 802.11y. Here high-powered stations can be used for backhaul Wi-Fi links in data networks. Table 9.7 gives the 3.6 GHz Wi-Fi band.

9.6.3 5 GHz Wi-Fi Channels and Frequencies

As the 2.4 GHz band becomes more crowded, many users are opting to use the 5 GHz ISM band. This provides more spectrum, but it is not as widely used for other appliances including items such as microwave ovens, etc.– microwave ovens work best around 2.4 GHz because of the absorption of the radiation by the food peaks around 2.4 GHz. Accordingly, 5GHz Wi-Fi generally encounters less interference. Table 9.8 shows the 5 GHz Wi-Fi channels and frequencies.

Many Wi-Fi routers provide the option for dual band Wi-Fi operation using this band and 2.4 GHz.

It will be seen that many of the 5 GHz Wi-Fi channels fall outside the accepted ISM unlicensed band, and as a result, various restrictions are placed on operation at these frequencies.

Table 9.7   3.6 GHz Wi-Fi band

Channel Number	Frequency (MHz)	5 MHz Bandwidth	10 MHz Bandwidth	20 MHz Bandwidth
131	3657.5	✓		
132	36622.5	✓		
132	3660.0		✓	
133	3667.5	✓		
133	3665.0			✓
134	3672.5	✓		
134	3670.0		✓	
135	3677.5	✓		
136	3682.5	✓		
136	3680.0		✓	
137	3687.5	✓		
137	3685.0			✓
138	3689.5	✓		
138	3690.0		✓	

Note: The channel center frequency depends upon the bandwidth used. This accounts for the fact that the center frequency for various channels is different if different signal bandwidths are used.

9.6.4 Additional Bands and Frequencies

In addition to the more established forms of Wi-Fi, new formats are being developed that will use new frequencies and bands. Technologies employing white space usage and also new standards using bands that are well into the microwave region will deliver gigabit Wi-Fi. These technologies will require the use of new spectrum for Wi-Fi, Table 9.9.

As Wi-Fi technology's use has increased out of all proportion and the data transfer speeds have risen significantly, so too has the way in which the bands are used.

Wi-Fi is available in many areas, in the home, office, and coffee shops. Wi-Fi access points are widely available, often providing dual band Wi-Fi – both 2.4 GHz and 5 GHz Wi-Fi to enable fast operation at all times.

Originally, the 2.4 GHz band was favored for Wi-Fi, but as the technology for the 5 GHz band fell it came into much greater use in view of its wider channel bandwidth capability.

Table 9.8 5-GHz Wi-Fi channels and frequencies

Channel Number	Frequency (MHz)	Europe (ETSI)	North America (FCC)	Japan
36	5180	Indoors	✓	✓
40	5200	Indoors	✓	✓
44	5220	Indoors	✓	✓
48	5240	Indoors	✓	✓
52	5260	Indoors / DFS / TPC	DFS	DFS / TPC
56	5280	Indoors / DFS / TPC	DFS	DFS / TPC
60	5300	Indoors / DFS / TPC	DFS	DFS / TPC
64	5320	Indoors / DFS / TPC	DFS	DFS / TPC
100	5500	DFS / TPC	DFS	DFS / TPC
104	5520	DFS / TPC	DFS	DFS / TPC
108	5540	DFS / TPC	DFS	DFS / TPC
112	5560	DFS / TPC	DFS	DFS / TPC
116	5580	DFS / TPC	DFS	DFS / TPC
120	5600	DFS / TPC	No access	DFS / TPC
124	5620	DFS / TPC	No access	DFS / TPC
128	5640	DFS / TPC	No access	DFS / TPC
132	5660	DFS / TPC	DFS	DFS / TPC
136	5680	DFS / TPC	DFS	DFS / TPC
140	5700	DFS / TPC	DFS	DFS / TPC
149	5745	SRD	✓	No access
153	5765	SRD	✓	No access
157	5785	SRD	✓	No access
161	5805	SRD	✓	No access
165	5825	SRD	✓	No access

Note 1: there are additional regional variations for countries including Australia, Brazil, China, Israel, Korea, Singapore, South Africa, Turkey, etc. Additionally, Japan has access to some channels below 5180 MHz.

Note 2: DFS = Dynamic Frequency Selection; TPC = Transmit Power Control; SRD = Short Range Devices 25 mW max power.

Table 9.9   Additional Wi-Fi bands and frequencies

Wi-Fi Technology	Standard	Frequencies Bands
White-Fi	802.11af	470–710 MHz
Microwave Wi-Fi	802.11ad	57.0–64.0 GHz ISM band (Regional variations apply) Channels: 58,32, 60.48, 62.64, and 64.80 GHz

As other Wi-Fi technologies come to the fore, many other frequencies are being used. Other unlicensed bands that are below 1 GHz as well as white space for White-Fi using the unused TV spectrum and also now increasingly higher frequencies into the microwave region where even greater bandwidths are available, but at the cost of shorter distance. We remind the reader that higher frequencies mean less coverage range at the same time allow data to be transmitted faster than lower frequencies. Shorter coverage because higher frequencies cannot penetrate solid objects, such as walls and floors. So the primary differences between the frequencies 2.4 GHz, 5 GHz and less than 1 GHz are the range (coverage) and bandwidth (speed) that the bands provide. The 2.4 GHz band provides coverage at a longer range but transmits data at slower speeds. The 5 GHz band provides less coverage but transmits data at faster speeds.

Each Wi-Fi technology has its own frequencies or bands and sometimes a different use of the channels

9.6.5 Wi-Fi – Radio Modulation

Wi-Fi systems use two primary radio transmission techniques.

- **802.11b (<=11 Mbps)**: The 802.11b radio link uses a direct sequence spread spectrum technique called **complementary coded keying** (CCK). The bit stream is processed with a special coding and then modulated using Quadrature Phase Shift Keying (QPSK).
- **802.11a and g (=54 Mbps)** – The 802.11a and g systems use 64-channel orthogonal frequency division multiplexing (OFDM). In an OFDM modulation system, the available radio band is divided into a number of sub-channels and some of the bits are sent on each. The transmitter encodes the bit streams on the 64 subcarriers using Binary Phase Shift Keying (BPSK), Quadrature Phase Shift Keying (QPSK), or one of two levels of Quadrature Amplitude Modulation (16, or 64-QAM). Some of the transmitted information is redundant, so the receiver does not have to receive all of the subcarriers to reconstruct the information.

The original 802.11 specifications also included an option for frequency **hopping spread spectrum** (FHSS), but that has largely been abandoned.

Adaptive Modulation: Wi-Fi uses adaptive modulation and varying levels of forward error correction to optimize transmission rate and error performance. As a radio signal loses power or encounters interference, the error rate will increase. Adaptive modulation means that the transmitter will automatically shift to a more robust, although less efficient, modulation technique in those adverse conditions.

9.6.6 Wi-Fi – Access Protocols

IEEE 802.11 wireless LANs use a media access control protocol called Carrier Sense Multiple Access with Collision Avoidance (CSMA/CA). While the name is similar to Ethernet's Carrier Sense Multiple Access with Collision Detection (CSMA/CD), the operating concept is totally different.

Wi-Fi systems are the half-duplex shared media configurations, where all stations transmit and receive on the same radio channel. The fundamental problem of a radio system is that a station cannot *hear* while it is sending, and hence, it is impossible to detect a collision. Because of this, the developers of the 802.11 specifications came up with a collision avoidance mechanism called the **Distributed Control Function** (DCF).

According to DCF, a Wi-Fi station will transmit only when the channel is clear. All transmissions are acknowledged, so if a station does not receive an acknowledgement, it assumes a collision occurred and retries after a random waiting interval.

The incidence of collisions will increase as the traffic increases or in situations where mobile stations cannot hear each other.

9.6.7 Wi-Fi – Quality of Service (QoS)

There are plans to incorporate quality of service (QoS) capabilities in Wi-Fi technology with the adoption of the IEEE 802.11e standard. The 802.11e standard will include two operating modes, either of which can be used to improve service for voice:

- Wi-Fi Multimedia Extensions (WME) – Mandatory
- Wi-Fi Scheduled Multimedia (WSM) – Optional

Wi-Fi Multimedia Extensions (WME): Wi-Fi Multimedia Extensions use a protocol called Enhanced Multimedia Distributed Control Access (EDCA), which is an extension of an enhanced version of the Distributed Control Function (DCF) defined in the original 802.11 MAC.

The *enhanced* part is that EDCA will define eight levels of access priority to the shared wireless channel. Like the original DCF, the EDCA access is a contention-based protocol that employs a set of waiting intervals and back-off timers designed to avoid collisions. However, with DCF all stations use the same values and hence have the same priority for transmitting on the channel.

With EDCA, each of the different access priorities is assigned a different range of waiting intervals and back-off counters. Transmissions with higher access priority are assigned shorter intervals. The standard also includes a packet-bursting mode that allows an access point or a mobile station to reserve the channel and send 3- to 5-packets in a sequence.

Wi-Fi Scheduled Multimedia (WSM): True consistent delay services can be provided with the optional Wi-Fi Scheduled Multimedia (WSM). WSM operates like the little used Point Control Function (PCF) defined with the original 802.11 MAC.

In WSM, the access point periodically broadcasts a control message that forces all stations to treat the channel as busy and not attempt to transmit. During that period, the access point polls each station that is defined for time sensitive service.

To use the WSM option, devices need to send a traffic profile describing bandwidth, latency, and jitter requirements. If the access point does not have sufficient resources to meet the traffic profile, it will return a *busy signal*.

9.7 Wi-Fi Key Topics and Issues

Wi-Fi is now an essential part of the connectivity system working alongside mobile communications, local area wired connectivity and much more. With the growing use of various forms of wireless connectivity for devices like smartphone and laptops as well as connected televisions, security system and a host more, the use of Wi-Fi will only grow. In fact with the Internet of Things now being a reality and its use increasing, the use of Wi-Fi will also continue to grow.

As new standards are developed, its performance will improve, both for office, local hotspots, and home Wi-Fi. For the future, not only will speeds improve, with the introduction of aspects like Gigabit Wi-Fi, but also the methods of use and its flexibility. In this way, Wi-Fi will remain a chosen technology for short-range connectivity.

With such increase in the use of Wi-Fi, any engineer working or using Wi-Fi must have enough knowledge in some essential topics and issues related to Wi-Fi. Some of the topics are theoretical and some practical.

The aim of this chapter is to give the reader the needed knowledge that covers such topics. Next is given a short brief upon such issues and topics.

- ***Wi-Fi variants and standards:*** As mentioned in Section 9.4, there are several different forms of Wi-Fi. Each variant is optimum for some applications. The first variant that was widely available were IEEE802.11a and 802.11b. They have long been superseded with a variety of variants offering much higher speeds and generally better levels of connectivity. There are many different Wi-Fi standard which have been used, each one with different levels of performance. IEEE 802.11a, 802.11b, g, n, 802.11ac, 802.11ad Gigabit Wi-Fi, 11af White-Fi, ah, ax, etc. It is important for the designer of Wi-Fi system to select the variant that is the optimum of this application. This topic is covered in Section 9.4.

- **Security problems:** Security concerns have held back Wi-Fi adoption in the corporate world. Hackers and security consultants have demonstrated how easy it can be to crack the current security technology known as wired equivalent privacy (WEP) used in most Wi-Fi connections. A hacker can break into a Wi-Fi network using readily available materials and software. This shows how important to study the topic of "Wi-Fi security" and how to take measures to increase the security of the Wi-Fi network.

- ***Using hotspots securely:*** This topic is completing the previous one: Security of the network. Wi-Fi hotspots are everywhere, and they are very convenient to use providing cheap access to data services. But public Wi-Fi hotspots are not particularly secure – some are very open and can open up the unwary user to having credentials and other secure details being obtained or computers hacked, etc. When using public Wi-Fi, great care must be taken and several rules should be followed to ensure that the malicious users do not take advantage. Wi-Fi security is always a major issue.

  When using a Wi-Fi link that could be monitored by someone close by, for example, when in a coffee shop, etc., it is important to make sure that the link is secure along with the website being browsed, that is, only visit https sites. It is always wise not to expose credit card details or login passwords, when on a public Wi-Fi link, even if the Wi-Fi link is secure. It is all too easy for details to be gathered and saved for use later.

If using a smartphone, it is far, far safer to use the mobile network itself. If necessary when using a laptop or tablet, link this to the smartphone as personal hotspot as this will have a password (remember to choose a safe one) and this is much less likely to be hacked. Wi-Fi security is the subject of Section 9.9.

- **Compatibility and interoperability:** One of the major problems with Wi-Fi is its compatibility and interoperability. For example, 802.11a products are not compatible with 802.11b products. Due to different operating frequencies, 802.11a hotspots would not help an 802.11b client. Due to lack of standardization, harmonization, and certification, different vendors come out with products that do not work with each other.

- *Positioning a Wi-Fi router:* The performance of a Wi-Fi router can be very dependent upon its location. When placing the router in bad (wrong) place, it cannot perform well. By locating a router in the best position, much better performance can be gained. The location of the Wi-Fi access point or router is key to providing good performance. Locating it in the right position can enable it to give much better service over more of the intended area. This will be discussed in Section 9.8.

- **Billing Issues** – Wi-Fi vendors are also looking for ways to solve the problem of back-end integration and billing, which have dogged the roll-out of commercial Wi-Fi hotspots. Some of the ideas under consideration for Wi-Fi billing such as per day, per hour, and unlimited monthly connection fees.

9.8 Wi-Fi Router Location: 802.11 Signal Coverage

The coverage of an IEEE 802.11 Wi-Fi router depends on many elements, and routers in some locations perform better than others.

By optimizing the location of the router, it will be possible to improve the Wi-Fi coverage, and this will result in improved speeds in the areas where it is needed and also coverage in more areas of the premises, whether they are a domestic residence, office or other premises.

For example, the choice of location for a Wi-Fi router in the home, office, or even a larger area such as a shopping center can make a major difference to the coverage and overall performance.

In the home, changing the location of a Wi-Fi, IEEE 802.11 router can significantly improve its performance, allowing much faster download speeds and better connectivity to be provided where it is needed.

9.8.1 Factors Affecting Wi-Fi Propagation and Coverage

There are many factors to be considered when looking at IEEE 802.11 Wi-Fi propagation and coverage. The Wi-Fi router placement must be chosen for best propagation coverage.

The environment in which routers are located is often far from ideal. The Wi-Fi signals suffer from the interaction with many objects that are within the environment and form part of the construction: walls, structural elements, furniture, windows, and ornaments, and in general, anything which is within the environment will have some form of effect.

As the home or office environment is so full of objects and structures, propagation for Wi-Fi signals is notoriously difficult to predict. They are affected in a number of ways:

- Free space path loss
- Reflection
- Absorption
- Diffraction
- Refraction

Some of these factors are considered here.

a. Wi-Fi propagation: path loss

Like all radio signals, Wi-Fi propagation is subject to the same laws of physics including those of path loss. The Wi-Fi coverage will be limited to some extent by the distance alone, although many other factors come in to play, Figure 9.10.

Under normal free space conditions the signal level is inversely proportional to the square of the distance from the transmitter.

b. Wi-Fi propagation: reflection

With many objects appearing in the signal path in the home, office or commercial / industrial environment, signals will be reflected by many surfaces, and this will have an impact on the Wi-Fi coverage, everything from walls to metal objects like desks, domestic appliances, etc.

These reflections give rise to multiple paths for the signal. Using an antenna technology known as MIMO (Multiple Input Multiple Output), Wi-Fi is now able to make use of these multiple paths to send data at a faster rate. However in the past, it would result in interference and reduction in data rates.

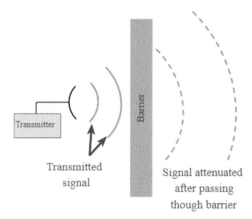

Figure 9.10   Attenuation resulting from barrier in signal propagation path.

Table 9.10   Relative absorption of various materials

Material	Relative Signal Absorption	Examples of Use of Material
Air	None	Free space outside, etc.
Wood	Low	Wooden furniture, doors, wooden partitioning.
Plastic	Low	Some partitioning, many plastic items
Glass	Low	Plain windows, glass used in partitions
Tinted glass	Medium	Tinted windows
Water	Medium	Water tanks, aquariums
Plaster	Medium	Partitioning, walls, plasterboard walls
Bricks	Medium to high	Brick walls
Ceramic	High	Tiles
Concrete	High	Concrete floors, pillars
Metal	High	Metal structures

c. Wi-Fi propagation: absorption

Attenuation resulting from barrier in signal propagation path

The type of medium through which the Wi-Fi signal passes will affect the level of attenuation. It is not possible to give exact figures for the different media, as specifications, thickness, and many other factors determine the overall level of attenuation. However, a table (Table 9.10 Relative absorption of various materials) of many common substances will give a guide to the likely affect.

9.8.2 Locating Router for Optimum Wi-Fi Coverage

From the above discussion about the factors affecting the Wi-Fi coverage, the following simple rules can help optimize the Wi-Fi coverage and signal propagation.

- *Ensure signal does not pass through thick walls:* Dense walls, especially those using concrete will significantly reduce the signal. For optimum Wi-Fi propagation and coverage, make sure the signal does not need to pass through thick walls to reach any high usage areas.
- *Locate router above desk height:* Most phones, laptops and tablets using a local Wi-Fi signal will be located above desk height and to gain the optimum signal path with the minimum of obstacles, it is best to locate the router above the desk height. Desks and other furniture will attenuate any signal, as will the wires and other metalwork associated with them.
- *Do not locate the router next to many other wires:* Although locating a router, for example, in a cupboard next to the fuse box etc. may be convenient, the wiring is likely to shield the signal from many areas.
- *Keep router as close to main areas as possible:* The Wi-Fi coverage and signal propagation will benefit from being as close to the main areas of use as possible. Path loss from the distance the signal travels will be minimized and in addition to this the signal will need to travel through fewer obstacles.

Other issues governing Wi-Fi coverage

The following factors also help Wi-Fi coverage and router performance. Some of these may need to be found by experimentation.

- *Locate router away from interference:* Locate equipment away from sources of possible interference. Obviously items like televisions that may be streaming data from a router will need to be able to receive a strong interference free signal. However the router also needs to be away from interference because it also receives management data from the remote equipment. If this suffers interference then data rates will be reduced. Sources of interference in the home are widespread. Microwave ovens transmit in the 2.4 GHz band and will cause interference, but other items like motors, fans, vacuum cleaners, fluorescent lights and many more create interference.

- ***Band selection:*** Wi-Fi is currently supported in both the 2.4 GHz and 5 GHz ISM bands. 2.4 GHz will provide better coverage as signals in this band will penetrate walls and floors, etc. better. But against this 5 GHz offers a wider bandwidth and will be able to achieve a higher throughput under ideal conditions.

- *Use Wi-Fi coverage planning tools:*

  When planning Wi-Fi coverage for a major area like a shopping mall or conference center, a more rigorous approach is needed. Software planning tools are used along with a comprehensive site survey using the plans for the area.

  These tools look at the Wi-Fi propagation characteristics and then calculate Wi-Fi coverage.

  These tools are not normally viable or accessible for domestic installations and small offices. It is for these instances where the guidelines above some in useful.

  Even when automated software tools are used, some practical input helps feed in the relevant data correctly and then adds value to the output from them.

9.8.3 Extending the Wi-Fi Coverage: Use of Wi-Fi Boosters, Range Extenders & Repeaters

It is possible to extend the Wi-Fi coverage by using a range extender, booster or repeater could be the solution.

Although Wi-Fi routers are now more effective than ever before, there can still be issues with dead-spots and poor coverage in the home or office.

One of the solutions to this issue is to use what can be termed, range extenders, boosters or repeaters.

These extenders, boosters and repeaters perform basically the same function, although there are some different types. The terms are used interchangeably: one person may use one term to talk about one type of item, whereas others will use the same term to talk about another. Next we will try to explain what these items are, how they work, and their advantages and disadvantages.

9.8.3.1 Wi-Fi boosters or range extenders

The term wireless booster indicates that the equipment boosts the Wi-Fi signal and extends the range of the router. Often these Wi-Fi boosters are plugged directly into the main router, typically via the Ethernet port and they transit a stronger signal.

Usually boosters refer to an upgrade at the Wi-Fi router itself making the Wi-Fi signal stronger. It can be an amplifier for the transmitted signal, or in some cases it may be a better antenna.

However, it is important to mention here that a booster is not a total solution of range extension. To have good coverage, it is necessary to be able to communicate in both directions. The client, that is, the item connected to the Wi-Fi router needs to be able to send data back, and the booster, if it is an amplifier will not affect this. So a booster on its own is unlikely to make much of a difference in many instances. Better antennas, though will make a difference as these will normally work in both directions.

9.8.3.2 Wi-Fi repeaters

Wi-Fi repeaters are another item used for extending the range and coverage. Often, they are also called range extenders, so make sure you know what you are buying.

As the name implies, a Wi-Fi repeater is a device that repeats the wireless signal from the main router to expand its coverage. It receives the Wi-Fi signal from the main router and then re-broadcasts it. In this way, it is able to extend the range of the main router signal, allowing coverage of areas that are not covered by the main Wi-Fi router.

The repeater has the same SSID (Service Set IDentifier) as the main access point, normally making connection easy.

The Wi-Fi repeater receives a frame from the main router, and then re-broadcasts it. This naturally takes twice as long (if not more) to transmit the frame to the final receiver. As such the performance of the wireless network will be slowed to at best half the rate of the original network.

Repeaters are a very good solution when coverage of a large residence is required, but care needs to be taken if large amounts data, for example, by video streaming, are to be used as the performance will be less than normal. With the speeds available today, this may be more than adequate, but it is best to be aware of the issue. Some more advanced repeaters make use of dual-band technology to enable the best operation to be achieved.

9.8.3.3 Powerline/homeplug extenders

One of the options available is to use what is termed a powerline extender. These typically use a standard known as Homeplug which transmits the data signal over the mains power lines within the home. The powerline Wi-Fi extender has the advantage that it does not degrade the performance of the Wi-Fi network.

9.9 Wi-Fi Security

Wi-Fi network security is an issue of importance to all Wi-Fi users. It is defined under the IEEE standard 802.11i and security schemes like as WEP, WPA, WPA2 and WPA3 are widely mentioned, with keys or codes being provided for the various Wi-Fi hotspots in use.

Wi-Fi security is of significant importance because huge number of people who are using it: at home, in the office, and when they are on the move. As the wireless signal can be picked up by non-authorized users, it is imperative to ensure that they cannot access the system.

An attacker can gain access to Wi-Fi network from the weak points of the network. The router and the hotspots are some of such windows that the attacker can use to get information from the system.

For this reason, when discussing Wi-Fi security, it is necessary to discuss the security of routers and hotspots, the matter that will be covered in this section and the following sections.

Wi-Fi Security Types: Since the invention of Wi-Fi in the 1990s, wireless networks have used several different security protocols. Each new standard provided greater security, and each promised to be easier to configure than those that came before. In spite of that, all the variants retain some inherent vulnerabilities. In addition, as each new protocol was released some systems were upgraded, and some were not. As a result, today there are a number of different security protocols in use. Some of these provide a pretty good level of protection, while some don't. There are three main security protocols in use today – WEP, WPA, and WPA2 – and one that is yet to be rolled out (WPA3).

WEP (Wired Equivalency Privacy): Wired Equivalent Privacy (WEP) was the first mainstream Wi-Fi security standard and was approved for use way back in 1999.

The aim for this key was to make wireless networks such as Wi-Fi as safe as wired communications. Unfortunately, this form of Wi-Fi network security did not live up to its name because it was soon hacked, and now there are many open source applications that can easily break into it in a matter of seconds.

In terms of its operation, the Wi-Fi WEP key uses a clear text message sent from the client. This is then encrypted and returned using a pre-shared key. A WEP comes in different key sizes. The common key lengths are normally 128 or 256 bits.

The security of the WEP system is seriously flawed. Primarily, it does not address the issue of key management and this is a primary consideration to any security system. Normally keys are distributed manually or via another secure route. The Wi-Fi WEP system uses shared keys – that is, the access point uses the same key for all clients, and therefore, this means that if the key is accessed, then all users are compromised. It only takes listening to the returned authentication frames to be able to determine the key.

Obviously, Wi-Fi WEP is better than nothing because not all people listening to a Wi-Fi access point will be hackers. It is still widely used and provides some level of security. However, if it is used, then higher layer encryption (SSL, TLS, etc.) should also be used when possible. WEP was officially abandoned by the Wi-Fi Alliance in 2004.

WPA (Wi-Fi Protected Access): In order to provide a workable improvement to the flawed WEP system, the WPA access methodology was devised. The scheme was developed under the auspices of the Wi-Fi Alliance and utilized a portion of the IEEE 802.11i security standard – in turn the IEEE 802.11i standard had been developed to replace the WEP protocol.

It increased security by using a pair of security keys: a pre-shared key (PSK), most often referred to as WPA Personal, and the Temporal Key Integrity Protocol (or TKIP) for encryption. TKIP is part of the IEEE802.11i standard and operates by performing per-packet key mixing with re-keying.

Although WPA represented a significant upgrade over WEP, it was also designed so that it could be rolled out across the ageing (and vulnerable) hardware designed for WEP. That meant that it inherited some of the well-known security vulnerabilities of the earlier system.

As a matter of fact, WPA, just like WEP, after being put through proof-of-concept and applied public demonstrations turned out to be pretty vulnerable to intrusion. The attacks that posed the most threat to the protocol were however not the direct ones, but those that were made on Wi-Fi Protected Setup (WPS) – auxiliary system developed to simplify the linking of devices to modern access points.

WPA (Wi-Fi Protected Access) scheme provides optional support for AES-CCMP algorithm. This provides a significantly improved level of security.

WPA2 (Wi-Fi Protected Access II): The WPA2 scheme for Wi-Fi network security has now superseded the basic WPA or WPAv1 scheme.

WPA2 implements the mandatory elements of IEEE 802.11i. In particular, it introduces CCMP, a new AES-based encryption mode with strong security.

WPA2 was developed in 2004 as the first truly new security protocol since the invention of Wi-Fi. The major advance made by WPA2 was the usage of the Advanced Encryption System (AES), a system used by the US government for encrypting Top Secret information. At the moment, WPA2 combined with AES represents the highest level of security typically used in home Wi-Fi networks, although there remain a number of known security vulnerabilities even in this system.

The main vulnerability to a WPA2 system is when the attacker already has access to a secured Wi-Fi network and can gain access to certain keys to perform an attack on other devices on the network. This being said, the security suggestions for the known WPA2 vulnerabilities are mostly significant to the networks of enterprise levels and not really relevant for small home networks.

Unfortunately, the possibility of attacks via the Wi-Fi Protected Setup (WPS) is still high in the current WPA2-capable access points, which is the issue with WPA too. And even though breaking into a WPA/WPA2 secured network through this hole will take anywhere around 2 to 14 hours, it is still a real security issue and WPS should be disabled and it would be good if the access point firmware could be reset to a distribution not supporting WPS to entirely exclude this attack vector.

Certification for WPA2 began in September 2004 and now it is mandatory for all new devices that bear the Wi-Fi trademark.

WPA3 (Wi-Fi Protected Access III): In 2018, the Wi-Fi Alliance announced the release of a new standard "WPA3" that will gradually replace WPA2. This new protocol has yet to be rolled out, but promises significant improvements over earlier systems. Devices compatible with the new standard are already being produced.

Update: It is hardly been a year since the launch of WPA3, and several Wi-Fi security vulnerabilities have already been unveiled, which could enable attackers to steal Wi-Fi passwords. The next-generation Wi-Fi security protocol relies on Dragonfly, an improved handshake that aims to protect against offline dictionary attacks.

However, security researchers discovered weaknesses in WPA3-Personal that allow an attacker to retrieve and recover passwords of Wi-Fi networks by abusing cache or timing-based side-channel leaks. The researchers identified two flaws in the WPA3 protocol. The first is associated with downgrade attacks, while the second leads to side-channel leaks. Since WPA2 is widely used by billions of devices worldwide, the universal adoption of WPA3 is expected to take a while. As such, most networks will support both WPA3 and WPA2 connections via WPA3's "transitional mode."

The transitional mode can be leveraged to carry out downgrade attacks by setting up a rogue access point that only supports the WPA2 protocol, forcing WPA3 devices to connect with WPA2's insecure 4-way handshake.

Researchers also found that the two side-channel attacks against the password encoding method of Dragonfly allow attackers to obtain Wi-Fi passwords by performing a password partitioning attack.

What WPA3 will introduce to improve security?
a. Password Protection

A fundamental weakness of WPA2 is that it lets hackers deploy a so-called offline dictionary attack to guess the user's password. An attacker can take as many shots as they want at guessing the user credentials without being on the same network, cycling through the entire dictionary – and beyond – in relatively short order.

WPA3 will protect against dictionary attacks by implementing a new key exchange protocol. WPA2 used an imperfect four-way handshake between clients and access points to enable encrypted connections; it is what was behind the notorious KRACK vulnerability that impacted basically every connected device (see latter about KRACK vulnerability). WPA3 will ditch that in favor of the more secure – and widely vetted – Simultaneous Authentication of Equals handshake (see Section 9.9.2).

The other benefit comes in the event that your password gets compromised nonetheless. With this new handshake, WPA3 supports forward secrecy, meaning that any traffic that came across your transom before an outsider gained access will remain encrypted. With WPA2, they can decrypt old traffic as well.

b. Safer Connections

When WPA2 came along in 2004, the Internet of Things (IoT) had not yet become anything close to the all-consuming security horror that is its present-day hallmark. No wonder, then, that WPA2 offered no streamlined way to safely onboard these devices to an existing Wi-Fi network. And in fact, the predominant method by which that process happens today – Wi-Fi Protected Setup – has had known vulnerabilities since 2011. WPA3 provides a fix.

Wi-Fi Easy Connect, as the Wi-Fi Alliance calls it, makes it easier to get wireless devices that have no (or limited) screen or input mechanism onto your network. When enabled, you will simply use your smartphone to scan a QR code on your router, then scan a QR code on your printer or speaker or other IoT device, and you are set – they are securely connected. With the QR code method, you're using public key-based encryption to onboard devices that currently largely lack a simple, secure method to do so.

That trend plays out also with Wi-Fi Enhanced Open, which the Wi-Fi Alliance detailed before. The reader has probably heard that it is necessary to avoid doing any sensitive browsing or data entry on public Wi-Fi networks. That is because with WPA2, anyone on the same public network as you can observe your activity, and target you with intrusions like man-in-the-middle attacks or traffic sniffing. On WPA3? Not so much.

When a person log onto a coffee shop's WPA3 Wi-Fi with a WPA3 device, his/her connection will automatically be encrypted without the need for additional credentials. It does so using an established standard called Opportunistic Wireless Encryption.

As with the password protections, WPA3's expanded encryption for public networks also keeps Wi-Fi users safe from a vulnerability they may not realize exists in the first place. In fact, if anything it might make Wi-Fi users feel too secure.

WEP versus WPA versus WPA2: Which Wi-Fi protocol is most secure?

When it comes to security, Wi-Fi networks are always going to be less secure than wired networks. In a wired network, data is sent via a physical cable, and this makes it very hard to listen in to network traffic. Wi-Fi networks are different. By design, they broadcast data across a wide area, and so network traffic can potentially be picked up by anyone listening in.

All modern Wi-Fi security protocols therefore make use of two main techniques: authentication protocols that identify machines seeking to connect to the network; and encryption, which ensures that if an attacker is listening in to network traffic they will not be able to access important data.

The way in which the three main Wi-Fi security protocols implement these tools is different (Table 9.11).

Without getting into the complicated details of each system, what this means is that different Wi-Fi security protocols offer different levels of protection. Each new protocol has improved security over those that came

Table 9.11 WEP versus WPA versus WPA2

	WEP	WPA	WPA2
Purpose	Making Wi-Fi networks as secure as wired networks (this didn't work!)	Implementation of IEEE802.1 li standards on WEP hardware	Complete implementation of IEEE802.1 li standards using new hardware
Data privacy (Encryption)	Rivest Cipher 4 (RC4)	Temporal Key Integrity Protocol (TKIP)	CCMP and AES
Authentication	WEP-Open and WEP-Shared	WPA-PSK and WPA-Enterprise	WPA-Personal and WPA-Enterprise
Data integrity	CRC-32	Message Integrity Code	Cipher block chaining message authentication code (CBC-MAC)
Key management	Not provided	4-way handshaking	4-way handshaking
Hardware compatibility	All hardware	All hardware	Older network interface cards are not supported (only newer than 2006)
Vulnerabilities	Highly vulnerable: susceptible to Chopchop, fragmentation, and DoS attacks	Better, but still vulnerable: Chopchop, fragmentation, WPA-PSK, and DoS attacks	The least vulnerable, although still susceptible to DoS attacks
Configuration	Easy to configure	Harder to configure	WPA-Personal is easy to configure, WPA-Enterprise less so
Replay attack protection	No protection	Sequence counter for replay protection	48-bit datagram / package number protects against replay attacks

before, and so, the basic rating from best to worst of the modern Wi-Fi security methods available on modern (after 2006) routers is like this:

- WPA2 + AES
- WPA + AES
- WPA + TKIP/AES (TKIP is there as a fallback method)

- WPA + TKIP
- WEP
- Open Network (no security at all)

9.9.1 Wi-Fi Vulnerability: KRACK Wi-Fi Vulnerability

There is a serious weakness in Wi-Fi Protected Access II (WPA2) proto-col that let Wi-Fi vulnerable for attacks: KRACK Wi-Fi vulnerability. The weakness is the result of the wireless encryption standard used by just about every Wi-Fi device in the world. The security hole enables hackers to steal unencrypted data transmitted via the Wi-Fi network, even if the network is password-protected. The exploit doesn't actually crack the Wi-Fi password of the victim, but rather replaces the encryption key to decrypt traffic on the network.

Because this is a vulnerability in WPA2 itself, nearly all internet-enabled devices are susceptible to the KRACK attack, regardless of the software they're running on. This vulnerability also exists in the earlier Wi-Fi Protected Access (WPA) as well as any cipher suite such as GCMP, WPA-TKIP and AES-CCMP.

The KRACK vulnerability has literally cracked Wi-Fi connection and security, for the worse. The vulnerability or exploit has left billions of devices and users across the world completely exposed, and has opened new doors for hackers and data snoopers. While the real impact and fallout of this exploit will make itself visible in the coming days, for now, internet-enabled devices, especially IoT (Internet of Things) devices, remain vulnerable more than ever before.

Till now, the best way to protect ourselves from the KRACK Wi-Fi flaw is by making sure we always use HTTPS. Moreover, using a VPN is a good idea as it encrypts all the traffic. Keep in mind that the attack doesn't retrieve the password of your Wi-Fi network, and therefore changing it wouldn't prove to be much of a help.

Which Security protocol is more secure?

Based on the above discussion of Wi-Fi vulnerability, the following question may arise for any user of Wi-Fi: **Which type of wireless security protocol is best for Wi-Fi?**

The key point here is this: the most secure Wi-Fi setup you can have today is WPA2 combined with AES. It will not always be possible to use this standard, though.

Table 9.12 Summary of Wi-Fi security protocols

Encryption Standard	Summary	How It Works	Should I Use It?
WEP	First 802.11 security standard: easy to hack.	Uses RC4 cipher.	No
WPA	Interim standard to address major security flaws in WEP.	Uses RC4 cipher, but adds longer (256-bit) keys.	Only if WPA2 is not available
WPA2	Current standard. With modern hardware increased encryption doesn't affect performance.	Replaces RC4 cipher with CCMP and AES for stronger authentication and encryption.	Yes

It might be, for instance, that your hardware does not support WPA2 or AES. This is a problem that can be overcome by upgrading your hardware.

The only disadvantage of using WPA2 and AES is that the military-grade encryption it uses can sometimes slow down your connection. This issue, though, mainly affects older routers that were released before WPA2 and only support WPA2 via a firmware upgrade. Any modern router will not suffer from this problem.

Another bigger problem is that all users are forced to use public Wi-Fi connections from time to time, and in some cases the level of security offered on them is poor. The best approach is therefore to be aware of the level of security offered on the networks you connect to, and to avoid sending passwords (or other important information) across poorly secured networks. All of this can be summed up in Table 9.12.

9.9.2 Wireless Security: Wi-Fi Authentication Modes

This sub-section gives brief discussion on the possible authentication schemes that are used in the wireless deployments. They are: Open Authentication and Pre-Shared Key (PSK)-based authentication. The former one is based on EAP frames to derive dynamic keys.

Open Authentication: The term Open Authentication is itself very misleading. It suggests that some kind of authentication is in place, but in fact, the authentication process in this scheme is more like formal step, rather than authentication mechanism. The process looks like how it is shown in Figure 9.11.

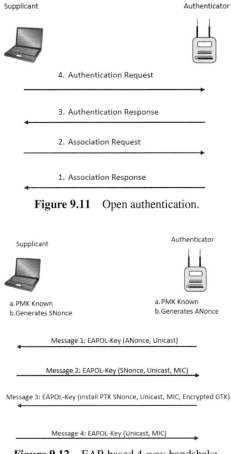

Figure 9.11 Open authentication.

Figure 9.12 EAP-based 4-way handshake.

The figure shows clearly that open authentication is not secure at all: it simply allows any client to authenticate to the network, without the right security check. This is why open authentication should never be used.

EAP-based 4-way handshake (with WPA/WPA2): When a wireless client authenticates to the AP, both of them go through the 4-step authentication process called **4**-way handshake (Figure 9.12). During those message exchanges, the shared password is derived between AP and wireless client, without being transmitted in any of those EAP messages.

The Pairwise Master Key (PMK) is something a hacker would like to collect, in order to break the network encryption scheme. PMK is only known to the Supplicant and Authenticator, but is not shared anywhere in transit.

However, the session keys are, and a combination of, ANonce, SNonce, PMK, MAC addresses of Supplicant and Authenticator. We may write that relation as the mathematical formula:

Sessions_keys = f(ANonce, SNonce, PMK, A_MAC, S_MAC).

In order to derive a PMK from that equation, one would have to break AES/RC4 (depending whether WPA2 or WPA is used). It is not that easy as the only practical approach is to perform a brute-force or dictionary attack (assuming you have a really good dictionary).

It is definitely a recommended authentication approach to use and definitely safer than using Open Authentication.

Wi-Fi Chalking: Wi-Fi chalking was a very funny concept in the history of wireless LAN, mainly used in the USA. The main idea was to mark the places, where open authentication or WLANs with weak authentication were implemented. By doing that, everyone who finds out this sign somewhere on the wall or ground, written with a chalk, can log into the Wi-Fi system without authentication.

9.9.3 Some Recommendations to Increase the Security of Wi-Fi Network

There are some simple steps the user of Wi-Fi can take to make his/her wireless network more secure, whether he/she are working in a business environment or simply looking to improve security of his/her home network.

The steps are:

- Move the Router to a Physically Secure Location
- Change the Default Router Login Information
- Change the Network Name
- Update the Firmware and Software
- Use WPA2
- Turn Off WPS
- Limit or Disable DHCP

9.9.3.1 Move the router to a physically secure location

The physical location of the router plays an important role of the Wi-Fi network security. The physical location will define how much of the signal is "leaking" outside your home/business place. The leaked signal can be picked up by neighbors, people outside in the street, etc. This will open doors for attacks.

It is important, accordingly, to place your router in a position where you can get a good signal everywhere you need it, and no-one else can.

In a business environment, the physical security of your router is even more important. Attack vectors can be introduced by the simple act of someone pushing the reset button on your router. It is recommended to keep the wireless router in a locked cabinet or office, and even think about video surveillance systems that will allow you to monitor access to it.

9.9.3.2 Change the default router login information

When installing the router for first time it comes with "admin password". It is very important to change this password.

While choosing a new password and username, you should pay attention to the general guidelines on choosing strong passwords: your new password should be at least 15 characters long, and include a mix of letters, numbers, and special characters. You should also change your username and password settings on a regular basis.

9.9.3.3 Change the network name

Like their generic passwords and usernames, most wireless routers arrive with generic Service Set Identifiers (SSIDs), which is the name that identifies your Wi-Fi network. Typically, these are something like 'Linksys' or 'Netgear3060', which gives you information on the make and model of your router. This is great during initial setup, because it allows you to find your new router.

The problem is that these names also give everyone, who can pick up your wireless signal a very useful piece of information: the make and model of your router. Knowing that there are lists online that detail the hardware and software vulnerabilities of almost every router out there, a potential attacker can quickly find out the best way to compromise your network.

This is a particular problem if you have not changed the default login information on your router (see above), because an attacker can then simply log into your router as an admin, and cause havoc.

9.9.3.4 Update your firmware and software

It is known that we should keep our software up to date in order to limit security vulnerabilities. The same is valid for the software and firmware of the router.

Firmware updates are particularly important, because firmware is the most basic code used by the router. New vulnerabilities in Wi-Fi router firmware are identified all the time, and with access to the firmware level of your router there is no end to the mischief that an attacker can cause.

Typically, firmware updates are released to patch specific security vulnerabilities, and will self-install after you download them. This makes them a simple step in securing your wireless network.

9.9.3.5 Use WPA2

It is important to use the most secure wireless network protocol that you can, that is the use of WPA2 combined with AES.

Most modern routers have the option to run several different types of Wi-Fi security protocol, in order to make them compatible with as wide a range of hardware as possible. This means that your router may be configured to use an outdated protocol out of the box.

If you are using an older router, it might be that it is not compatible with WPA2, or with AES. If this is the case, there are few options. First, you should check for a firmware upgrade that will allow your router to use WPA: since WPA was designed to be compatible with older WEP routers, many now have this functionality.

If you cannot find a firmware upgrade, then you must think about upgrading your hardware.

9.9.3.6 Turn Off WPS

Although WPA2 is far more secure than the protocols that came before it, it retains a number of specific security vulnerabilities that we discussed before. Some of these are caused by a feature of WPA2 that was designed to make setting up the wireless network easier: WPS.

Wi-Fi Protected Setup (WPS) means that connecting a device to your Wi-Fi network for the first time is as easy as pushing a button. If you leave WPS enabled, anyone who can physically access your router can gain a foothold in your network.

Turning off WPS is easy enough: login to your router as an admin user, and you should see an option to disable it.

9.9.3.7 Limit or disable DHCP

Disabling the Dynamic Host Configuration Protocol (DHCP) server that the router uses, results in more security for the network. This system automatically assigns IP addresses to every device connected to your router, allowing additional devices to connect to your wireless network easily. The problem is that it will give anyone connected to your network an IP address, including someone seeking to gain unauthorized access.

There are two approaches that you can take to combat this potential vulnerability. The first is to limit the DHCP range that your router uses, which has the effect of limiting the number of machines it can connect to. The second approach is to disable DHCP entirely. This means you will have to manually assign every device an IP address every time it connects to your network.

Whether these approaches are suitable for your network will depend on how you use it. If you commonly connect and reconnect multiple devices to your router, it can become very time consuming to manually assign each an IP address. On the other hand, if the number of devices you want to connect is limited and predictable, disabling DHCP gives you a lot of control over who is connected to your network.

9.10 Wi-Fi Hotspot

Hotspots are access points generally enabled by one device to connect other devices and give them access to the internet. By accessing a hotspot, it is possible to connect laptop, smartphone or any other device with wireless connectivity to the internet.

Hotspots can be open, password protected, free or paid. However, connecting to an open public hotspot is not always a correct decision as they accompany certain security threats. Thus, we do not recommend our readers to connect to any open public hotspot without taking a few necessary precautions.

9.10.1 Differences between Wi-Fi and Mobile Hotspot

Before discussing the hotspot security concerns and the measures that can be taken to avoid them. Let us explore the types of hotspot technology.

Mobile Hotspot: This type of hotspot refers to the internet access points that could be created using devices that are mobile and can be carried from one place to another. Mobile hotspots can be further classified into two categories: personal hotspots and portable hotspots.

Personal Hotspot: With the advancement in technology, every cellular device has become a hotspot host. At first, hotspot devices could only be located at public places. Now, every individual carries a device that can turn into a hotspot access point. This is often referred to as a personal hotspot.

When using smartphone (Android or iOS), tablet or an iPad, it is possible to share your internet connection with other devices in the vicinity by creating a hotspot, allowing other devices to connect via Wi-Fi network. This hotspot can also be password protected, limiting it from any unwanted access.

With respect to hotspot security, personal hotspots encompass tier 2 security. Despite the fact that these hotspots are created on personal devices, host device could be infected with malware. This can compromise the security of hotspot network and put connected devices at risk.

Portable Hotspot: As discussed before, technological advancements have enabled smartphones to become hotspot hosts; however, there is a limit to which smartphones can be used as hotspots.

First, to become a hotspot access point, you need to know if your cellular network allows tethering, allowing your device to share its data network.

Second, sharing the data with multiple devices connected to your smartphone can quickly drain your battery. Third, it is better to be safe than sorry; one must not casually allow others to share data using a personal device.

To address the discrepancies above, we have a solution called portable hotspot devices. These are mini devices that serve as hotspot on-the-go, offering a wide range of data plans. They also double as a dedicated hotspot to multiple devices. AT&T Velocity, Verizon Jetpack, Karma Go and MCD-4800 are few examples of portable hotspots.

Portable hotspots can be classified as tier 1 in terms of hotspot security. Upon suspicion of network security being compromised, a user can either opt password reset or flash the device with hard reset.

Since these are dedicated hotspot devices backed by the support of telecom giants, chances of a hacker penetrating into their security are quite thin. Thus, if you are a frequent traveler and a security enthusiast, then a mobile hotspot device is a must-have.

Wi-Fi Hotspot: In contrast to mobile hotspots that use cellular data to enable data sharing, Wi-Fi hotspots allow people to get internet access via Wi-Fi technology. Using a router connecting to an ISP, a Wi-Fi hotspot cannot be mobile. A Wi-Fi hotspot can either be open or closed as per the host's preferences.

Open Public Wi-Fi: A Wi-Fi router connected with an ISP that has intentionally or unintentionally turned off its authentication requirements could be called open public Wi-Fi. Any device within the range of that router can share its internet access without any limitation. The host of an open public Wi-Fi often has no control over bandwidth allocation or cap over its usage. Open public Wi-Fi hotspots are usually free, though, risky to connect to.

In terms of hotspot security, open public Wi-Fi networks are least secure in comparison with other substitutes. Lack of authorization at the time of establishing the connection is what makes these networks least secure. Anyone can connect to the network and infect its security. Since worms can infect devices via nodes, security of any device connected to the network is compromised.

Further, devices opting to connect to an open public Wi-Fi network are more susceptible to be connected to a fake network. There are multiple ways through which security of such hotspot networks can be exploited by hackers. Thus, we do not recommend connecting to such networks.

Closed Public Wi-Fi: Unlike open public Wi-Fi hotspot, closed public hotspots usually involve some management and control. Closed public hotspots have authorization access enabled. Hence, only users with credentials or privileges can access the network. The closed public Wi-Fi hotspot hosts often manage bandwidth allocation, usage, upload and download limit or users along with access control – allowing only specific external devices to the Internet. Such access points are usually paid, but, sometimes free.

In comparison with open public Wi-Fi hotspots, closed public Wi-Fi hotspots are more steadfast in terms of hotspot security. These networks are closely administered and require authentication at the time of connection. These hotspots are likely secure from man-in-the-middle attacks. Moreover, due to authentication, chances for a user to connect to a fake network are quite slim.

9.10.2 Main Threats to Hotspot Security?

Connecting to an open Wi-Fi hotspot is never secure without necessary precautions. Though there are many security threats out there, let us explore a few significant ones.

Fake Networks – Rogue APs: As the name says itself, fake networks, also known as evil twin hotspots are rogue access points that can infect your device once connected. This network can infect any device that has a tethering ability, be it a smartphone or a laptop. Copying the same name along with similar security credentials, these fake hotspots are created to hack into the targeted devices when connected.

Another trick hackers use to increase the possibility for a device to connect to a fake network rather the original one is by making the fake network available in the proximity of the device. By default, the device catches the stronger signal or the one that is physically closer and gets connected to the fake network.

Further, the hackers can also take the original network out of the equation with a simple denial of service attack – leaving insecure devices no other choice but to connect to a fake and malicious network.

Wi-Fi Pineapple: Wi-Fi Pineapple was never meant to be used for exploiting devices. The core purpose of creating the device was to pen test system vulnerabilities and find out loopholes in network or hotspot security. It was mainly used by ethical hackers to execute network penetration tests.

These penetration tests were ethical since the system proprietor was aware and had given his consent for them. In short, the access was authorized. The method gained popularity because it minimized the complexities for performing network penetration tests which required specialized software and OS. Nevertheless, the usage of this method is not confined to ethical hacking only.

If you have forgotten to turn off Wi-Fi on your device, the Pineapple will intercept your device's signal – connecting it instantly to a honeypot hotspot. It then initiates man-in-the-middle attack (explained later) by exploiting network SSID that is recognized by your device. Even though Wi-Fi Pineapple is connected to the network, the internet connection is not disrupted. Albeit the security of the network is sabotaged, a user has no idea of what is happening.

Man-In-The- Middle Attack: As the name clearly states, man-in-the-middle attack refers to an unauthorized network interception. It happens when a hacker has successfully intercepted your network signal and has now access to the information you share or receive. Think of it as someone who is overhearing everything you are saying. What you speak, he listens. What you write, he reads. Such attacks are usually backed by the motives of unauthorized access to information and identity theft. The hacker now has access to your messages, emails, and information you send over the internet.

The hacker will also gain access to your credit card information and bank details if you intend to do online shopping while your connection is compromised. Further, they can even access the information that you have saved during previous transactions. Even though the websites you visit are Hyper Text Transfer Protocol Secure, the hacker can quickly get around the encryption by either routing you to the fake version of a real site or use a couple of tricks up his sleeve to remove that HTTPS encryption altogether.

Cookie "Sidejacking": Cookie side-jacking is another form of a man-in-the-middle attack and is also known as session hijacking. During this hacking attempt, the hacker gets access to a victim's online account(s). Whenever you log in to your account, be it a social media account, online banking or any website requiring sign-in credentials, the system identifies your credentials and the server grants you requested access.

Something called a session cookie facilitates this process. This cookie is stored into your device as long as you are logged in. The moment you sign out, the server nullifies session token, requiring you to re-enter credentials the next time you sign in. Cookie side-jacking refers to the situation where a hacker steals your session token and uses it to grant himself unauthorized access to your accounts.

Worms – Hotspot: A worm is a malicious computer program that infects vulnerable networks to spread into the connected devices. The core objective of this malicious program is to replicate; thus, the moment it infects a device, it starts spreading. It usually targets devices with security vulnerabilities or loopholes and can spread through nodes.

Hence, it is safe to assume that worms can infect a device is connected to a compromised hotspot. Worms not only lower your device's performance but, also make them susceptible to hacking attempts.

Wi-Fi – Summary: Wi-Fi is a universal wireless networking technology that utilizes radio frequencies to transfer data. Wi-Fi allows high-speed Internet connections without the use of cables.

The term Wi-Fi is a contraction of "wireless fidelity" and commonly used to refer to wireless networking technology. The Wi-Fi Alliance claims rights in its uses as a certification mark for equipment certified to 802.11x standards.

Wi-Fi is a freedom – freedom from wires. It allows you to connect to the Internet from just about anywhere – a coffee shop, a hotel room, or a conference room at work. What's more – it is almost 10 times faster than a regular dial-up connection. Wi-Fi networks operate in the unlicensed 2.4 radio bands, with an 11 Mbps (802.11b) or 54 Mbps (802.11a) data rate, respectively.

To access Wi-Fi, you need Wi-Fi enabled devices (laptops or PDAs). These devices can send and receive data wirelessly in any location equipped with Wi-Fi access.

References

Useful Links on Wi-Fi

- Wi-Fi Aliance - Official website of Wi-Fi Alliance, the best resource for Wi-Fi information.
- https://www.netspotapp.com/wifi-encryption-and-security.html
- https://www.electronics-notes.com/
- WiMAX Forum – Official website of WiMAX Forum, the best resource for WiMAX information.
- IEEE 8092.16 Specification – The IEEE 802.16 Working Group on Broadband Wireless Access Standards.
- WiMAX Forum White Papers – Read more about WiMAX wireless technology and how it addresses market problems through these white papers.
- WiMAX Industry – Find latest news and market trends about WiMAX.
- WiMAX.com – Latest WiMAX news, market trends, WiMAX Forums, etc.

10

WiMAX

Summary

WiMAX is:

- Acronym for **Worldwide Interoperability for Microwave Access**.
- Based on Wireless MAN technology.
- A wireless technology optimized for the delivery of IP centric services over a wide area.
- A scalable wireless platform for constructing alternative and complementary broadband networks.
- A certification that denotes interoperability of equipment built to the IEEE 802.16 or compatible standard. The IEEE 802.16 Working Group develops standards that address two types of usage models –

  - A fixed usage model (IEEE 802.16-2004).
  - A portable usage model (IEEE 802.16e).

10.1 Introduction

In Section 9.1 of Chapter 9, an introduction to wireless technologies was given. One of the widely used wireless technologies, the Wi-Fi, was the subject of Chapter 9. The current chapter, Chapter 10, introduces another wireless technology, WiMAX.

WiMAX is one of the hottest broadband wireless technologies around today. WiMAX technology is a wireless broadband communications technology based around the IEE 802.16 standard providing high-speed data over a wide area.

The letters of WiMAX stand for **W**orldwide **I**nteroperability for **M**icrowave Access (**AX**ess), and it is a technology for point-to-multipoint wireless networking.

WiMAX technology is able to meet the needs of a large variety of users from those in developed nations wanting to install a new high-speed data network very cheaply without the cost and time required to install a wired network to those in rural areas needing fast access where wired solutions may not be viable because of the distances and costs involved, effectively providing WiMAX broadband. Additionally, it is being used for mobile applications, providing high-speed data to users on the move.

It is possible to say that WiMAX is a standardized wireless version of Ethernet intended primarily as an alternative to wire technologies (such as Cable Modems, DSL, and T1/E1 links) to provide broadband access to customer premises.

More strictly, WiMAX is an industry trade organization formed by leading communications, component, and equipment companies to promote and certify compatibility and interoperability of broadband wireless access equipment that conforms to the IEEE 802.16 and ETSI HIPERMAN standards.

WiMAX would operate similar to Wi-Fi, but at higher speeds over greater distances and for a greater number of users. WiMAX has the ability to provide service even in areas that are difficult for wired infrastructure to reach and the ability to overcome the physical limitations of traditional wired infrastructure.

WiMAX was formed in April 2001, in anticipation of the publication of the original 10.66 GHz IEEE 802.16 specifications. WiMAX is one of the family members of 802.16 as the Wi-Fi Alliance is a member of 802.11.

10.2 WiMAX and Other Wireless Technologies

The competition with WiMAX, 802.16 depends upon the type or version being used. Although initially it was thought that there could be significant competition with Wi-Fi, there are other areas to which WiMAX is posing a threat.

- **DSL cable lines:** WiMAX is able to provide high-speed data links to users, and in this way, it can pose a threat to DSL cable operators.
- **Cell phone operators:** As LTE was being developed and the initial roll-outs were taking place, cell phone operators saw the mobile version of WiMAX as a significant threat. It was even considered for adoption as the IMT 4G standard, but LTE was adopted as the standard, leaving WiMAX for fixed WiMAX broadband, last mile links and a variety of other point-to-point applications.

WiMAX technology has been deployed in many areas. Although initially seen as a candidate for 4G, its use is decreasing, although it is used as WiMAX broadband and also for last mile links.

10.3 Basics of WiMAX Technology

The standard for WiMAX technology is a standard for Wireless Metropolitan Area Networks (WMANs) that has been developed by working group number 16 of IEEE 802, specializing in point-to-multipoint broadband wireless access. Initially, 802.16a was developed and launched, but now it has been further refined. 802.16d or 802.16-2004 was released as a refined version of the 802.16a standard aimed at fixed applications. Another version of the standard, 802.16e or 802.16-2005, was also released and aimed at the roaming and mobile markets.

WiMAX broadband technology uses some key technologies to enable it to provide high-speed data rates:

- *OFDM (Orthogonal Frequency Division Multiplex):* OFDM has been incorporated into WiMAX technology to enable it to provide high-speed data without the selective fading and other issues of other forms of signal format.

  Orthogonal Frequency Division Multiplex (OFDM) is a form of signal format that uses a large number of close-spaced carriers that are each modulated with low-rate data stream. The close-spaced signals would normally be expected to interfere with each other, but by making the signals orthogonal to each other there is no mutual interference. The data to be transmitted are shared across all the carriers, and this provides resilience against selective fading from multi-path effects (see Section 10.11.7 for more information).

- *MIMO (Multiple Input Multiple Output):* WiMAX technology makes use of multipath propagation using MIMO. By utilizing the multiple signal paths that exist, the use of MIMO either enables operation with lower signal strength levels, or it allows for higher data rates.

  MIMO is a form of antenna technology that uses multiple antennas to enable signals travelling via different paths as a result of reflections to be separated and their capability used to improve the data throughput and/or the signal to noise ratio, thereby improving system performance.

The design of WiMAX network is based on the following major principles:

- **Spectrum** – able to be deployed in both licensed and unlicensed spectra.
- **Topology** – supports different Radio Access Network (RAN) topologies.
- **Interworking** – independent RAN architecture to enable seamless integration and interworking with Wi-Fi, 3GPP and 3GPP2 networks and existing IP operator core network.
- **IP connectivity** – supports a mix of IPv4 and IPv6 network interconnects in clients and application servers.
- **Mobility management** – possibility to extend the fixed access to mobility and broadband multimedia services delivery.

WiMAX has defined two MAC system profiles: the basic ATM and the basic IP. They have also defined two primary PHY system profiles, the 25-MHz-wide channel for use in (US deployments) the 10.66 GHz range, and the 28-MHz-wide channel for use in (European deployments) the 10.66 GHz range.

The WiMAX technical working group is defining MAC and PHY system profiles for IEEE 802.16a and HiperMan standards. The MAC profile includes an IP-based version for both wireless MAN (licensed) and wireless HUMAN (license-exempt). WiMAX Physical and MAC Layers are explained latter.

IEEE Standard 802.16 was designed to evolve as a set of air interfaces standards for WMAN based on a common MAC protocol, but with physical layer specifications dependent on the spectrum of use and the associated regulations.

The WiMAX framework is based on several core principles:

- Support for different RAN topologies.
- Well-defined interfaces to enable 802.16 RAN architecture independence while enabling seamless integration and interworking with Wi-Fi, 3GPP3 and 3GPP2 networks.
- Leverage and open, IETF-defined IP technologies to build scalable all-IP 802.16 access networks using common off the shelf (COTS) equipment.
- Support for IPv4 and IPv6 clients and application servers, recommending use of IPv6 in the infrastructure.
- Functional extensibility to support future migration to full mobility and delivery of rich broadband multimedia.

10.4 WiMAX History

The history of WiMAX starts back in the 1990s with the realization that there would be a significant increase in data traffic over telecommunications networks. With wired telecommunications networks being very expensive, especially in outlying areas and not installed in many countries, methods of providing wireless broadband were investigated.

WiMAX history started with these investigations into what was termed the last mile connectivity – methods of delivering high speed data to a large number of users who may have no existing wired connection.

The possibility of low cost last mile connectivity along with the possibility of a system that could handle backhaul over a wireless link proved to be a compelling argument to develop a new wireless data link system.

The next major phase in WiMAX history was the development of the standards by the IEEE.

The 802.16 standards working group was set up by the IEEE in 1999 under the IEEE 802 LAN/MAN Standards Committee. The first 802.16 standard was approved in December 2001, and this was followed by two amendments to the basic 802.16 standard. These amendments addressed issues of radio spectrum and interoperability and came under the designations 802.16a and 802.16c.

The first 802.16 standard, of December 2001, was designed to specialize point-to-multipoint broadband wireless transmission in the 10.66 GHz spectrum with only a light-of-sight (LOS) capability. But with the lack of support for non-line-of-sight (NLOS) operation, this standard is not suitable for low-frequency applications. In 2003, the IEEE 802.16a standard was published to accommodate this requirement.

In September 2003, a major revision project was commenced. This had the aim of aligning the standard with the European/ETSI HIPERMAN standard. It was also intended to incorporate conformance test specifications within the overall standard. The project was completed in 2004, and the standard was released as 802.16d, although it is often referred to as 802.16-2004 in view of the release date. With the release of the 802.16-2004 standard, the previous 802.16 documents, including a, b, and c amendments, were withdrawn.

These standards define the BWA for stationary and nomadic use which means that end devices cannot move between base stations (BS) but they can enter the network at different locations. In 2005, an amendment to 802.14-2004, the IEEE 802.16e was released to address the mobility which enables

Table 10.1   Summary of IEEE 802.16 family of standards

Standard	802.16	802.16a/802.16REVD	802.16e
Spectrum	10–66 GHz	<11 GHz	< 6 GHz
Channel conditions	Line-of-sight only	Non-line-of-sight	Non-line-of-sight
Speed (bit rate)	32 to 134 Mbps	75 Mbps max, 20-MHz channelization	15 Mbps max, 5-MHz channelization
Modulation	QPSK, 16-QAM, 64-QAM	OFDM 256 subcarrier QPSK, 16-QAM, 64-QAM	Same as 802.16a
Mobility	Fixed	Fixed	Pedestrian mobility, regional roaming
Channel bandwidths	20, 25, and 28 MHz	Selectable between 1.25 and 20 MHz	Same as 802.16a with sub-channels
Typical cell radius	1–3 miles	3–5 miles (up to 30 miles, depending on tower height, antenna gain, and transmit power)	1–3 miles

mobile stations (MB) to handover between BSs while communicating. This standard is often called "Mobile WiMAX7#8221." Table 10.1 provides a summary of the IEEE 802.16 family of standards.

10.5 WiMAX Versions

Since its initial conception, new applications for WiMAX have been developed, and as a result, there are two types of WiMAX technology that are available: *802.16d and 802.16e.*

The two types of WiMAX broadband technology are used for different applications, and although they are based on the same standard, the implementation of each has been optimized to suit its particular application.

- *802.16d – DSL replacement:* The 802.16d version is often referred to as 802.16-2004, and it is closer to what may be termed the original version of WiMAX defined under 802.16a. It is aimed at fixed applications, providing a wireless equivalent of DSL broadband data – often called WiMAX broadband. In fact, the WiMAX Forum describes the technology as "a standards-based technology enabling the delivery of last mile wireless broadband access as an alternative to cable and DSL."

- 802.16d is able to provide data rates of up to 75 Mbps, and as a result, it is ideal for fixed, DSL replacement applications as WiMAX broadband. It may also be used for backhaul where the final data may be distributed further to individual users. Cell radii are typically up to 75 km.
- *802.16e – Nomadic/Mobile:* While 802.16/WiMAX was originally envisaged as being a fixed only technology, with the need for people on the move requiring high-speed data at a cost less than that provided by cellular services and opportunity for a mobile version was seen and 802.16e was developed. This standard is also widely known as 802.16-2005. It currently provides the ability for users to connect to a WiMAX cell from a variety of locations, and there are future enhancements to provide cell handover. 802.16e is able to provide data rates up to 15 Mbps, and the cell radius distances are typically between 2 and 4 km.

10.6 Benefits of Using WiMAX

- WiMAX can satisfy a variety of access needs. Potential applications include extending broadband capabilities to bring them closer to subscribers, filling gaps in cable, DSL and T1 services, Wi-Fi, and cellular backhaul, providing last 100 m access from fiber to the curb and giving service providers another cost-effective option for supporting broadband services.
- WiMAX can support very high bandwidth solutions where large spectrum deployments (i.e., >10 MHz) are desired using existing infrastructure keeping costs down while delivering the bandwidth needed to support a full range of high-value multimedia services.
- WiMAX can help service providers meet many of the challenges they face due to increasing customer demands without discarding their existing infrastructure investments because it has the ability to seamlessly interoperate across various network types.
- WiMAX can provide wide area coverage and quality of service capabilities for applications ranging from real-time delay-sensitive voice-over-IP (VoIP) to real-time streaming video and non-real-time downloads, ensuring that subscribers obtain the performance they expect for all types of communications.
- WiMAX, which is an IP-based wireless broadband technology, can be integrated into both wide-area third-generation (3G) mobile and wireless and wireline networks allowing it to become part of a seamless anytime, anywhere broadband access solution.

- Ultimately, WiMAX is intended to serve as the next step in the evolution of 3G mobile phones, via a potential combination of WiMAX and CDMA standards called 4G.

10.7 WiMAX Features

WiMAX is a wireless broadband solution that offers a rich set of features with a lot of flexibility in terms of deployment options and potential service offerings. Some of the more features that deserve highlighting are as follows:

a. Two Types of Services

WiMAX can provide two forms of wireless service:

- **Non-line-of-sight** service is a Wi-Fi sort of service. Here a small antenna on your computer connects to the WiMAX tower. In this mode, WiMAX uses a lower frequency range – 2 to 11 GHz (similar to Wi-Fi).
- **Line-of-sight** service is where a fixed dish antenna points straight at the WiMAX tower from a rooftop or pole. The line-of-sight connection is stronger and more stable, so it's able to send a lot of data with fewer errors. Line-of-sight transmissions use higher frequencies, with ranges reaching a possible 66 GHz.

b. OFDM-based Physical Layer

The WiMAX physical layer (PHY) is based on orthogonal frequency division multiplexing (OFDM), a scheme that offers good resistance to multipath, and allows WiMAX to operate in NLOS conditions.

Mobile WiMAX uses orthogonal frequency division multiple access (OFDM) as a multiple-access technique, whereby different users can be allocated different subsets of the OFDM tones.

c. Very High Peak Data Rates

WiMAX is capable of supporting very high peak data rates. In fact, the peak PHY data rate can be as high as 74 Mbps when operating using a 20 MHz wide spectrum.

More typically, using a 10 MHz spectrum operating using TDD scheme with a 3:1 downlink-to-uplink ratio, the peak PHY data rate is about 25 Mbps and 6.7 Mbps for the downlink and the uplink, respectively.

d. Scalable Bandwidth and Data Rate Support

WiMAX has a scalable physical-layer architecture that allows for the data rate to scale easily with available channel bandwidth.

For example, a WiMAX system may use 128-, 512-, or 1,048-bit FFTs (Fast Fourier Transforms) based on whether the channel bandwidth is 1.25, 5, or 10 MHz, respectively. This scaling may be done dynamically to support user roaming across different networks that may have different bandwidth allocations.

e. Adaptive Modulation and Coding (AMC)

WiMAX supports a number of modulation and forward error correction (FEC) coding schemes and allows the scheme to be changed as per user and per frame basis, based on channel conditions.

AMC is an effective mechanism to maximize throughput in a time-varying channel.

f. Link-layer Retransmissions

WiMAX supports automatic retransmission requests (ARQ) at the link layer for connections that require enhanced reliability. ARQ-enabled connections require each transmitted packet to be acknowledged by the receiver; unacknowledged packets are assumed to be lost and are retransmitted.

g. Support for TDD and FDD

IEEE 802.16-2004 and IEEE 802.16e-2005 support both time division duplexing and frequency division duplexing, as well as a half-duplex FDD, which allows for a low-cost system implementation.

h. Flexible and Dynamic per User Resource Allocation

Both uplink and downlink resource allocation are controlled by a scheduler in the base station. Capacity is shared among multiple users on a demand basis, using a burst TDM scheme.

i. Support for Advanced Antenna Techniques

The WiMAX solution has a number of hooks built into the physical-layer design, which allows for the use of multiple-antenna techniques, such as beam-forming, space-time coding, and spatial multiplexing.

j. Quality-of-service Support

The WiMAX MAC layer has a connection-oriented architecture that is designed to support a variety of applications, including voice and multimedia services.

WiMAX system offers support for constant bit rate, variable bit rate, real-time, and non-real-time traffic flows, in addition to best-effort data traffic.

WiMAX MAC is designed to support a large number of users, with multiple connections per terminal, each with its own QoS requirement.

k. Robust Security

WiMAX supports strong encryption, using Advanced Encryption Standard (AES), and has a robust privacy and key-management protocol.

The system also offers a very flexible authentication architecture based on **Extensible Authentication Protocol (EAP)**, which allows for a variety of user credentials, including username/password, digital certificates, and smart cards.

l. Support for Mobility

The mobile WiMAX variant of the system has mechanisms to support secure seamless handovers for delay-tolerant full-mobility applications, such as VoIP.

m. IP-based Architecture

The WiMAX Forum has defined a reference network architecture that is based on an all-IP platform. All end-to-end services are delivered over an IP architecture relying on IP-based protocols for end-to-end transport, QoS, session management, security, and mobility.

10.8 WiMAX – Building Blocks

A WiMAX system consists of two major parts:

- A WiMAX base station
- A WiMAX receiver

10.8.1 WiMAX Base Station

A WiMAX base station consists of indoor electronics and a WiMAX tower similar in concept to a cell-phone tower. A WiMAX base station can provide coverage to a very large area up to a radius of 6 miles. Any wireless device within the coverage area would be able to access the Internet.

The WiMAX base stations would use the MAC layer defined in the standard, a common interface that makes the networks interoperable and would allocate uplink and downlink bandwidth to subscribers according to their needs, on an essentially real-time basis.

Each base station provides wireless coverage over an area called a cell. Theoretically, the maximum radius of a cell is 50 km or 30 miles; however, practical considerations limit it to about 10 km or 6 miles.

10.8.2 WiMAX Receiver

A WiMAX receiver may have a separate antenna or could be a stand-alone box or a PCMCIA card sitting in your laptop or computer or any other device. This is also referred as customer premise equipment (CPE).

WiMAX base station is similar to accessing a wireless access point in a Wi-Fi network, but the coverage is greater.

10.8.3 Backhaul

A WiMAX tower station can connect directly to the Internet using a high-bandwidth, wired connection (e.g., a T3 line). It can also connect to another WiMAX tower using a line-of-sight microwave link.

Backhaul refers both to the connection from the access point back to the base station and to the connection from the base station to the core network.

It is possible to connect several base stations to one another using high-speed backhaul microwave links. This would also allow for roaming by a WiMAX subscriber from one base station coverage area to another, similar to the roaming enabled by cell phones.

10.9 WiMAX – Reference Network Model

The IEEE 802.16e-2005 standard provides the air interface for WiMAX, but does not define the full end-to-end WiMAX network. The WiMAX Forum's Network Working Group (NWG) is responsible for developing the end-to-end network requirements, architecture, and protocols for WiMAX, using IEEE 802.16e-2005 as the air interface.

The standard now used is available from the WiMAX Forum as WiMAX Forum Network Architecture, document: WMF – T32-002-R010v04, and it is dated February 03, 2009.

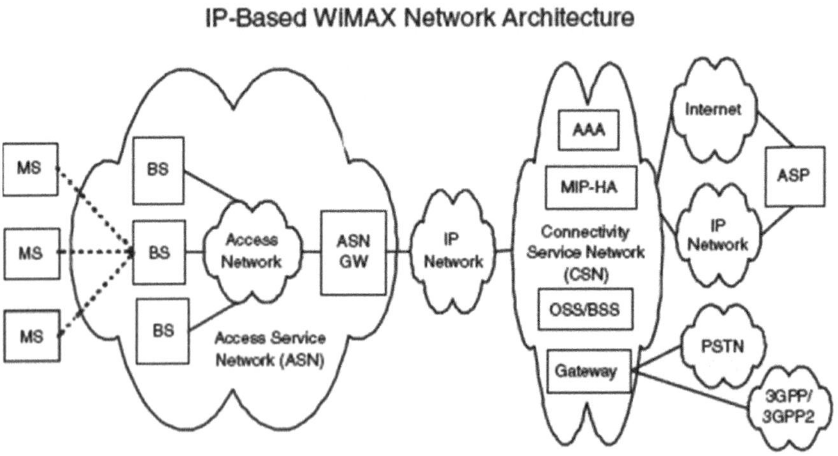

IP-Based WiMAX Network Architecture

Figure 10.1   IP-based WiMAX network architecture.

The WiMAX NWG has developed a network reference model to serve as an architecture framework for WiMAX deployments and to ensure interoperability among various WiMAX equipment and operators. The WiMAX network architecture is based upon an all-IP model.

The network reference model envisions a unified network architecture for supporting fixed, nomadic, and mobile deployments and is based on an IP service model. Below is a simplified illustration of an IP-based WiMAX network architecture.

10.9.1 WiMAX Network Architecture Major Elements or Areas

The network reference model developed by the WiMAX Forum NWG defines a number of functional entities and interfaces between those entities. Figure 10.1 shows some of the more important functional entities.

The overall network architecture comprises three major elements or areas:

- **Base station (BS):** The BS is responsible for providing the air interface to the MS. Additional functions that may be part of the BS are micro-mobility management functions, such as handoff triggering and tunnel establishment, radio resource management, QoS policy enforcement, traffic classification, DHCP (Dynamic Host Control Protocol) proxy, key management, session management, and multicast group management.
- **Access service network gateway (ASN-GW):** The ASN gateway typically acts as a layer 2 traffic aggregation point within an ASN.

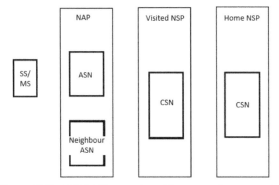

Figure 10.2 WiMAX network architecture reference model.

Additional functions that may be part of the ASN gateway include intra-ASN location management and paging, radio resource management, and admission control, caching of subscriber profiles, and encryption keys, AAA client functionality, establishment, and management of mobility tunnel with base stations, QoS and policy enforcement, foreign agent functionality for mobile IP, and routing to the selected CSN.

- **Connectivity service network (CSN):** The CSN provides connectivity to the Internet, ASP, other public networks, and corporate networks. The CSN is owned by the NSP and includes AAA servers that support authentication for the devices, users, and specific services. The CSN also provides per user policy management of QoS and security. The CSN is also responsible for IP address management, support for roaming between different NSPs, location management between ASNs, and mobility and roaming between ASNs.

The WiMAX architecture framework allows for the flexible decomposition and/or combination of functional entities when building the physical entities. For example, the ASN may be decomposed into base station transceivers (BST), base station controllers (BSC), and an ASNGW analogous to the GSM model of BTS, BSC, and Serving GPRS Support Node (SGSN).

10.9.2 WiMAX Network Architecture Major Entities

The WiMAX architecture developed by the WiMAX Forum Support is a unified network architecture that supports fixed, nomadic, and mobile operations. The WiMAX network architecture is based upon an all-IP model.

The overall WiMAX network comprises a number of different entities that make up the different major areas described above. These include the following entities:

- ***Subscriber Station (SS)/Mobile Station (MS):*** The Subscriber station (SS) may often be referred to as the Customer Premises Equipment (CPE). These take a variety of forms and these may be termed "indoor CPE" or "outdoor CPE" – the terminology is self-explanatory. The outdoor CPE has the advantage that it provides better performance as a result of the better position of the antenna, whereas the indoor CPE can be installed by the user. Mobile Stations may also be used. These are often in the form of a dongle for a laptop.
- ***Base Station (BS):*** The base-station forms an essential element of the WiMAX network. It is responsible for providing the air interface to the subscriber and mobile stations. It provides additional functionality in terms of micro-mobility management functions, such as handoff triggering and tunnel establishment, radio resource management, QoS policy enforcement, traffic classification, DHCP (Dynamic Host Control Protocol) proxy, key management, session management, and multicast group management.
- ***ASN Gateway (ASN-GW):*** The ASN gateway within the WiMAX network architecture typically acts as a layer 2 traffic aggregation point within the overall ASN.

The ASN-GW may also provide additional functions that include intra-ASN location management and paging, radio resource management and admission control, caching of subscriber profiles and encryption keys. The ASN-GW may also include the AAA client functionality (see below), establishment and management of mobility tunnel with base stations, QoS and policy enforcement, foreign agent functionality for mobile IP, and routing to the selected CSN.

- ***Home Agent (HA):*** The Home Agent within the WiMAX network is located within the CSN. With Mobile-IP forming a key element within WiMAX technology, the Home Agent works in conjunction with a "Foreign Agent," such as the ASN Gateway, to provide an efficient end-to-end Mobile IP solution. The Home Agent serves as an anchor point for subscribers, providing secure roaming with QOS capabilities.
- ***Authentication, Authorization, and Accounting Server (AAA):*** As with any communications or wireless system requiring subscription services,

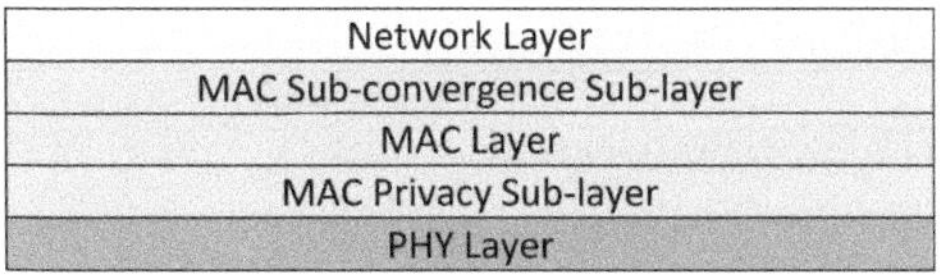

| Network Layer |
| MAC Sub-convergence Sub-layer |
| MAC Layer |
| MAC Privacy Sub-layer |
| PHY Layer |

Figure 10.3 The IEEE 802.16 protocol structure.

an Authentication, Authorization and Accounting server is used. This is included within the CSN.

10.10 IEEE 802.16 Protocol Architecture

The IEEE 802.16 protocol architecture is structured into two main layers: the Medium Access Control (MAC) layer and the Physical (PHY) layer besides the Network Layer, as shown in Figure 10.3.

MAC layer consists of three sub-layers. The first sub-layer is Service-Specific Convergence Sub-layer (CS), which maps higher level data services to MAC layer service flow and connections. The second sub-layer is Common Part Sub-layer (CPS), which is the core of the standard and is tightly integrated with the security sub-layer (MAC Privacy Sub-layer). This layer defines the rules and mechanisms for system access, bandwidth allocation, and connection management. The MAC protocol data units are constructed in this sub-layer. The last sub-layer of MAC layer is the Security Sub-layer or Privacy Sub-layer, which lies between the MAC CPS and the PHY layer, addressing the authentication, key establishment and exchange, encryption and decryption of data exchanged between MAC and PHY layers.

The PHY layer provides a two-way mapping between MAC protocol data units and the PHY layer frames received and transmitted through coding and modulation of radio frequency signals.

10.11 WiMAX – Physical Layer

The WiMAX physical layer is based on orthogonal frequency division multiplexing. OFDM is the transmission scheme of choice to enable high-speed data, video, and multimedia communications and is used by a variety of commercial broadband systems, including DSL, Wi-Fi, Digital Video Broadcast-Handheld (DVB-H), and MediaFLO, besides WiMAX.

OFDM is an elegant and efficient scheme for high data rate transmission in a non-line-of-sight or multipath radio environment.

Table 10.2 Modulation and coding schemes supported by WiMAX

	Downlink	Uplink
Modulation	BPSK, QPSK, 16 QAM, 64 QAM; BPSK optional for OFDMA-PHY	BPSK, QPSK, 16 QAM; 64 QAM optional
Coding	Mandatory: convolutional codes at rate 1/2, 2/3, 3/4, 5/6 Optional: convolutional turbo codes at rate 1/2, 2/3, 3/4, 5/6; repetition codes at rate 1/2, 1/3, 1/6, LDPC, RS-Codes for OFDM-PHY	Mandatory: convolutional codes at rate 1/2, 2/3, 3/4, 5/6 Optional: convolutional turbo codes at rate 1/2, 2/3, 3/4, 5/6; repetition codes at rate 1/2, 1/3, 1/6, LDPC

10.11.1 Adaptive Modulation and Coding in WiMAX

WiMAX modulation and coding is adaptive, enabling it to vary these parameters according to prevailing conditions. WiMAX modulation and coding can be changed on a burst by burst basis per link, depending on channel conditions. Using the channel quality feedback indicator, the mobile can provide the base station with feedback on the downlink channel quality. For the uplink, the base station can estimate the channel quality, based on the received signal quality.

Table 10.2 provides a list of the various modulation and coding schemes supported by WiMAX.

10.11.2 PHY-layer Data Rates

Because the physical layer of WiMAX is quite flexible, data rate performance varies based on the operating parameters. Parameters that have a significant impact on the physical-layer data rate are channel bandwidth and the modulation and coding scheme used. Other parameters, such as number of sub-channels, OFDM guard time, and oversampling rate, also have an impact.

Table 10.3 is the PHY-layer data rate at various channel bandwidths, as well as modulation and coding schemes.

10.11.3 WiMAX Frequencies and Spectrum Allocations

The IEEE 802.16 WiMAX standard allows data transmission using multiple wireless broadband frequency ranges. The original 802.16a standard specified transmissions in the range 10–66 GHz, but 802.16d allowed lower frequencies in the range 2 to 11 GHz. The lower frequencies used in the later specifications means that the signals suffer less from attenuation and therefore they provide improved range and better coverage within buildings.

Table 10.3   PHY-layer channel bandwidths, modulation, and coding schemes

Channel Bandwidth	3.5 MHz		1.25 MHz		5 MHz		10 MHz	
PHY mode	256 OFDM		128 OFDM		512 OFDM		1,024 OFDM	
Oversampling	8/7		28/25		28/25		28/25	
Modulation and code rate	PHY-layer data rate (Kbps)							
	DL	UL	DL	UL	DL	UL	DL	UL
BPSK, 1/2	946	326			Not applicable			
QPSK, 1/2	1,882	653	504	154	2,520	653	5,040	1,344
QPSK, Â¿	2,822	979	756	230	3,780	979	7,560	2,016
16 QAM, 1/2	3,763	1,306	1,008	307	5,040	1,306	10,080	2,688
16 QAM 3/4	5,645	1,958	1,512	461	7,560	1,958	15,120	4,032
64 QAM, 1/2	5,645	1,958	1,512	461	7,560	1,958	15,120	4,032
64 QAM, 2/3	7,526	2,611	2,016	614	10,080	2,611	20,160	5,376
64 QAM, 3/4	8,467	2,938	2,268	691	11,340	2,938	22,680	6,048
64 QAM, 5/6	9,408	3,264	2,520	768	12,600	3,264	25,200	6,720

This brings many benefits to those using these data links within buildings and means that external antennas are not required.

Different bands are available for WiMAX applications in different parts of the world. The frequencies commonly used are 3.5 and 5.8 GHz for 802.16d and 2.3, 2.5, and 3.5 GHz for 802.16e but the use depends upon the countries as shown in Table 10.4.

10.11.4 WiMAX Speed and Range

WiMAX is expected to offer initially up to about 40 Mbps capacity per wireless channel for both fixed and portable applications, depending on the particular technical configuration chosen, enough to support hundreds of businesses with T-1 speed connectivity and thousands of residences with DSL speed connectivity. WiMAX can support voice and video as well as Internet data.

WiMAX developed to provide wireless broadband access to buildings, either in competition to existing wired networks or alone in currently unserved rural or thinly populated areas. It can also be used to connect WLAN hotspots to the Internet. WiMAX is also intended to provide broadband connectivity to mobile devices. It would not be as fast as in these fixed applications, but expectations are for about 15 Mbps capacity in a 3 km cell coverage area.

Table 10.4   WiMAX frequency bands at different countries

Region	Frequency bands (GHz)	Comments
Canada	2.3 2.5 3.5 5.8	
USA	2.3 2.5 5.8	
Central and South America	2.5 3.5 5.8	The spectrum is very fragmented and allocations vary from country to country
Europe	2.5 3.5 3.5	The spectrum is very fragmented and varies from country to country. The 2.5 GHz allocation is currently allocated to IMT 2000. 5.8 GHz is also not available in most European countries.
Middle East and Africa	2.5 5.8	The spectrum is very fragmented.
Russia	2.5 3.5 5.8	The 2.5 GHz allocation is currently allocated to IMT 2000.
Asia Pacific (China, India, Australia, etc.)	2.3 2.5 3.3 3.5 5.8	The spectrum is very fragmented and varies between countries.

With WiMAX, users could really cut free from today's Internet access arrangements and be able to go online at broadband speeds, almost wherever they like from within a MetroZone.

WiMAX could potentially be deployed in a variety of spectrum bands: 2.3, 2.5, 3.5, and 5.8 GHz

10.11.5 WiMAX Data Structure

Although WiMAX can be deployed as TDD (Time Division Duplex), FDD (Frequency Division Duplex), and half-duplex FDD, the most common arrangement is the TDD mode. This allows for a greater efficiency in spectrum usage than FDD mode.

Using TDD mode the WiMAX base station and the end users transmit on the same frequency, but to enable them not to interfere with each other their transmissions are separated in time. In order to achieve this, the base station first transmits a subframe, and this is followed by a short gap which is called the transmit/receive transition gap (TTG). After this gap, the users or remote stations are able to transmit their subframes. The timing of these "uplink" subframes needs to be accurately controlled and synchronized so that they do not overlap whatever distance they are from the base station. Once all the uplink subframes have been transmitted, another short gap known as the receive/transmit transition gap (RTG) is left before the base station transmits again.

There are slight differences between the WiMAX subframes transmitted on the uplink and downlink. The downlink subframe begins with a preamble, after which a header is transmitted and this is followed by one or more bursts of data. The modulation within a subframe may change, but it remains the same within an individual burst. Nevertheless it is possible for the modulation type to change from one burst to the next. The first bursts to be transmitted use the more resilient forms of modulation such as BPSK and QPSK. Later bursts may use the less resilient forms of modulation such as 16 QAM and 64 QAM that enable more data to be carried.

Using this RF interface, WiMAX is able to provide a very effective form of wireless broadband system that can be sued in many areas.

10.11.6 WiMAX – IEEE Standards

The IEEE 802.16, the *Air Interface for Fixed Broadband Wireless Access Systems*, also known as the IEEE WirelessMAN air interface, is an emerging suite of standards for fixed, portable, and mobile BWA in MAN.

These standards are issued by IEEE 802.16 work group that originally covered the wireless local loop (WLL) technologies in the 10.66 GHz radio spectrum, which were later extended through amendment projects to include both licensed and unlicensed spectra from 2 to 11 GHz.

The WiMAX umbrella currently includes 802.16-2004 and 802.16e. 802.16-2004 utilizes OFDM to serve multiple users in a time division fashion in a sort of a round-robin technique, but done extremely quickly so that users have the perception that they are always transmitting/receiving. 802.16e utilizes OFDMA and can serve multiple users simultaneously by allocating sets of *tones* to each user.

Table 10.5 is the chart of various IEEE 802.16 Standards related to WiMAX.

Note: The IEEE 802.16 standards for BWA provide the possibility for inter-operability between equipment from different vendors, which is in contrast to the previous BWA industry, where proprietary products with high prices are dominant in the market.

Table 10.5   Various IEEE 802.16 standard related to WiMAX

	802.16	802.16a	802.16e
Spectrum	10–66 GHz	2–11 GHz	< 6 GHz
Configuration	Line-of-sight	Non-line-of-sight	Non-line-of-sight
Bit rate	32 to 134 Mbps (28 MHz Channel)	= 70 or 100 Mbps (20 MHz channel)	Up to 15 Mbps
Modulation	QPSK, 16-QAM. 64-QAM	256 sub-carrier OFDM using QPSK, 16-QAM, 64-QAM, 256-QAM	
Mobility	Fixed	Fixed	= 75 MPH
Channel bandwidth	20, 25, 28 MHz	Selectable 1.25 to 20 MHz	5 MHz (Planned)
Typical cell radius	1–3 miles	3–5 miles	1–3 miles
Completed	Dec., 2001	Jan., 2003	2nd half of 2005

10.11.7 OFDM Basics

OFDM belongs to a family of transmission schemes called multicarrier modulation, which is based on the idea of dividing a given high-bit-rate data stream into several parallel lower bit-rate streams and modulating each stream on separate carriers, often called sub-carriers or tones.

Multicarrier modulation schemes eliminate or minimize inter-symbol interference (ISI) by making the symbol time large enough so that the channel-induced delays are an insignificant (typically, <10%) fraction of the symbol duration.

Therefore, in high-data-rate systems in which the symbol duration is small, being inversely proportional to the data rate, splitting the data stream into many parallel streams increases the symbol duration of each stream such that the delay spread is only a small fraction of the symbol duration.

OFDM is a spectrally efficient version of multicarrier modulation, where the subcarriers are selected such that they are all orthogonal to one another over the symbol duration, thereby avoiding the need to have non-overlapping subcarrier channels to eliminate inter-carrier interference.

In order to completely eliminate ISI, guard intervals are used between OFDM symbols. By making the guard interval larger than the expected multipath delay spread, ISI can be completely eliminated. Adding a guard interval, however, implies power wastage and a decrease in bandwidth efficiency.

10.12 WiMAX – MAC Layer

The WiMAX MAC layer or IEEE 802.16 MAC is an essential elements within the overall WiMAX software stack. These elements enable WiMAX to perform as an effective wireless broadband system.

The WiMAX MAC layer is a form of MAC used for the WiMAX system.

A MAC layer or Media Access Control data communication protocol sub-layer may also be known as a Medium Access Control layer.

A MAC layer is a sub-layer of the Data Link Layer. This is defined in the standard seven-layer OSI model as layer 2. The MAC layer provides addressing and channel access control mechanisms that make it possible for several terminals or network nodes to communicate within a multi-point network, typically a local area network (LAN) or metropolitan area network (MAN).

The 802.16 MAC is designed for point-to-multipoint (PMP) applications and is based on collision sense multiple access with collision avoidance (CSMA/CA).

The primary task of the WiMAX MAC layer is to provide an interface between the higher transport layers and the physical layer.

The MAC layer takes packets from the upper layer, these packets are called MAC service data units (MSDUs) and organizes them into MAC protocol data units (MPDUs) for transmission over the air. For received transmissions, the MAC layer does the reverse.

The IEEE 802.16-2004 and IEEE 802.16e-2005 MAC designs include a convergence sublayer that can interface with a variety of higher-layer protocols, such as ATM TDM Voice, Ethernet, IP, and any unknown future protocol.

10.12.1 Requirements Needed from MAC Layer to Meet

The WiMAX MAC layer has to meet a number of requirements:

- *Point to multipoint:* One of the main requirements for WiMAX is that it must be possible for a base station to communicate with a number of different outlying users, either fixed or mobile. To achieve this, the IEEE 802.16, WiMAX MAC layer is based on collision sense multiple access with collision avoidance, CSMA/CA to provide the point to multipoint, PMP capability.
- *Connection orientated*
- *Supports communication in all conditions:* The WiMAX MAC layer must be able to support a large number of users along with high data

rates. As the traffic is packet data orientated it must be able to support both continuous and "bursty" traffic. Most data traffic is "bursty" in nature having short times of high data rates then remaining dormant for a short while.

- *Efficient spectrum use:* The WiMAX MAC must be capable of supporting methods that enable very efficient use of the spectrum.
- *Variety of QoS options:* To provide the support for different forms of traffic from voice data to Internet surfing, etc., a variety of different classes and forms of QoS support are needed. Support for QoS is a fundamental part of the WiMAX MAC-layer. The WiMAX MAC utilizes some of the concepts that are embedded in the DOCSIS cable modem standard.
- *Multiple WiMAX/IEEE 802.16 physical layers:* With different variants, the WiMAX MAC layer must be able to provide support for the different PHYs.

10.12.2 Some Features of MAC Layer

The MAC incorporates several features suitable for a broad range of applications at different mobility rates, such as the following:

- Privacy key management (PKM) for MAC layer security. PKM version 2 incorporates support for extensible authentication protocol (EAP).
- Broadcast and multicast support.
- Manageability primitives.
- High-speed handover and mobility management primitives.
- Three power management levels, normal operation, sleep, and idle.
- Header suppression, packing and fragmentation for an efficient use of spectrum.
- Five service classes, unsolicited grant service (UGS), real-time polling service (rtPS), non-real-time polling service (nrtPS), best effort (BE), and Extended real-time variable rate (ERT-VR) service.

These features combined with the inherent benefits of scalable OFDMA make 802.16 suitable for high-speed data and bursty or isochronous IP multimedia applications.

Support for QoS is a fundamental part of the WiMAX MAC-layer design. WiMAX borrows some of the basic ideas behind its QoS design from the DOCSIS cable modem standard.

Strong QoS control is achieved by using a connection-oriented MAC architecture, where all downlink and uplink connections are controlled by the serving BS.

WiMAX also defines a concept of a service flow. A service flow is a unidirectional flow of packets with a particular set of QoS parameters and is identified by a *service flow identifier* (SFID).

10.12.3 WiMAX MAC Layer Operation

As shown in Figure 10.3 the WiMAX MAC layer is primarily an adaptation layer between the physical layer and the upper layers within the overall stack.

One of the main tasks of the WiMAX MAC layer is to transfer data between the various layers.

- Transmission of data – reception of MAC Service Data Units, MSDUs from the layer above. It then aggregates and encapsulates them into MAC Protocol Data Units, MPDUs, before passing them to the physical layer, PHY for transmission.
- Reception of data – the WiMAX MAC layer takes MPDUs from the physical layer. It decapsulates and reorganizes them into MSDUs, and then passes them on to the upper-layer protocols.

For the different formats: IEEE 802.16-2004 and IEEE 802.16e-2005, the WiMAX MAC design includes a convergence sublayer. This is used to interface with a variety of higher-layer protocols, such as ATM, Ethernet, IP, TDM Voice, and other future protocols that may arise.

WiMAX defines a concept of a service flow and has an accompanying Service Flow Identifier, SFID. The service flow is a unidirectional flow of packets with a particular set of QoS parameters, and the identifier is used to identify the flow to enable operation.

There is an additional layer between the WiMAX MAC itself and the upper layers. This is called the Convergence Sublayer. For the upper protocol layers, the convergence sublayer acts as an interface to the WiMAX MAC. Currently the convergence sublayer only supports IP and Ethernet, although other protocols can be supported by encapsulating the data.

The WiMAX MAC layer provides for a flexible allocation of capacity to different users. It is possible to use variably sized MPDUs from different flows – these can be included into one data burst before being handed over to the PHY layer for transmission. Also, multiple small MSDUs can be aggregated into one larger MPDU. Conversely, one big MSDU can be fragmented into multiple small ones in order to further enhance system performance. This level of flexibility gives significant improvements in overall efficiency.

We are going to discuss again the layers of the stack while discussing Protocol Architecture and Security Solutions.

10.12.4 WiMAX MAC Connection Identifier

Before any data is transferred over a WiMAX link, the user equipment or mobile station and the base station must create a connection between the WiMAX MAC layers of the two stations. To achieve this, an identifier known as a Connection Identifier, CID, is generated and assigned to each uplink/downlink connection. The CID serves as an intermediate address for the data packets transmitted over the WiMAX link.

There is another identifier used within the WiMAX MAC layer. This identifier is known as the Service Flow Identifier, SFID and is assigned to unidirectional packet data traffic by the base station. It is worth noting that the base station WiMAX MAC layer also handles the mapping of the SFIDs to CIDs to provide the required quality of service.

The WiMAX MAC layer also incorporates a number of other features including power-management techniques and security features.

The WiMAX MAC layer has been developed to provide the functionality required for a point to multipoint system and to provide wireless broadband. The WiMAX MAC layer is also able to provide support for the different physical layers needed for the different flavors of WiMAX that are in use.

10.13 WiMAX – Mobility Support

WiMAX envisions four mobility-related usage scenarios:

- **Nomadic:** The user is allowed to take a fixed subscriber station and reconnect from a different point of attachment.
- **Portable:** Nomadic access is provided to a portable device, such as a PC card, with expectation of a best-effort handover.
- **Simple mobility:** The subscriber may move at speeds up to 60 kmph with brief interruptions (less than 1 sec) during handoff.
- **Full mobility:** Up to 120 kmph mobility and seamless handoff (less than 50 ms latency and $< 1\%$ packet loss) is supported.

It is likely that WiMAX networks will initially be deployed for fixed and nomadic applications and then evolve to support portability to full mobility over time.

The IEEE 802.16e-2005 standard defines a framework for supporting mobility management. In particular, the standard defines signaling mechanisms for tracking subscriber stations as they move from the coverage range of one base station to another when active or as they move from one paging group to another when idle.

The standard also has protocols to enable a seamless handover of ongoing connections from one base station to another.

The standard also has protocols to enable a seamless handover of ongoing connections from one base station to another. The WiMAX Forum has used the framework defined in IEEE 802.16e-2005, to further develop mobility management within an end-to-end network architecture framework. The architecture also supports IP-layer mobility using mobile IP.

10.14 WiMAX: Protocol Architecture And Security Solutions

In order to understand WiMAX security issues it is needed at first to understand WiMAX architecture and how securities specifications are addressed in WiMAX. WiMAX architecture was given in Section 10.9.

We start this section by considering the architecture of the two main layers of WiMAX: Physical layer and MAC layer.

The reason is the fact that WiMAX gives securities specifications of each sub-layer.

10.14.1 IEEE 802.16 Protocol Architecture

The IEEE 802.16 protocol architecture is structured into two main layers: the Medium Access Control (MAC) layer and the Physical (PHY) layer, as shown in Figure 10.4.

From the figure and as mentioned before, MAC layer consists of three sub-layers:

- Service Specific Convergence Sub-layer (CS), which maps higher level data services to MAC layer service flow and connections.

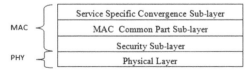

Figure 10.4 The IEEE 802.16 protocol structure.

- Common Part Sub-layer (CPS), which is the core of the standard and is tightly integrated with the security sub-layer. This layer defines the rules and mechanisms for system access, bandwidth allocation and connection management.
- Security (Privacy) Sub-layer, which lies between the MAC CPS and the PHY layer, addressing security issues as authentication, key establishment and exchange, encryption and decryption of data exchanged between MAC and PHY layers.

The PHY layer provides a two-way mapping between MAC protocol data units and the PHY layer frames received and transmitted through coding and modulation of radio frequency signals.

10.14.2 WiMAX Security Solutions

By adopting the best technologies available today, the WiMAX, based on the IEEE 802.16e standard, provides strong support for authentication, key management, encryption and decryption, control and management of plain text protection and security protocol optimization. In WiMAX, most of security issues are addressed and handled in the MAC security sub-layer as described in Figure 10.5:

Two main entities in WiMAX, including Base Station (BS) and Subscriber Station (SS), are protected by the following WiMAX security features:

a. **Security association:** A security association (SA) is a set of security information parameters that a BS and one or more of its client SSs share. Each SA has its own identifier (SAID) and also contains a cryptographic suite identifier (for selected algorithms), traffic encryption keys (TEKs) and initialization vectors.

b. **Public key infrastructure:** WiMAX uses the Privacy and Key Management Protocol (PKM) for secure key management, transfer and

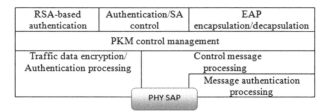

Figure 10.5   MAC security sub-layer.

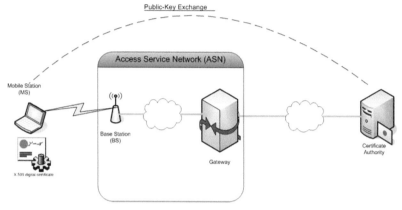

Figure 10.6 Public key infrastructure in WiMAX [Image from http://2008.telfor.rs/files/rad ovi/02_32.pdf].

exchange between mobile stations. This protocol also authenticates an SS to a BS. The PKM protocol uses X.509 digital certificates, RSA (Rivest–Shamir–Adleman) public-key algorithm and a strong encryption algorithm [Advanced Encryption Standard (AES)]. The initial draft version of WiMAX uses PKMv1 which is a one-way authentication method and has a risk for Man-in-the-middle (MITM) attack. To deal with this issue, in the later version (802.16e), the PKMv2 was used to provide two-way authentication mechanism. Figure 10.6 provides an overview of public key infrastructure in WiMAX.

c. **Device/user authentication:** Generally, WiMAX supports three types of authentication which are handled in the security sub-layer. The three authentication types are as follows:

    i. Type 1 – RSA-based Authentication: The first type is RSA-based authentication which applies X.509 certificates together with RSA encryption. The X.509 certificate is issued by the SS manufacturer and contains the SS's public key (PK) and its MAC address. When requesting an Authorization Key (AK), the SS sends its digital certificate to the BS, the BS validates the certificate, and then uses the verified PK to encrypt an AK and pass it to the SS.

    ii. Type 2 – EAP-based Authentication: The second type is EAP (Extensive Authentication Protocol) based authentication in which the SS is authenticated by an X.509 certificate or by a unique operator-issued credential such as a SIM, USIM or even by username/password, Figure 10.7. The network operator can choose

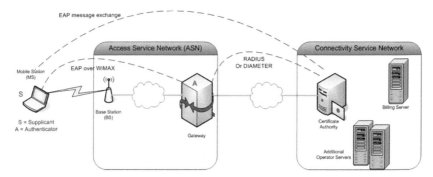

Figure 10.7   EAP-based authentication [Image from http://2008.telfor.rs/files/radovi/02_32.pdf].

one of three types of EAP: EAP-AKA (Authentication and Key Agreement), EAP-TLS (Transport Layer Security) and EAP-TTLS MS-CHAP v2 (Tunneled Transport Layer Security with Microsoft Challenge-Handshake Authentication Protocol version 2).

   iii. Type 3 – RSA-based authentication followed by EAP authentication: The third type of authentication that the security sublayer supports is the RSA-based authentication followed by EAP authentication.

d. Authorization: Following the authentication process is the authorization process in which SS requests for an AK and a SAID from BS by sending an Authorization Request message. This message contains SS's X.509 certificate, encryption algorithms and cryptographic ID. The BS then interacts with an AAA (Authentication, Authorization and Accounting) server to validate the request from the SS and sends back an Authorization Reply which includes the AK encrypted with the SS's public key, a lifetime key and an SAIS.

e. Data privacy and integrity: Data privacy and integrity is achieved, generally, by encryption. WiMAX adopts the AES algorithm for encryption. "The AES cipher is specified as a number of repetitions of transformation rounds that convert the input plain-text into the final output of cipher-text. Each round consists of several processing steps, including one that depends on the encryption key. A set of reverse rounds are applied to transform cipher-text back into the original plaintext using the same encryption key". Since DES is no more secure enough and double DES is in use in many application, but AES is recommended in WiMAX with many supported block cipher operation modes: CCM-Mode and Electronic Code Book mode (ECB-Mode) (in IEEE 802.16-2004), Cipher Block Chaining mode (CBC-Mode), Counter mode (CTR-Mode), Cipher Feedback mode (CFB –MODE), AND AES-Key-Wrap.

WiMAX has been designed carefully with security concerns but it is still vulnerable to various attacks. The following section will present these security issues in WiMAX.

10.14.3 WiMAX Security Vulnerabilities and Countermeasures

WiMAX has security vulnerabilities in both PHY and MAC layers, exposing to various classes of wireless attack including interception, fabrication, modification, and replay attacks. Some vulnerabilities of WiMAX originate from flaws of IEEE 802.16 on which WiMAX is based. A lot of problems and flaws have been fixed in the enhanced version but WiMAX still has some exposes.

Some possible threats or vulnerabilities are reviewed next and some solutions will be discussed.

10.14.3.1 Threats to the PHY layer

As described in Section 10.14.2, WiMAX security is implemented in the security sub-layer which is above the PHY layer. Therefore, the PHY is unsecure, and it is not protected from attacks targeting at the inherent vulnerability of wireless links such as jamming, scrambling, or water torture attack. WiMAX supports mobility, and thus it is more vulnerable to these attacks because the attackers do not need to reside in a fixed place and the monitoring solutions presented below will be more difficult.

10.14.3.1.1 Jamming attack

In literature, **jamming attack is described as** "the transmission of radio signals that disrupt communications by decreasing the Signal-to-Inference-plus-Noise ratio (SINR). SINR is the ratio of the signal power to the sum of the interference power from other interfering signals and noise power". Also, it is described as an attack "achieved by introducing a source of noise strong enough to significantly reduce the capacity of the channel". Jamming can be either intentional or unintentional. It is not difficult to perform a jamming attack because necessary information and equipment are easy to acquire.

We note here that jamming represents the most serious security threat in the field of Wireless Sensor Networks (WSNs), as it can easily put out of order even WSNs that utilize strong high-layer security mechanisms, simply because it is often ignored in the initial WSN design. For this reason, we are giving next some of the countermeasures of jamming attack.

Jamming attack can be prevented by many means, and some of them are as follows:

- Increasing the power of signals
- Increasing the bandwidth of signals using spreading techniques such as frequency spread spectrum (FHSS)
- Using direct sequence spread spectrum (DSS)
- Using Radio Direction Finding tools, the sources of jamming are easy to be located. It is easy to detect jamming using radio spectrum monitoring equipment.

In the following, some details of these techniques are given taking into consideration the case of WSN case:

A. Regulated Transmitted Power: The use of low transmission power decreases the discovery probability from an attacker (an attacker must locate first the target before transmitting jamming signal). Higher transmitted power implies higher resistance against jamming because a stronger jamming signal is needed to overcome the original signal. In some wireless networks especially Wireless Sensor Network, the nodes possess the capability to change the output power of their transmitter.

B. Frequency-Hopping Spread Spectrum (FHSS): Frequency-Hopping Spread Spectrum (FHSS) is a spread-spectrum method of transmitting radio signals by rapidly switching a carrier among many frequency channels, using a shared algorithm known both to the transmitter and the receiver. In case of WSN, FHSS brings forward many advantages in WSN environments:

- It minimizes unauthorized interception and jamming of radio transmission between the nodes.
- The SNR required for the carrier, relative to the back-ground, decreases as a wider range of frequencies is used for transmission.
- It deals effectively with the multipath effect.
- Multiple WSNs can coexist in the same area without causing interference problems.

One of the main drawbacks of frequency-hopping is that the overall bandwidth required is much wider than that required to transmit the same data using a single carrier frequency.

However, transmission in each frequency lasts for a very limited period of time so the frequency is not occupied for long.

C. Direct Sequence Spread Spectrum (DSSS): Direct Sequence Spread Spectrum (DSSS) transmissions are performed by multiplying the data (RF carrier) being transmitted and a Pseudo-Noise (PN) digital signal. This PN digital signal is a pseudorandom sequence of 1 and -1 values, at a frequency (chip rate) much higher than that of the original signal. This process causes the RF signal to be replaced with a very wide bandwidth signal with the spectral equivalent of a noise signal; however, this noise can be filtered out at the receiving end to recover the original data, through multiplying the incoming RF signal with the same PN modulated carrier.

The first three of the above-mentioned FHSS advantages also apply to DSSS. Furthermore, the processing applied to the original signal by DSSS makes it difficult to the attacker to descramble the transmitted RF carrier and recover the original signal. Also since the transmitted signal of DSSS resembles white noise, radio direction finding of the transmitting source is a difficult task.

We should note that although 802.15.4 standard uses DSSS modulation, that does not make it invulnerable to jamming. On the contrary due to the limited supported chip rate (2 Mchip/s) and the restricted transmission power of sensor nodes (typically 0 dBm) the network is very likely to collapse under a jamming attack.

D. Hybrid FHSS/DSSS: Hybrid FHSS/DSSS communication between WSN nodes represents a promising anti-jamming measure. In general terms direct-sequence systems achieve their processing gains through interference attenuation using a wider bandwidth for signal transmission, while frequency hopping systems through interference avoidance. Consequently using both these two modulations, resistance to jamming may be highly increased. Also Hybrid FHSS/DSSS compared to standard FHSS or DSSS modulation provides better Low-Probability-of-Detection/Low-Probability-of-Interception (LPD/LPI) properties. Fairly specialized interception equipment is required to mirror the frequency changes uninvited. It is stressed though that both the frequency sequence and the PN code of DSSS should be known to recover the original signal. Thus Hybrid FHSS/DSSS improves the ability to combat the near-far problem which arises in DSSS communications schemes. Another welcome feature is the ability to adapt to a variety of channel problems.

E. Ultra Wide Band Technology: Ultra Wide Band (UWB) technology is a modulation technique based on transmitting very short pulses on a large spectrum of a frequency band simultaneously. This renders the transmitted signal very hard to be intercepted/jammed and also resistant to multipath effects. In the context of WSNs, UWB can provide many advantages. UWB-based sensor networks guarantee more accurate localization and prolonged battery lifetime. The IEEE standard for UWB, 802.15.3.a, is under development.

10.14.3.1.2 Scrambling attack

Scrambling is a kind of jamming but only provoked for short intervals of time and targeted to specific WiMAX frames or parts of frames at the PHY layer. Attackers can selectively scramble control or management information in order to affect the normal operation of the network. Slots of data traffic belonging to the targeted SSs can be scrambled selectively, forcing them to retransmit. It is more difficult to perform an scrambling attack than to perform a jamming attack due to "the need, by the attacker, to interpret control information and to send noise during specific intervals".

Since scrambling is intermittent, it is more difficult to detect scrambling than jamming. Fortunately, we can use anomalies monitoring beyond performance norm (or criteria) to detect scrambling and scramblers.

10.14.3.1.3 Water torture attack

This is another typical attack in which an attacker forces an SS to drain its battery or consume computing resources by sending a series of bogus frames. This kind of attack is considered even more destructive than a typical Denial-of-Service (DoS) attack since the SS which is usually a portable device is likely to have limited resources.

To prevent this kind of attack, a sophisticated mechanism is necessary to discard bogus frames, thus avoiding running out of battery or computational resources.

10.14.3.1.4 Other threats

In addition to threats from jamming, scrambling and water torture attacks, 802.16 is also vulnerable to other attacks such as forgery attacks in which an attacker with an adequate radio transmitter can write to a wireless channel. In mesh mode, 802.16 is also vulnerable to replay attacks in which an attacker resends valid frames that the attacker has intercepted in the middle of forwarding (relaying) process.

WiMAX has fixed the security flaw of 802.16 by providing mutual authentication to defend these kinds of attacks.

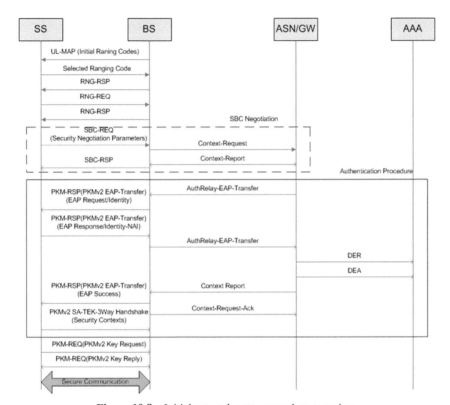

Figure 10.8   Initial network entry procedure overview.

10.14.3.2 Threats to the MAC layers

There are many defects or flaws in WiMAX security solutions at the MAC layer. The vulnerabilities with MAC management messages are presented first in section 10.14.3.2.1 and Section 10.14.3.2.2. Then, vulnerabilities in authentication mechanism and some specific attacks are discussed.

10.14.3.2.1 Threats to Mac Management Message in Initial Network Entry

The initial network entry procedure is very important since it is the first gate to establish a connection to Mobile WiMAX by performing several steps including: initial Ranging process, SS Basic Capability (SSBC) negotiation, PKM authentication and registration process as depicted in Figure 10.8.

   The vulnerability of using Ranging Request-Response (RNG-REQ, RNG-RSP) messages: This message is used in the initial ranging process.

The RNG-REQ message is sent by a SS trying to join a network to propose a request for transmission timing, power, frequency and burst profile information. Then, the BS responds by sending a RNG-RSP message to fine-tune the setting of transmission link. After that, the RNG-RSP can be used to change the uplink and downlink channel of the SS. There are several threats related to these messages. For instance, an attacker can intercept the RNG-REQ to change the most preferred burst profile of SS to the least effective one, thus downgrading the service. An attacker can also spoof or modify ranging messages to attack or interrupt regular network activities. This vulnerability can lead to a DoS attack which will be presented in details in Section 10.13.3.2.4.

Other initial network entry vulnerability: There is a more general vulnerability of initial network entry. During the initial network entry process, many important physical parameters, performance factors, and security contexts between SS and BS, specifically the SBS negotiation parameters and PKM security contexts, play a major role.. Although the security schemes offered WiMAX include a message authentication scheme using HMAC/CMAC codes and traffic encryption scheme using AES based on PKMv2, these schemes are applied only to normal data traffic after initial network entry process. Subsequently, the parameters exchanged during this process are not securely protected, bringing a possible exposure to malicious users to attack.

One possible solution to this vulnerability is by using Diffie–Hellman key agreement scheme as depicted in Figure 10.9.

In this approach, the Diffie-Hellman key agreement scheme will be used for SS and BS to generate a shared common key called "pre-TEK" separately and establish a secret communication channels in the initial ranging procedure. After that, the SBC security parameters and PKM security contexts can be exchanged securely.

10.14.3.2.2 Threats to Access Network Security

There is also vulnerability in access network security in WiMAX. In order to accommodate the requirements of WiMAX End-to-End Network Systems Architecture for mobile WiMAX network, the WiMAX forum defined network Reference Model (NRM) which consists of the following entities (Figure 10.2): Subscriber Station (SS), Access Service Network (ASN), and Connectivity Service Network (CSN). ASN consists of at least one BS and one ASN Gateway (ASN/GW) forming a complete set of network functions necessary to provide radio access to mobile subscribers. CSN consists of

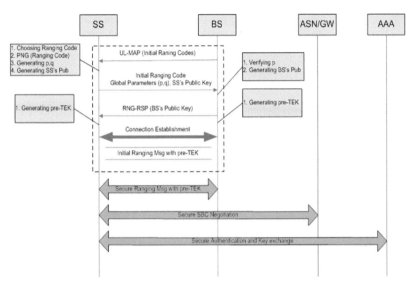

Figure 10.9 Proposed network initial entry approach.

AAA Proxy/Server, Policy, Billing, and Roaming Entities forming a set of network functions to provide IP connectivity services to subscribers. This AAA-architecture-based model is illustrated in Figure 10.10.

This model is divided into three insecure domains and one secure domain. The only secure domain covered by encryption and authentication schemes in 802.16 standard is the data communications between SS and BS. The initial network entry which is examined in the 3b section belongs to domain A. Domain B and C are considered insecure because the Network Working Group in WiMAX forum just assumes that domain B is in a trusted network without proposing any protection and just suggests a possibility of applying an IPSec tunnel between ASN and AAA in domain C.

Some researchers proposed a countermeasure for this problem by using a simple and efficient key exchange method based on PKI. Their method is described in Figure 10.11.

In this approach, all network devices have their certificate and a certificate chain for verification. The PKI structure is used as a method to obtain correspondent's public keys and verify the certificates, thus enabling entities to create a shared secret key for establishing a secure connection.

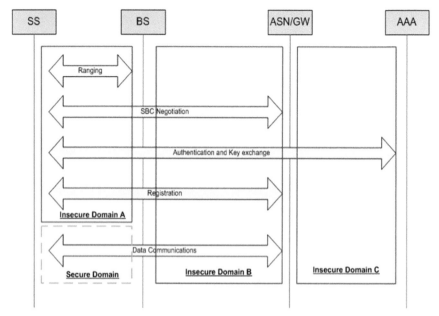

Figure 10.10   Access network security overview.

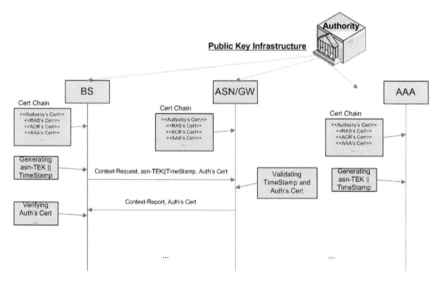

Figure 10.11   Proposed access network approach.

10.14.3.2.3 Threats to Authentication

Many serious threats also arise from the WiMAX's authentication scheme in which masquerading and attacks on the authentication protocol of PKM are the most considerable.

Masquerading threat: Masquerade attack is a type of attack in which one system assumes the identity of another, that is, one entity pretends to be a different entity. WiMAX supports unilateral device level authentication which is a RSA/X.509 certificate based authentication. The certificate can be programmed in a device by the manufacturer. Therefore sniffing and spoofing can make a masquerade attack possible. Specifically, there are two techniques to perform this attack: identity theft and rogue BS attack.

- *Identity theft:* An attacker reprograms a device with the hardware address of another device. The address can be stolen by interfering the management messages.
- *Rogue BS attack*: SS can be compromised by a forged BS which imitates a legitimate BS. The rogue BS makes the SSs believing that they are connected to the legitimate BS, thus it can intercept SSs' whole information. In IEEE 802.16 using PKMv1, the lack of mutual authentication prevents confirming the authentication of BS and makes Man-In-The-Middle (MITM) attack through rogue BS possible by sniffing Auth-related message from SS. However, it is difficult to successfully perform this kind of attack in WiMAX which supports mutual authentication by using PKMv2.

Attacks on the authentication protocols of basic PKM in 802.16 and its later version-PKMv2: By adopting new version of PKM, WiMAX fixes many flaws in PKMv1 such as vulnerability to MITM due to the lack of mutual authentication. However, the newly proposed PKMv2 has been found to be also vulnerable to new attacks.

- *Attacks on basic PKM authentication protocol:* Attacker can intercept and save the messages sent by a legal SS and then perform a replay attack against the BS. The SS also might face with this kind of attack. In the worst case, since mutual authentication is not supported in basic PKM, BS is not authenticated. Therefore malicious BS can perform a MITM attack by making its own Auth-Reply message and gain the control of the communication of victim SS. As a fact, the Basic PKM has many flaws such that it provides almost no guarantees to SS about the AK. These problems have been fixed in the Intel Nonce version of PKM.

- *Attacks on Intel Nonce Version PKM:* In this version, nonce is a possible alternative to timestamp in authentication protocol. This approach does not protect a BS from a replay attack.
- *Attacks on PKMv2:* This version provides a three-way authentication with a confirmation message from SS to BS. There are two possible attacks as follows. First, a replay attack can be performed if there is no signature by SS. Second, even with the signature form SS, an interleaving attack is still possible.

10.14.3.2.4 Other threats

Some serious attacks can exploit vulnerabilities in many aspect of the MAC layers. Two of the most destructive attacks can be man-in-the-middle (MITM) and denial of service (DoS) attacks.

Man in the middle attack (MITM): Although WiMAX can prevent MITM attack through rogue BS by using PKMv2, it is still vulnerable to MITM attack. This possibility is due to the vulnerabilities in initial network entry procedure which is already presented in Section 10.14.3.2.2 of this chapter. In Section 10.14.3.2.2, it is known that WiMAX standard does not provide any security mechanism for the SSBC negotiation parameters. Some researchers show that through intercepting and capturing message in the SSBC negotiation procedure, an attacker can imitate a legitimate SS and send tamped SSBC response message to the BS while interrupting the communication between them. The spoof message would inform the BS that the SS only supports low security capabilities or has no security capability. If the BS still accepts, then the communication between the SS and the BS will not have a strong protection. Under these circumstances, the attacker is able to wiretap and tamper all the information transmitted. A solution based on Diffie–Hellman key exchange protocol was proposed to stop this kind attack.

Denial of Service attack (DoS): Denial of Service attack, in general, prevents or inhibits the normal use or management of communication facilities. There are many vulnerabilities exposing IEEE 802.16e networks to DoS attacks such as unprotected network entry, unencrypted management communication, unprotected management frame, weak key sharing mechanism in multicast and broadcast operations, and Reset-Command message).

Some of noticeable DoS attacks may include the following:

- *DoS attacks based on Ranging Request/Response (RNG-REG/RNG-RSP) messages:* An attacker can forge a RNG-RSP message to minimize the power level of SS to make SS hardly transmit to BS, thus triggering initial ranging procedure repeatedly. An attacker can also perform a water torture DoS by maximizing the power level of SS, effectively draining the SS's battery.

- *DoS attacks based on Mobile Neighbor Advertisement (MOB_NBR_ADV) message:* MOB_NBR_ADV message is sent from serving BS to publicize the characteristics of neighbor base stations to SSs searching for possible handovers. This message is not authenticated. Thus it can be forged by an attacker in order to prevent the SSs from efficient handovers downgrading the performance or even denying the legitimate service.
- *DoS attacks based on Fast Power Control (FPC) message:* FPC message is sent from BS to ask a SS to adjust its transmission power. This is also one of the management messages which are not protected. An attacker can intercept and use FPC message to prevent a SS from correctly adjusting transmission power and communicating with the BS. The attacker can also use this message to perform a water torture DoS attack to drain the SS's battery.
- *DoS attacks based on Authorization-invalid (Auth-invalid) message:* The Auth-invalid is sent from a BS to a SS when AK shared between BS and SS expires or BS is unable to verify the HMAC/CMAC properly. This message is not protected by HMAC and it has PKM identifier equal to zero. Thus, it can be used as DoS tool to invalidate legitimate SS.
- *DoS attacks based on Reset Command (RES-CMD) message:* This message is sent to request a SS to reinitialize its MAC state machine, allowing a BS to reset a non-responsive or malfunction SS. This message is protected by HMAC but is still potential to be used to perform a DoS attacks.

In order to prevent DoS attacks, it is necessary to fix the vulnerabilities in the initial network entry. This is discussed in Section 10.14.3.2.1. Some researchers also suggest that the authentication mechanism should be extended to as many management frame as possible. They also suggest using digital signatures as an authentication method.

10.15 WiMAX – Security Functions

In Section 10.14, we discussed the vulnerability and the possible attacks on WiMAX. In this section, we introduce the security functions that WiMAX is providing. WiMAX systems were designed at the outset with robust security in mind. The standard includes state-of-the-art methods for ensuring user data privacy and preventing unauthorized access with additional protocol optimization for mobility.

As mentioned in the above section, security is handled by a privacy sublayer within the WiMAX MAC. The key aspects of WiMAX security are as follows:

Support for privacy: User data is encrypted using cryptographic schemes of proven robustness to provide privacy. Both AES (Advanced Encryption Standard) and 3DES (Triple Data Encryption Standard) are supported. The 128-bit or 256-bit key used for deriving the cipher is generated during the authentication phase and is periodically refreshed for additional protection.

Device/user authentication: WiMAX provides a flexible means for authenticating subscriber stations and users to prevent unauthorized use. The authentication framework is based on the Internet Engineering Task Force (IETF) EAP, which supports a variety of credentials, such as username/password, digital certificates, and smart cards.

WiMAX terminal devices come with built-in X.509 digital certificates that contain their public key and MAC address. WiMAX operators can use the certificates for device authentication and use a username/password or smart card authentication on top of it for user authentication.

Flexible key management protocol: The Privacy and Key Management Protocol Version 2 (PKMv2) is used for securely transferring keying material from the base station to the mobile station, periodically re-authorizing and refreshing the keys.

Protection of control messages: The integrity of over-the-air control messages is protected by using message digest schemes, such as AES-based CMAC or MD5-based HMAC.

Support for fast handover: To support fast handovers, WiMAX allows the MS to use pre-authentication with a particular target BS to facilitate accelerated re-entry. A three-way handshake scheme, discussed before, is supported to optimize the re-authentication mechanisms for supporting fast handovers, while simultaneously preventing any man-in-the-middle attacks.

10.16 WiMAX and Wi-Fi Comparison

WiMAX is similar to Wi-Fi, but on a much larger scale and at faster speeds. A nomadic version would keep WiMAX-enabled devices connected over large areas, much like today's cell phones. We can compare it with Wi-Fi based on the following factors:

a. **IEEE Standards:** Wi-Fi is based on IEEE 802.11 standard whereas WiMAX is based on IEEE 802.16. However, both are IEEE standards. Wi-Fi is based on IEEE 802.11 standard whereas WiMAX is based on IEEE 802.16. However, both are IEEE standards.

b. **Range:** Wi-Fi typically provides local network access for a few hundred feet with the speed of up to 54 Mbps, and a single WiMAX antenna is expected to have a range of up to 40 miles with the speed of 70 Mbps or more. As such, WiMAX can bring the underlying Internet connection needed to service local Wi-Fi networks.

c. **Scalability:** Wi-Fi is intended for LAN applications, users scale from one to tens with one subscriber for each CPE device. Fixed channel sizes (20MHz). WiMAX is designed to efficiently support from one to hundreds of Consumer premises equipment (CPE), with unlimited subscribers behind each CPE. Flexible channel sizes from 1.5 MHz to 20 MHz.

d. **Bit rate:** Wi-Fi works at 2.7 bps/Hz and can peak up to 54 Mbps in 20 MHz channel. WiMAX works at 5 bps/Hz and can peak up to 100 Mbps in a 20 MHz channel.

e. **Quality of Service:** Wi-Fi does not guarantee any QoS but WiMAX provides several level of QoS.

As such, WiMAX can bring the underlying Internet connection needed to service local Wi-Fi networks. Wi-Fi does not provide ubiquitous broadband while WiMAX does.

Table 10.6 gives a summary of the comparison.

Reference and Resources

The following resources are used while writing this chapter. The resources contain additional information on WiMAX. The reader can use them to get more in-depth knowledge on this topic.

Useful Links on WiMAX

- Quantum WiMAX – Quantum Networks, A WiMAX Company.
- WiMAX Technology – Good site to give latest information on Wi-Fi and WiMAX.

Table 10.6   Wi-Fi versus WiMAX

Feature	WiMAX (802.16a)	Wi-Fi (802.11b)	Wi-Fi (802.11a/g)
Primary application	Broadband Wireless Access	Wireless LAN	Wireless LAN
Frequency band	Licensed/unlicensed 2 to 11 GHz	2.4 GHz ISM	2.4 GHz ISM (g) 5 GHz U-NII (a)
Channel bandwidth	Adjustable 1.25 M to 20 MHz	25 MHz	20 MHz
Half-/full-duplex	Full	Half	Half
Radio technology	OFDM (256-channels)	Direct sequence spread spectrum	OFDM (64-channels)
Bandwidth efficiency	<=5 bps/Hz	<=0.44 bps/Hz	<=2.7 bps/Hz
Modulation	BPSK, QPSK, 16-, 64-, 256-QAM	QPSK	BPSK, QPSK, 16-, 64-QAM
FEC	Convolutional Code Reed-Solomon	None	Convolutional Code
Encryption	Mandatory-3DES Optional-AES	Optional-RC4 (AES in 802.11i)	Optional-RC4 (AES in 802.11i)
Mobility	Mobile WiMAX (802.16e)	In development	In development
Mesh	Yes	Vendor proprietary	Vendor proprietary
Access protocol	Request/Grant	CSMA/CA	CSMA/CA

- WiMAX Forum – Official website of WiMAX Forum, the best resource for WiMAX information.
- IEEE 8092.16 Specification – The IEEE 802.16 Working Group on Broadband Wireless Access Standards.
- WiMAX Forum White Papers – Read more about WiMAX wireless technology and how it addresses market problems through these white papers.
- WiMAX Industry – Find latest news and market trends about WiMAX.
- WiMAX.com – Latest WiMAX news, market trends, WiMAX Forums, etc.
- Wi-Fi Aliance – Official website of Wi-Fi Alliance. The best resource for WiMAX information.
- Wi-Fi.com – Wi-Fi.com is a wireless Internet portal offered by Salient Properties, LLC.

References

[1] http://en.wikipedia.org/wiki/802.16

[2] http://en.wikipedia.org/wiki/WiMAX

[3] http://www.wimaxforum.org/about

[4] http://en.wikipedia.org/wiki/Advanced_Encryption_Standard

[5] http://www.cse.wustl.edu/~jain/cse574-08/

[6] David Johnson and Jesse Walker, "Overview of IEEE 802.16 Security", Intel Corp, IEEE Security and Privacy, 2004, http://portal.acm.org/citation.cfm?id=1009288

[7] Michel Barbeau, "WiMAX/802.16 Threat Analysis", Proceedings of the 1st ACM international workshop on Quality of service & security in wireless and mobile networks, Quebec, Canada 2005. http://portal.acm.org/citation.cfm?id=1089761.1089764

[8] Mahmoud Narsreldin, Heba Aslan, Magdy El-Hennawy, Adel El-Hennawy, "WiMAX security", 22nd International Conference on Advanced Information Networking and Applications, 2008. http://portal.acm.org/citation.cfm?id=1395554

[9] Andreas Deininger, Shinsaku Kiyomoto, Jun Kurihara, Toshiaki Tanaka, "Security Vulnerabilities and Solutions in Mobile WiMAX", International Journal of Computer Science and Network Security, VOL.7 No.11, November 2007. http://paper.ijcsns.org/07_book/200711/20071102.pdf

[10] Abdelrahman Elleithy, Alaa Abuzaghleh, Abdelshakour Abuzneid, "A new mechanism to solve IEEE 802.16 authentication vulnerabilities", Computer Science and Engineering Department University of Bridgeport, Bridgeport, CT. http://www.asee.org/activities/organizations/zones/proceedings/zone1/2008/Professional/ASEE12008_0022_paper.pdf

[11] Tao Han, Ning Zhang, Kaiming Liu, Bihua Tang, Yuan'an Liu, "Analysis of Mobile WiMAX Security: Vulnerabilities and Solutions", Mobile Ad Hoc and Sensor Systems, 2008. MASS 2008, http://ieeexplore.ieee.org/document/4660134/

[12] Sen Xu, Chin-Tser Huang, "Attacks on PKM Protocols of IEEE 802.16 and Its Later Versions", 3rd International Symposium on Wireless Communication Systems, ISWCS 2006. http://ieeexplore.ieee.org/document/4362284/

[13] Taeshik Shon, Wook Choi, "An Analysis of Mobile WiMAX Security: Vulnerabilities and Solutions", Lecture notes in computer science, Springer, 2007. http://www.springerlink.com/content/d03p14w7720x842l/

[14] Sheraz Naseer, Dr. Muhammad Younus, Attiq Ahmed, "Vulnerabilities Exposing IEEE 802.16e Networks To DoS Attacks: A Survey", Proceedings of the 2008 Ninth ACIS International Conference on Software Engineering, Artificial Intelligence, Networking, and Parallel/Distributed Computing, 2008. http://ieeexplore.ieee.org/document/4617395/

[15] R. Poisel, "Modern Communications Jamming Principles and Techinques", Artech House Publishers, 2003. Click here to order from Amazon.com

[16] Ayesha Altaf, Rabia Sirhindi, Attiq Ahmed, "A Novel Approach against DoS Attacks in WiMAX Authentication using Visual Cryptography", The Second International Conference on Emerging Security Information, Systems and Technologies, SECURWARE, Cap Esterel, France 2008, http://ieeexplore.ieee.org/document/4622589/

[17] D.W. Park, "A Study of Packet Analysis regarding a DoS Attack in WiBro Environments", IJCSNS International Journal of Computer Science and Network Security, vol. 8, no. 12, December 2008, http://paper.ijcsns.org/07_book/200812/20081257.pdf

[18] Mitko Bogdanoski, Pero Latkoski, Aleksandar Risteski, Borislav Popovski, "IEEE 802.16 Security Issues: A Survey", Faculty of Electrical Engineering and Information Technologies, Ss. Cyril and Methodius University, Skopje, Macedonia. http://2008.telfor.rs/files/radovi/02_32.pdf

11

INSTEON Technology

11.1 Introduction to INSTEON: Overview

In our home of the near future, artificial-intelligence-enabled appliances will make our lives simpler. These innovations will include our smart microwave/convection ovens downloading new cooking recipes, a thermostat that automatically changes to its energy saving set point when the security system is enabled, bathroom floors and towel racks that heat up when the bath runs, and our mobile phones alerting us when someone has approached our house to give us the option to allow trusted friends/family inside. The market for possible solutions for home automation is broken into two distinct camps: wired and wireless. The wired camp includes Ethernet, power line, and Home PNA (IEEE 1901). The wireless camp consists of 802.11 abgn/ac, (IEEE 802.15.1) Bluetooth, ZigBee, and X10 as well as a multitude of others.

INSTEON is the most unique protocol on the list of smart home protocols because it incorporates wireless and powerline (PL) technologies into a single network. This gives the flexibility to installing accessories anywhere in the home as long as there is either an outlet or a wireless signal in the house. Devices communicate with each other using the INSTEON protocol over the air via radio frequency (RF) and over the powerline (PL) as illustrated in Figure 11.1.

INSTEON is a home automation technology that enables simple, low-cost devices developed for peoples' comfort and home security like light switches, lights, thermostats, leak sensors, remote controls, motion sensors, dimmers, HVAC System, LED bulbs, low voltage controllers, on the security side devices like sprinkler control, garage door monitor, hidden door sensor, motion sensor detector and transmitter, smoke detector, energy monitoring device, camera and other electrically powered devices to interoperate through power lines, radio frequency (RF) communications, or both. Nowadays, the security system consists of sprinklers, fire detectors, smoke detectors, cameras, etc., but all are not interconnected and do not use traditional technology. INSTEON concept is that it connects all devices to form a mesh network. And the most important concept that attracts all engineers is its dual band technology. This technology includes transmission of data through powerline and RF communication. All the devices form peer-to-peer network such that any device in the network can transmit, repeat, or receive.

Some devices follow only RF communication or communicate via powerline and some devices use both to communicate and are called dual-band devices. Although it uses both RF and powerline, the networking concept is very simple and not complicated that includes routing table, addressing, easily comprehensible data packet, simple error check mechanism and other cumbersome tasks. All of these together make working of the network without any impediment.

It employs a dual-mesh networking topology in which all devices are peers meaning that any device can transmit, receive, or repeat messages independently without requiring a master controller or complex routing software. Like other home automation systems, it has been associated with the Internet of Things (IoT).

One of the main features of INSTEON is its robustness. An INSTEON network becomes more robust and reliable as it is expanded because INSTEON devices repeat messages received from other INSTEON devices. INSTEON devices repeat each other's messages by simulcasting them at precisely the same time, so with more devices the INSTEON signal gets stronger. Dual-mesh communications using both the powerline and the airwaves ensure that there are multiple pathways for messages to travel. INSTEON signals automatically jump from one layer to the other, and back. Each layer works around road-blocks in the other. It is estimated that the error rate for a dual-band technology might be 100 times smaller than that of a single-band technology.

INSTEON-based products were launched in 2005 by Smartlabs, the company which holds the trademark for INSTEON. A Smartlabs subsidiary, also named INSTEON, was created to market the technology. There are now millions of INSTEON nodes in use around the world in many different products, including dimmers and switches, handheld and tabletop remotes, thermostats, sprinkler controllers, energy monitoring devices, sensors and low-voltage input/output interfaces. INSTEON networks easily connect to other, larger networks, such as LANs and Wi-Fi. Many different user interfaces, including smartphones, PCs, and third-party controllers, can manage an INSTEON network. Applications running on such devices not only can talk to INSTEON

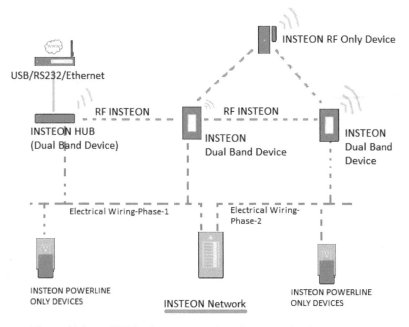

Figure 11.1   INSTEON home automation (Courtesy of INSTEON Inc.).

products but also can interoperate with cameras, sensors, and other devices using different networking protocols, including TCP/IP, ZigBee, and Z-Wave.

Devices communicate with each other using the INSTEON protocol over the air via radio frequency (RF) and over the powerline (PL) as illustrated in Figure 11.1.

11.1.1 The Need for INSTEON

For more than three decades, X10 technology which works only using existing powerlines was in use for home automation. During this long period of time, X10 technology did not develop itself to satisfy the need of more equipped home automation system. Reliability, response time, and efficiency of the X10 system is not as expected. A system which overcomes all of these is required. This led to the invention of INSTEON network protocol in June 2004. INSTEON network protocol uses dual band, both RF and powerline, and this improves reliability. INSTEON is proved 30 times faster than X10, and this promotes response time. Both of these together increases overall efficiency of the system. Hence, INSTEON is a successful home network management protocol and now the authority in the same domain. Due to the network's rapidness, the name of the protocol INSTEON is derived from the phrase **"Instant on."**

11.2 Characteristics of INSTEON Technology

Some of the important characteristics of INSTEON Technology are discussed here:

Reliability: INSTEON's dual band nature of communication helps sender to send data to nodes in different routes, such that if one fails the other successfully transmits the command. This holds good because of another concept of repeating messages. The function of a node as a repeater is to check the received packet and take actions accordingly. Once the node receives the packet it checks for the "to address," if the packet's to address is not equal to the address of the node that received the packet, the node sends out the packet for retransmission instead of dropping it. These features thus aid reliable transmission of the message and proper outcome is achieved.

Installation: The users can install their INSTEON product by themselves without requirement of any specialized trained engineers. The INSTEON nodes are pre-addressed at the time of manufacturing process. Therefore there is no initialization process to occur once the nodes are powered on. Second, INSTEON setup demands no separate wiring to form network, but uses the existing powerline to communicate within the devices. The installation is said to be called as "Plug and Tap" process, which has two simple steps:

- First is to plug a device like lamp or a bulb or any other device that has to be controlled automatically into the INSTEON module, which is plugged to the power line via power socket.
- Second step is to press the set buttons in the module and in the mobile phone.

INSTEON application, which is used to control the device automatically, is well explained and wide range of products are shown in the website. Therefore, installation of INSTEON home networking devices is easy.

Utilization: Controlling and using INSTEON devices is simple, because it does not require complicated programming or functions to make the network operational by the user. Just pressing buttons and switches is involved. Therefore INSTEON products can be termed as user friendly devices.

Response Time: Due to dual band feature and due to high speed of the travelling signal, the network does not experience any delay in the transmission of commands. INSTEON system is build such that a function can be executed in 0.04 s. Hence, the name INSTEON is derived from "Instant On" as explained earlier.

Affordability: The network is not made to engage into performing cumbersome algorithms to determine the route to destination and the software used is simple too. These together make the networking protocol affordable to users.

Backward Compatibility: X10 technology mostly uses existing powerline to communicate with devices. Also X10 was the recent technology before INSTEON was evolved. INSTEON technology can work on both RF and powerline. INSTEON can function with some X10 devices, therefore INSTEON is called as backward compatible technology.

11.3 How INSTEON Works

There are three fundamental differences between INSTEON and all of other command and control networks.

- INSTEON is a dual-mesh network (two uncorrelated media).
- INSTEON propagates messages by simulcasting.
- INSTEON utilizes Statelink.

11.3.1 INSTEON Is a Dual-Mesh Network

There are 115 million households in the United States. Virtually, all have powerline wiring, and the ISM (industrial, scientific, medical) radio bands are freely available for unlicensed home use; thus, utilizing communications over both the powerline and over the airwaves makes the most sense for command and control networking. Single media communication exclusively over powerline or radio bands is fraught with problems. The regulations require that radio devices using the ISM bands must be able to tolerate interference from other devices, and there are many ways in the home that radio signals can be attenuated and reflected. The powerline is notorious for electrical noise and phase bridging problems (see below) can prevent signals from half of the circuits in a house from reaching the other half.

INSTEON solves the single media signaling problems because it is a dual-mesh network. As shown in Figure 11.1, INSTEON devices can communicate with each other using both radio and the powerline, and this is called "Dual Band" technology. Wireless network use RF (Radio Frequency) to communicate, while the wired network devices use existing powerline to communicate or in other words, devices can communicate through wired and wireless medium. These devices are called as "dual band devices". These devises are RF compatible to transmit data wirelessly and uses powerline to transmit through a wire.

INSTEON dual-band devices solve a significant problem encountered by networking technologies that can only communicate via the powerline. Electrical power is most commonly distributed to homes in North America as split-phase 220 V alternating current (220 VAC). At the main electrical junction box to the home, the single three-wire 220 VAC powerline is split into a pair of two-wire 110 VAC powerlines, known as Phase 1 and Phase 2. Phase 1 wiring usually powers half the circuits in the home and Phase 2 powers the other half.

The problem is that powerline signals originating on one phase needing to reach a receiver on the other phase are severely attenuated, because in many cases there is no direct circuit connection for them to travel over.

A traditional solution to this problem is to connect a signal coupling device between the powerline phases, either by hardwiring it in at a junction box or by plugging it into a 220 VAC outlet. INSTEON automatically solves the powerline phase coupling problem through the use of dual-band INSTEON devices – INSTEON RF messages automatically couple the phases when received by a dual-band device on the "other" phase.

As shown in Figure 11.2, devices on INSTEON networks can also interface with the larger world. When suitably equipped with a dedicated serial interface, such as U.S.B, RS-232, Ethernet, or Wi-Fi, INSTEON devices can also interface with computers, smartphones, and other digital equipment. Serial communications can bridge networks of INSTEON devices to otherwise incompatible networks of devices in a home, to computers, to other nodes on a local-area network (LAN) or to the global Internet. Such connections to outside resources allow networks of INSTEON devices to exhibit complex, adaptive, people-pleasing behaviors.

11.3.1.1 Mesh topology or peer-to-peer topology

There are many topologies that a network can form: Star topology, Ring topology, Bus topology, Tree topology and Mesh topology. Each topology has its advantages and limitations. The best suitable topology for home networking protocol is Mesh topology considering the following features. Mesh topology is a network topology in which all the devices of the network can

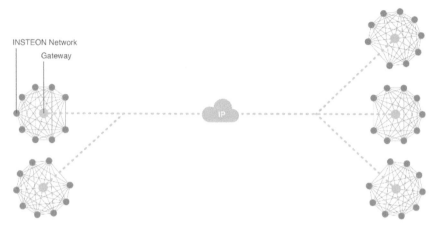

INSTEON Network
Gateway

Figure 11.2   INSTEON networks interface with the larger world.

communicate with each other, which means all the devices are interconnected with each other. The advantages of this topology are that more than one communication within the network is possible simultaneously and more than one routes are available between source and destination. Figure 11.3 shows a mesh topology, in which all the devices are interconnected.

Figure 11.3 demonstrates a basic INSTEON network, which consists of nodes that can communicate using a wired medium via powerline and nodes that can communicate wirelessly via RF communication. One node in the center of the network in the figure is called as a DB device, which supports communication wirelessly and on a wired medium.

11.3.1.2 INSTEON is a peer-to-peer network

All INSTEON devices are peers, meaning that any device can act as a controller (sending messages), responder (receiving messages), or repeater1 (relaying messages).

All the devices forming the network have equal rights to access the resource. The process is decentralized; there is no master to control the other devices as in server and client communication, where server is considered to be the master. Here, in INSTEON, any node can perform the duties of a sender or a receiver or an intermediate node that just passes the data it receives.

This relationship is illustrated in Figure 11.4, where INSTEON device 1, acting as a controller, sends messages to INSTEON devices 2, 3, and 4 acting as responders. INSTEON devices 5, 6, and 7 acting as controllers can also send messages to a single INSTEON device 3 acting as a responder.

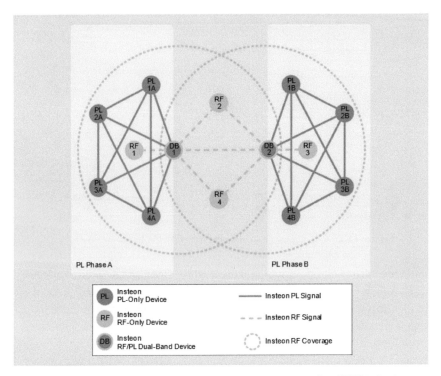

Figure 11.3   An INSTEON message repeating (courtesy of INSTEON Inc.).

Any non-battery-powered INSTEON device can repeat messages, as with device B in Figure 11.5, which is shown relaying a message from device A acting as a controller to device C acting as a responder.

11.3.2 INSTEON Simulcasts Repeated Messages

INSTEON devices repeat each another's INSTEON messages by simulcasting them in precise timeslots synchronized to the powerline zero crossing. To avoid runaway signals, each message is assigned a maximum number of times to be repeated. Each repeated message then contains how many repeats remain until it reaches zero. Presently, the maximum number of repeats can be set to 0, 1, 2, or 3. While all INSTEON devices can act as repeaters, in virtually all situations battery-powered devices do not to conserve battery power.

INSTEON devices automatically and immediately act as repeaters upon power-up – they do not need to be specially installed using some network

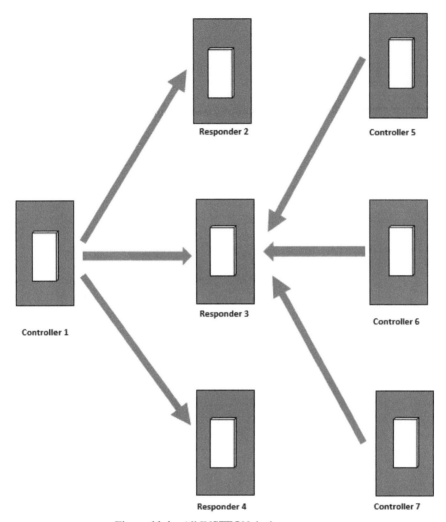

Figure 11.4 All INSTEON devices are peers.

setup procedure. Adding more devices not only increases the strength of the simulcast signal but also increases the number of available pathways for messages to travel. This path diversity results in highly reliable messaging, so the more devices in an INSTEON network, the better.

Simulcasting does not require routing of the message. This makes the networking infrastructure much simpler and less expensive than a routed network. Also, the simulcast propagation automatically produces simultaneous

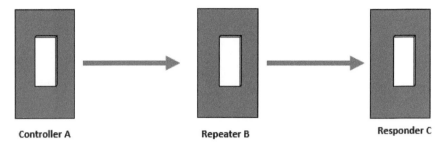

Controller A   **Repeater B**   **Responder C**

Figure 11.5   Non-battery powered node acts as repeater.

response at all responders. This means that responders don't fire at different times creating a "popcorn" effect.

11.3.3 INSTEON Utilizes Statelink

Statelink is a simple, yet powerful tool. Instead of using "commands" to get responders to their desired states, INSTEON simply signals the responder to "recall" the state at which it was at when added to the "scene." Therefore, a generic signal can be sent by any controller resulting in any number of responders recalling their scene states even though the products might vary greatly (e.g., thermostats, lights, pumps, clocks, and security systems).

Since these scene messages are generic in nature, all INSTEON products can both send and respond to them. This fundamentally guarantees that products are both forward and backward compatible – a virtually priceless advantage.

Another enormous advantage of Statelink is that when combined with simulcasting, they create an infrastructure that scales to very large installations. INSTEON installations with over 400 nodes are relatively common.

11.4 INSTEON Specifications

INSTEON, as mentioned before, is a true peer-to-peer dual-mesh network. Its most important property is its simplicity. INSTEON messages are fixed in length and synchronized to the AC powerline zero crossings. Because messages propagate by synchronous simulcasting, no network controllers or routing tables are necessary – a three-byte source and destination address in each message suffices. Optimized for home command and control, INSTEON allows infrastructure devices like light switches, clocks, thermostats, security sensors, and remote controls to be networked together at low cost. In turn, these devices can appear as nodes on larger networks, such as Wi-Fi LANs, the Internet, telephony, and broadband entertainment distribution systems, because INSTEON can connect them using internetworking devices. Table 11.1 shows the main features of INSTEON. The following section will discuss these features in detail.

11.5 The INSTEON Communication Protocols

Protocol is defined as a step-by-step procedure that a data message should follow starting from sending end device till it reaches the receiver. There are a various protocols that INSTEON packets should adhere for a successful transaction between the source and the destination.

1. All devices are capable of receiving a message and resending it. Therefore, all devices act as repeaters. A node receives a packet, checks for "to address," and, if it is destined to it, decodes the information. If the packet is destined to some other node, then the device sends the packet out without any alteration. Therefore, all devices follow two-way repeater communication.

2. Sender who sends a packet to the destination node should know whether it has reached the destination without any impediment or loss; therefore, all the messages that are received by the destination are acknowledged to the sender that the receiver got the packet successfully. In INSTEON, all the messages that are received are acknowledged back to the sender, except for the broadcast message. This procedure in communication helps successful transmission of data with high reliability and reduces loss of packets. In the case of no acknowledgement or ACK is received, the sender retransmits the packet.

3. The concept of Negative Acknowledgement or NACK Adds more reliability to data communication. Circumstances at which NACK should be sent out to the sender, informing about the data disruption are: When data is dropped or lost or in the intermediate node in the path of communication, NACK is to be sent to the sender that notifies the sender that the message has been dropped. Secondly, NACK is issued to the sender, when the data receiver erroneous packet. In all cases after NACK is received by the sender from the receiver that encounters packet loss and / or erroneous data, the packet is retransmitted by the sender.

Table 11.1   INSTEON characteristic summary

INSTEON Property	Specification	
Network	Dual-mesh (RF and powerline)	
	Peer-to-peer	
	Mesh topology	
	Unsupervised	
	No routing table	
Protocol	All non-battery-powered devices are two-way repeaters	
	Messages synchronized to powerline	
	Repeated messages are simulcast	
	Message error detection	
	Messages acknowledged	
	Retry if not acknowledged	
Data Rate	Instantaneous powerline	13,165 bits/sec
	Sustained powerline	2,880 bits/sec
	Instantaneous RF	38,400 bits/sec
Message types	Standard	10 bytes
	Extended	24 bytes
Message format	From address	3 bytes
	To address	3 bytes
	Flags	1 byte
	Command	2 bytes
	User data	14 bytes (extended format)
	Message integrity	1 byte
Devices support	Unique I.D.s	16,777,216
	Commands	65,536
	Members within a group	Limited only by memory
INSTEON Engine Memory requirements	RAM	80 bytes
	ROM	3K byte
Typical application (light switch, lamp dimmer)	RAM	256 bytes
	EEPROM	256 bytes
Memory Requirements	Flash	7K bytes

(*Continued*)

Table 11.1 Continued

INSTEON Property	Specification	
Device installation	Plug – in	
	Wire – in	
	Battery – operated	
Security	Physical device possession	
	Address masking	
	Encrypted message payloads	
X10 compatibility	INSTEON devices may send and receive X10 commands	
Powerline physical layer	Frequency	131.65 KHz
RF physical layer	Modulation	BPSK
	Min. transmit level	$3.16\ V_{PP}$ into 5 ohms
	Min receive level	10 mV
	Phase bridging	INSTEON RF or hardware
	Frequency	915.00 MHz in U.S.
		869.85 MHz in Europe
		921.00 MHz in Australia
	Modulation	FSK
	Sensitivity	-103 dbm
	Range	150 ft unobstructed line-of-sight

4. Above all discussed concepts, the protocol must have measures to ensure that the message contents reach the destination without any error. The receiver may receive erroneous message and respond wrong according to the message received. This could be fixed using the parity check method, wherein a bit field included in the message byte informs whether all the bits in the message hold even or odd parity. More effective mechanisms as Cryptographic hash function, Repetition codes, Checksum, Cyclic Redundancy Check, or CRC can be used. X10 Technology has no error detection mechanism, and this is a drawback for a protocol to be called as a reliable and well-defined protocol.

In INSTEON home networking technology, CRC Protocol is used. Every message that is sent out from a device has a field in it that stores a value. This value is called as Cyclic Redundancy Check or CRC value. The sender before sending the packet and after encoding the information into the packet subjects, the data bits in the packet to error control algorithm. The result of the process is saved into the CRC field. Now the packet is sent out of the sender, and the receiver receives the packet and runs the CRC algorithm with input to the algorithm as the bits in the packet received. The receiver obtains the value as the result of CRC algorithm and compares the value with the CRC field in the original packet it received. If the value is same, the packet is categorized as accurate packet and the receiver starts to decode the information in the packet. If the values are dissimilar, then the received packet has error and is dropped, and notifies the sender about the error in the packet by sending negative acknowledgement.

INSTEON as dual-mesh communication has two protocols: RF Protocol and Powerline Protocol. The two protocols have same message structure but have different "Packet" structure.

11.5.1 The INSTEON Messages

Message Size and Type: The INSTEON message is of less size, which aids faster transmission of data. Therefore commands can be executed rapidly in about 0.04 seconds. Users can expect instant switching on of lamps or lights and other household electronic products. There are two types of messages in INSTEON: standard message and extended message. An extended message can carry the information in a standard message, whereas a standard message is incapable of carrying all the information as in an extended message. For basic communication, like just to turn on the lights, a standard message is enough. And for functions beyond some basic communication, an extended message is required. The standard message is of size 10 Bytes and the extended message is of size 24 Bytes.

Message Format: Message format describes each and every field and its significance in the INSTEON message. Let us now look in to the fields of the two available types of message in detail.

1. Standard Message: This basic and simple message, which is of size 10 bytes has the following fields as presented in the Figure 11.6.

Table 11.2 presents the data size of each field, which is then followed by the description of each field (see also Table 11.1).

From Address	To Address	Flags	Command	CRC

Figure 11.6   Standard message structure.

Table 11.2 Standard message field size

Field	Size (Bytes)
From Address	3
To Address	3
Flags	1
Command	2
Error Detection	1

Table 11.3 Flag fields in a standard message

Flag Field	Size (Bytes)	Bit Position	Description
Message type	1	7	NACK/broadcast
	1	6	Group message
	1	5	ACK
	1	4	Extended/standard message
Message retransmission	2	2 to 3	Hops left
	2	0 to 1	Maximum Hops

From Address: This field is a 3 bytes or 24 bits field that holds source address. This field is used to determine the sender by the receiver node, and in the case of loss of data, negative data should be sent by the receiver; hence, it is a necessity for the receiver to know the sender address to send the negative acknowledgment.

To Address: To identify the message's destination, to address is encoded into the packet from the database. If a receiver receives a message, it first checks the "to address."

If the "to address" is same as its own receiver device address, then it is understood that the message is intended to this device that has received the message; otherwise, the message is retransmitted to the neighbors.

Flags: This field, which is of 1 byte size, describes the type of message and controls retransmission of the message at the device holding this message. Table 11.3 shows the flag fields in a standard message.

Message type field in a standard message: The message type field holds the information on the type of the message. From the fourth bit position, we can learn that whether the message is standard or extended type of message. If that field contains the bit value "1," then the message is an extended message; otherwise, the message is standard message, if it contains the bit value "0." The rest of the bits in the fifth, sixth, and seventh positions play very

Table 11.4 Details of message type field in a standard message

Bit position			
5	6	7	Significance
0	0	0	Direct message
0	0	1	ACK for direct message
0	1	0	Group cleanup direct message
0	1	1	ACK for group cleanup direct message
1	0	0	Broadcast message
1	0	1	NACK for direct message
1	1	0	Group broadcast message
1	1	1	NACK for group cleanup direct message

important role in the classification of the message. Group cleanup message has always a bit value "1" in the sixth position. For broadcast and NACK type of messages, the fifth bit position carries bit value "1." For the purpose of NACK and ACK message, the seventh bit position takes the bit value "1." Table 11.4 displays the significance of each bit of the field.

The details of the message type field in a standard message are as follows:

Direct Messages: The direct message is nothing but a normal message that carries command to the receiver. This message can also be called as point-to-point message.

Broadcast Message: This is not a point-to-point message, because the message is not sent to just one device, but sent to all the devices in the network; it can also be called as one-to-all message.

Group Message: This type of message is sent to a group of devices from one device, therefore can be called as one-to-many message.

Acknowledgement Message: This is a positive response sent by the receiver to the sender for the successful receipt of the packet received by the device from the sender. There is acknowledgment message for a direct message, group cleanup message, but there is no acknowledgement for broadcast message.

Negative Acknowledgement: This is the negative response to a garbled message or a message with error or garbled by the receiver. This is a kind of request for the retransmission of the message with proper computation by the source, this is one of the conditions for retransmission of message at the time of reception of a message with error. Outcome of the combination of these above type of messages gives us many more messages depending upon the purpose, serving many functions.

Message Retransmission: This is one of the most important concepts in INSTEON technology that serves to increase reliability. The two fields called Hops left and Hop count plays vital role in message retransmission technique.

Hop count is a two-bit information field that can store numbers from zero to three in binary form, because messages are encoded with binary digits for digital transmission of the packets. The purpose of this field is to store the value of the maximum number of transmission allowed in a single transaction of the packet between the sender and the receiver.

Hops left field is again a two-bit information field capable of storing numbers from zero to three. The value in the field decrements by one count for every retransmission of the packet. Retransmission is done within the limit of the value in the hop count field. The hop left field is initially loaded with the value in the hop count field.

After every retransmission, as the value decrements, at one stage the value turns to zero in the hops left field. At this condition the packet is dropped, and there is no more retransmission.

Command Field: There are a lot of predefined commands against each value of the two-byte command field. This field provides additional functionality to the message with the above-mentioned functions. Since there are two bytes of size, this field can hold around 65,536 commands for each transaction. The huge number states that 65,536 commands can be created, but depends on the application of INSTEON protocol in each residency. These commands are stored in a database, and the source fetches the command byte from the database every time the user tries to use the application.

Cyclic Redundancy Check: This is an important field that validates the correctness of the message. CRC involves a protocol in which first a polynomial is generated, which is called as the divisor. The message bits are called as the dividend.

As mentioned before, as the message with the CRC value which is of 1 byte is received by the receiver, the receiver again performs CRC process with the same message bits as dividend and divisor polynomial. The outcome of the process, which is the quotient is acquired, then checked with the value in the received message CRC field. The value should be the same, meaning the message is correct and can be used to decode the information. If not, the message has to be discarded.

Figure 11.7   Extended message structure.

Table 11.5   Extended message field size

Field	Size (Bytes)
From Address	3
To Address	3
Flags	1
Command	2
User Data	14
Error Detection Value	1

2. Extended Message

The Extended message, Figure 11.7, is same as the standard message except the additional user data field which is of size 14 bytes, making the size of the extended message equal to 24 bytes in total, whereas the standard message is only 10 bytes. The purpose of having the user field is for advanced applications.

Table 11.5 presents the data size of each field, which is then followed by the description of each field.

The user data field is placed in between the command and the CRC field. Although the user data field is 14 bytes and increases the overall size of the message, the speed is maintained, INSTEON is still instantaneous. All the fields except "User Data" are defined before.

User Data: The user data field is the only addition to the standard message that leads to the formation of extended message. This field is encrypted to render a secure and private connection between the source and destination. If the information encoded in the user field exceeds the 14 bytes data field, more INSTEON extended messages can be appended. The receiver has the ability to receive and put them together in order using packetizing process.

11.5.2 The INSTEON Packets

INSTEON messages are broken up to form packets. Packet structure is different for powerline communication and RF communication.

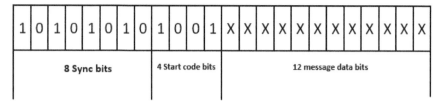

Figure 11.8 Start packet structure.

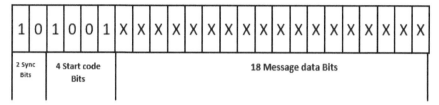

Figure 11.9 Body packet structure.

11.5.2.1 Powerline packet

Each packet consists of 24 bits. At the start of standard message packet, there exists a special type of packet called start packet (SP). And the source packet is followed by body packet (BP).

Start Packet: This packet has three subdivisions – Sync bits, Start code bits, and Message data bits. The start packet has initial eight sync bits, which indicates the start of packet and consists of alternate pattern of 1s and 0s as shown in Figure 11.8 and then follows four start code bits, which indicates the start of message data bits and takes the bit value 1001. Lastly, the start packet has the message data packet, which is of length 12 bits.

Body Packet: The body packet, Figure 11.9, follows the start packet.

The body packet has two sync bits to identify the start of the packet, which has the binary value 10. Four start code bits follow the sync bit and the trailer is the message data bits, which is of 18 bits size.

Powerline Standard Message Packet: The standard message packet, Figure 11.10, comprises of one start packet in the beginning, followed by four body packets. Since each packet is of 24 bits size, the total length of the packet is 120 bits. Start packet has 12 bits of message data and the body packet has 18 bits of message data, and therefore, ((1 SP * 12 bits) + (4 BP * 18 bits)) yields 84 bits. Considering 80 bits, which is 10 bytes gives us the size of one standard message. Therefore, as told earlier, transmission of one

Figure 11.10   Powerline standard message packet.

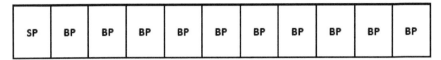

Figure 11.11   Powerline extended message packet.

standard message requires one start packet and four body packets. The total standard packet is of 120 bits or 15 bytes.

Powerline Extended Message Packet: The extended message packet, Figure 11.11, comprises of one start packet and 10 body packets. As we know, the start packet has 12 bits of data message and the body packet has 18 bits of data message, and therefore, ((1 SP * 12 bits) + (10 BP * 18 bits)) summing to 192 bits, which is equal to 24 bytes, which is exactly the size of extended message size. The total size of the extended packet is 264 bits or 33 bytes.

11.5.2.2 RF packet

Unlike powerline packet, RF packets are not broken down into start packet and body packet. Due to this reason, INSTEON packet sent via RF communication is faster than the data sent via powerline. The RF standard message packet, Figure 11.12, and RF extended message packet, Figure 11.13, are almost same, except the inclusion of data bits, which is 80 bits or 10 bytes in the case of standard data message and 192 data bits or 24 bytes in the case of extended data message.

RF Standard Message Packet: This single packet has two sync bytes that indicate the start of the RF packet, followed by start code byte, which is of one byte size, indicates the start of data message packet, then follows the 80 bits or 10 bytes of data message bits.

The trailer attached to this packet is the CRC redundant value for error detection mechanism.

The total packet size comes to 14 bytes.

RF Extended Message Packet: This again has two sync bytes, which specifies the start of a packet, followed by one start code byte that specifies the start of message data. The date message of 192 bits or 24 bytes follow and the packet ends with the trailer, which is of one byte that stores CRC redundant value.

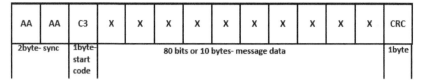

Figure 11.12 RF standard message packet.

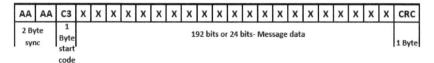

Figure 11.13 RF extended message packet.

The total size of the packet is 28 bytes, or in the terms of bits, the size of extended message packet is 224 bits.

11.5.3 INSTEON Data Rates

In the above, we discussed the two types of messages: Standard and Extended. Extended message carries 112 (14 bytes * 8 bits) bits of excess data when compared to the standard one, and sometimes multiple extended messages are sent. Therefore, it is necessary to set a data rate.

There must also be circumstances at which large amount of multiple data has to be transported spontaneously for certain applications, and to achieve this, we have Instantaneous data rate, which is 13,165 bits/s. With this data rate, it is possible to send around 49.8 powerline extended message packets in one second, and in the case of standard message, approximately 109.7 powerline standard message packets can be sent out in a second. Thus, this data rate called "**Instantaneous data rate**" is apt for applications that use heavy data messages and demands rapid transmission.

Previously, it was considered that communication via powerline is impossible due to heavy loss and noise. Nowadays, we have strong error detection and error correction mechanism and modulation techniques which support data transmission through powerline successfully, hence development in home networking protocol. There is a research going on to provide internet access to the users carried by existing powerline that would provide downloading speed of 14 Mbps and more research to attain the speed of 30–60 Mbps using powerline.

Now for RF packets, using instantaneous data rate of 13165 bits/s, the standard message packet which is of 112 bits or 14 bytes, with 10 bytes or

80 bits data message, the number of RF standard message packets that can be sent in one second is 117.5 RF standard message packets. For RF extended message packet which is of 224 bits or 28 bytes, with 192 bits or 24 bytes of data message, the number of RF extended message packet that can be sent is 58.7 message packets.

There is also another data rate provided in INSTEON home networking protocol, which is called as "**sustained data rate**" and the speed is 2880 bits/s. With this data rate, we can send around 10.9 powerline extended data message packets in one second. And in the case of standard message packet type, we can send 24 powerline standard message packets in one second.

With a sustained speed of 2880 bits/s, 25.7 RF standard message packets can be sent in one second. And in the case of RF extended message packet, 12.8 packets can be sent in one second.

From the above information, we can infer that using instantaneous data rate, we can send powerline extended data message packet 4.5 times the powerline extended data message sent with sustained data rate. And in the case of powerline standard message, we can send powerline standard message using instantaneous data rate again 4.5 times the standard message sent with the help of sustained data rate.

Considering RF standard message packets, with instantaneous speed, we can send RF standard message packet 4.5 times the RF standard message packet sent using sustained speed of 2880 bits/s. Similarly with RF extended message packet, with instantaneous speed, 4.5 times the RF extended message packet with sustained speed can be sent.

Instantaneous data rate is faster than the sustained data rate, although sustained data rate is actually fast. Thus, analyzing the data rate we can come to a conclusion that INSTEON Technology is faster and can operate and respond to applications much faster in about 0.04 seconds.

INSTEON Powerline Data Rates – More Consideration
(The Details are available at www.INSTEON.com)

Some more details about the data rate are considered here. INSTEON Standard messages contain 120 raw data bits and require 5 zero crossings or 41.667 milliseconds to send. Extended messages contain 264 raw data bits and require 11 zero crossings or 91.667 milliseconds to send. The raw sustained INSTEON bitrate is therefore 2880 bps (bits per second) for either kind of message.

Table 11.6 INSTEON data rates

Condition			Bits Per Second		
Max. Hops	ACK	Retries	Standard Message (Usable Data)	Extended Message (Usable Data)	Extended Message (User Data Only)
0	No	0	1440	1698	1034
1	No	0	720	849	517
2	No	0	480	566	345
3	No	0	360	425	259
0	Yes	0	720	849	517
1	Yes	0	360	425	259
2	Yes	0	240	283	173
3	Yes	0	180	213	130

However, the INSTEON protocol waits for one additional zero crossing after each Standard message and for two additional zero crossings after each Extended message to allow for transmitter "politeness" and possible RF message transmission. Therefore, the actual sustained bitrate is 2400 bps for Standard messages or 2437 bps for Extended messages, instead of the 2880 bps it would be without waiting for the extra zero crossings.

INSTEON Standard messages contain 9 bytes (72 bits) of usable data, not counting packet sync and start code bits, nor the message integrity byte. Extended messages contain 23 bytes (184 bits) of usable data using the same criteria. Therefore, the usable data bitrates are further reduced to 1440 bps for Standard messages and 1698 bps for Extended messages. Counting only the 14 bytes (112 bits) of user data in Extended messages, the user data bitrate is 1034 bps.

These data rates assume that messages are sent with max. hops set to zero and that there are no message retries. They also do not take into account the time it takes for a message to be acknowledged. Table 11.6 shows net data rates when multiple hops and message acknowledgement are taken into account. To account for retries, divide the given data rates by one plus the number of retries (up to a maximum of five possible retries).

11.5.4 INSTEON Devices

Device Function: Primarily the INSTEON devices can function as the following types: Controllers, Repeaters, and Responders. These simple device functions enable INSTEON home-control network protocol to work without any impediment.

Controllers: The function of a control is to transmit data, and therefore, it's an originator or a source or a sender. The controller's task is to get the destination device address from the database and apply the same into the to address field and simultaneously adding command from the database into the command field as per the request by the user. Before this process, the node adds its own device address into the from address field. Other formalities to build the data packet take place. Once the packet is formed, it is set out to reach the destination from the controller.

Responder: The responder is nothing but the destination host. It receives the data packet sent out from the controller and decodes the packet to get necessary information to operate specific application. The to address in the data packet is the device address of the responder.

Repeater: Repeater plays a vital role in the INSTEON protocol. It is an intermediate between the repeater and the responder. Sometimes, there might not be an involvement of Repeater during message transmission, and only the controller and the responder can carry out transmission of data packet. There is a limitation on the number of repeaters an INSTEON message can be passed. This condition limits data flooding and creation of loop in the network. In INSTEON network, any device can act as a controller or a repeater or a responder, which means the device can work as responder, repeater, and controller, irrespective of whether the device is a dual-band device or only RF device or only powerline device. Therefore, there is no master slave communication, whereas INSTEON technology exhibits peer-to-peer communication.

Figure 11.14 demonstrates simple direct message transmission between a controller and the responder. This type of communication is possible, when the responder is just one step away from the controller. Figure 11.15 consists of the intermediate device or the repeater. This type of condition occurs when the responder and the controller are faraway, hence require repeater to help transmission of the data.

***INSTEON Device Hardware Requirements*:** A normal INSTEON device or module that is capable of switching on or off appliances should have some memory to run the system. Therefore the RAM memory size is of 256 Bytes. The memory that stores operating system to run the system is Flash memory and is of size 7 Kbytes. The memory that stores configuration and loads, when the system switches on is EEPROM and is also of size 256 bytes.

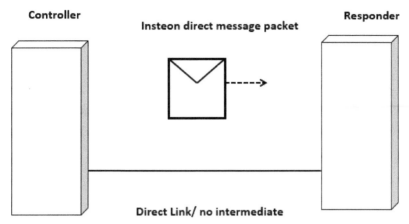

Figure 11.14 Simple INSTEON network device function without a repeater.

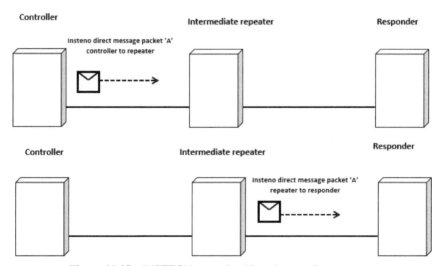

Figure 11.15 INSTEON network with an intermediate repeater.

Installation of INSTEON Devices: Installation is very easy, and no specialized engineer is required; users themselves can perform the installation process, because it is just plug-in and tap-in or wire-in process. After connecting the device to the ports, the set button is to be pressed and in seconds the device joins the network.

Powering INSTEON Devices: We know INSTEON devices comprise both wired and wireless devices. Wireless devices that communicate via RF are battery operated, whereas the devices that can communicate with the devices via powerline are powered with regular residential power supply, step down to required voltage.

Frequency of Operation and Modulation technique: If the device supports RF communication, to communicate devices wirelessly, the frequency of operation is 915 MHz. The modulation technique used in the communication is Frequency Shift Keying (FSK). In wired communication, the wired devices communicate over powerline using the frequency 131.65 KHz. And the modulation technique is Bipolar Phase Shift Keying (BPSK).

11.5.5 Security in INSTEON Devices

The security in INSTEON communication is carried out by encrypting INSTEON messages. Address masking is a concept of identifying by means of segregating data packet, and this is adopted by INSTEON technology, which provides some level of security.

11.5.6 Powerline Signaling / INSTEON Packet Timing

INSTEON Signaling: We know that INSTEON devices can communicate wirelessly or through wired medium, with the help of RF or powerline, respectively. In this section, we are going to study about signaling techniques used in RF and powerline communication. A signal is a physical parameter that carries information between sender and receiver. Therefore, the signal has to be generated depending upon the information to be passed or otherwise defined as encoding information in the form of bits into signal, which varies depending upon RF and powerline. INSTEON uses existing wired powerline connections to transmit data. The powerline voltage is at 110 V and at 60 Hz frequency. At every zero crossing of the powerline signal, INSTEON data are transferred. This is done by a signal to the powerline signal at zero crossing. The signal which is added is of peak-to-peak amplitude of 4.64 V and the carrier frequency used is 131.65 KHz with modulation technique Binary.

INSTEON Packet Timing: All INSTEON powerline packets contain 24 bits. Since a bit takes 10 cycles of 131.65 KHz carrier (see Figure 11.16), there are 240 cycles of carrier in an INSTEON packet. An INSTEON powerline packet therefore lasts 1.823 milliseconds.

The powerline environment is notorious for uncontrolled noise, especially high-amplitude spikes caused by motors, dimmers, and compact fluorescent

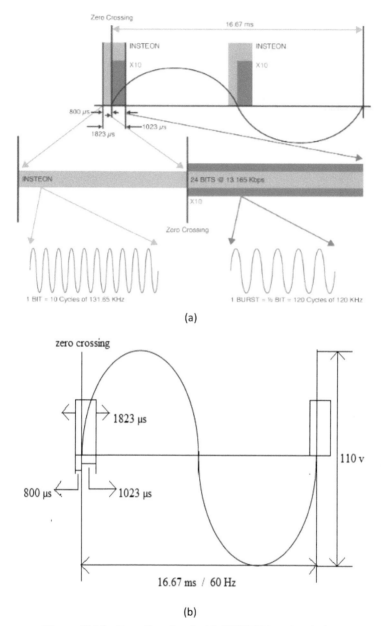

Figure 11.16   Powerline signal with INSTEON packet timing.

lighting. This noise is minimal during the time that the current on the powerline reverses direction, a time known as the powerline zero crossing. Therefore, INSTEON packets are transmitted during the zero crossing quiet time, as shown in Figure 11.16.

The top of Figure 11.16 a shows a single powerline cycle, which possesses two zero crossings. An INSTEON packet is shown at each zero crossing. INSTEON packets begin 800 microseconds before a zero crossing and last until 1023 microseconds after the zero crossing.

INSTEON message packets are sent in at the zero crossing of the powerline only, because at high amplitudes, there could be noise that could disrupt INSTEON packet. We know that the center frequency is 131.65 KHz, there are 24 bits in a single INSTEON packet, and each bit takes 10 cycles to reach the destination. Therefore, 240 cycles are required to send a single INSTEON packet. With the center frequency 131.65 KHz, the time period is 0.007 ms. And totally for 24 cycles to carry one INSTEON packet, it takes 1.823 ms or 10823.0 Âts (24 cycles * 0.007 ms).

From Figure 11.16, we can notice that the INSTEON packet starts 800 Âts before the powerline signal, and its transmission goes till 1023 Âts, therefore completing one full packet transmission of 1823 Âts. The most important fact that lies in this concept is X10 compatibility. X10 message packet takes 1023 Âts to be transmitted to destination. After the powerline signal reaches zero, both INSTEON packets and X10 packets take 1023 Âts to complete the transmission. Therefore, from this we can understand that both INSTEON and X10 technologies are compatible and INSTEON technology is actually backward compatible.

To transfer an INSTEON standard powerline data, we know that one start packet and four body packets, so totally five packets, are required. While sending the next data, which requires again five set of standard powerline message packets, single zero crossing of the powerline is not considered for transmission of the packet. Therefore, the next zero crossing is used to send the next set of packets, instead of taking immediate powerline zero crossing.

To transfer five packets of standard powerline data, total time period of 50.01 ms is required as shown in Figure 11.17. Totally, six zero powerline crossings are used to transmit the powerline standard packet.

Considering extended powerline message packets transmission, totally 11 extended packets of data are sent using 13 zero crossing, two zero crossings are left without any transmission intermediate of two packet transmission. The total time period required to transfer the extended powerline packet would be approximately around 100.02 ms, Figure 11.18,

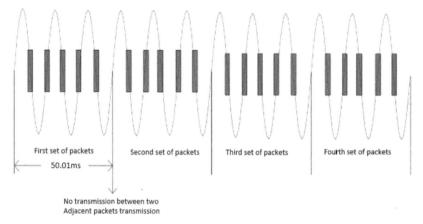

Figure 11.17 Standard powerline packet transmission.

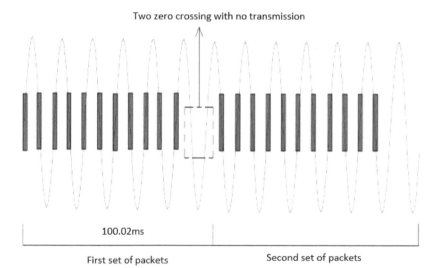

Figure 11.18 Extended powerline packet transmission.

11.6 INSTEON Communication

Powerline Communication: One of the means that INSTEON devices communicate each other is via powerline communication. The powerline cables are generally used to provide electricity to residential and commercial buildings and we also know there are many modulation techniques to transport useful data modulating it to suit the original signal of the medium.

Taking the advantage of these two concepts, the method of sending required data through a powerline implying certain type of modulation, thus found its application in many domains, and especially, this type of communication protocol is very much suitable method in home control networking protocol, because this does not demand for installation of a new wired infrastructure.

The powerline communication takes place by producing a signal called carrier that depends upon the information to be transported. This carrier signal should not go below 60 Hz, because the powerline signal operates with 60 Hz frequency.

Classification: Depending upon the frequency band, the powerline communication is grouped into two types:

- **Narrowband Powerline Communication:** The frequency range of operation is between 3 KHz and 500 KHz and speed in the range of 100 Kbps. In this type of powerline communication, long distance data transmission is possible with the use of repeaters.
- **Broadband Powerline Communication:** This type of communication takes place in short distance data transmission, while this method of communication takes the frequency range in MHz range and provides data rate within Mbps range.

Generally, Narrowband communication is implemented in almost all the applications. This is also used in INSTEON home-control networking protocol; although the system involves short distance communication, the data rate and the frequency of communication are well suitable for the application.

Communication Model: The basic communication process flow at the sender end is as shown in Figure 11.19.

Source: The source here should be in digital form. If the output of the source is analog, then the signal has to be digitized using converters and is sent to the next step.

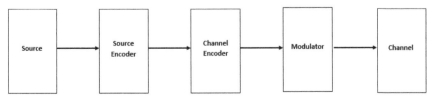

Figure 11.19 Powerline communication source.

Source Encoder

Here compression of the digital bits takes place. This avoids sending high volume of bits by removing redundant bits.

Channel Encoder

Here, error detection mechanism is applied to compute the useful redundant bits that carry the information of the result obtained from CRC process and used later at the receiver end.

Modulator

Now the digital processed signal is fed into the modulator to convert it into analog signal called carrier signal and made suitable for transmission over the powerline.

Some of the modulation techniques that can be used are BPSK, FSK, GFSK, etc. In INSTEON, as already mentioned earlier, BPSK is used, because of its simple method of generation of the signal.

Channel

This is the medium of communication. A channel could be wired or wireless, but in powerline communication the channel should be definitely a wired one, where the channel carries both powerline signal and INSTEON data.

The receiver has the blocks shown in Figure 11.20, which is the reverse of the source end process.

Demodulator: The demodulator is responsible to convert back the analog carrier signal to its digital equivalent and appropriate for further process.

Channel Decoder: Here error check process takes place. The result is compared with the redundant CRC bits from the received signal, and if both match the received bits and are error free, they gain entry to the next level

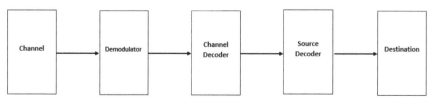

Figure 11.20 Powerline communication destination.

of the process. Otherwise, the received message is considered to have error and dropped and NACK is sent to the sender notifying the disrupted message containing error.

Source Decoder: The decoder decompresses the received signal to gain the original signal. Data compression should be performed such that there is extraction of lossless data.

Destination: This is the end point of powerline communication. The digital bits are delivered to device to decode necessary information.

Multihop Data Transmission in Powerline Communication: There are many multihop transmission in powerline communication in which some of them are explained below:

- **Incremental Redundancy (IR) Multihop:** If the powerline signal with data is transmitted, all the devices in the link will receive the message. After receiving the data, the nodes retransmit, and then all the adjacent nodes receive the message again. This happens multiple times, and each device receives multiple copies of the source data. Now it seems that more than once the same data is received simultaneously. This fact is actually advantageous, because by receiving more than once the same data, the device could apply various combining strategies, wherein the best suitable signal is transported, taking into account various factors. With the above multihop retransmission technique, there are a few ways to choose one relay technique, which is discussed as below:
- **Fixed (Fi) Selection:** In this method of choosing a relay route is by assigning dedicated nodes as retransmitting node. Only the assigned nodes in the multihop track can retransmit and others cannot. The retransmitting node should be placed in the way such that, its placement is between the source and destination path evenly, such that signal attenuation in minimized.
- **Channel Adaptive (CA) Selection:** In this process of determining retransmitting device, a metric value is calculated at each node. The link rate depends upon metric calculation. The node that possesses maximum metric value becomes the next retransmitting node. In this type of technique, different nodes take up the responsibility of retransmitting at different time. Here, there is no specific node selected to retransmit, which might be a burden of retransmitting to one single node as in the above method.

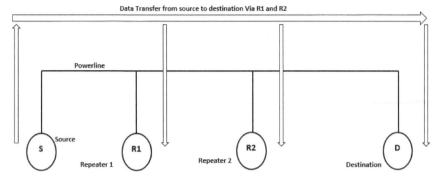

Figure 11.21 Simple multihop demonstration.

- **Automatic Repeat Request (ARQ):** This method is an automatic self-detection process. Here the node that first receives a source signal and that could decode information properly, requests to become the next retransmitting node. The node that successfully decodes the information tends to send the acknowledgement first and this is how the node takes the responsibility of becoming the retransmitting node.
- **Timeslot Synchronization:** This is the technique used in INSTEON home control networking. The time period to send all the packets in the powerline is called as timeslot. Therefore fixed timeslots exists in this protocol. Taking advantage of this concept, multihop transmission is introduced in this protocol, Figure 11.21. Each devices either transmits, retransmits or send acknowledgement at each timeslot. If one device starts sending a message all the other adjacent devices will just listen and do nothing. In the next timeslot, the immediate adjacent device goes into the retransmitting state. Similarly for every timeslot the function of a device varies and successful retransmission is possible. With this method the devices do not retransmit the message every time they receive a packet from any node. The advantage of this method is that we can avoid to certain extent the retransmission of the message every time slot by every devices, this in turn avoids flooding in the network.

The source tends to send information to the destination, in the path the packet has to traverse through the other two nodes in between, which are the repeaters. Initially, once the data leaves the source, the data are received by all the other nodes in the path, because the data use powerline communication to travel. Now if all the devices receive the message, the received message is repeated again, and there is no account of how many times the message is repeated; this situation leads to the flooding of message in the network. This problem is fixed with the introduction of timeslot synchronization concept to avoid flooding. Where at the first timeslot the source sends the data, later in the next, R1 retransmits the data and other devices, although they receive data, they do nothing and are in just listening state. In this way, the data propagates and reaches the destination.

Attenuation Characteristics of Carrier in Powerline Communication: The major attribute of the powerline communication is that the carrier amplitude faces serious distortion with respect to distance and frequency. To measure the attenuation characteristics of the carrier, the expression that can be used is:

$$\text{Attenuation} = 20 \log (Vr / Vt)$$

where Vr is the voltage value of the carrier received, Vt is the voltage value of the actual carrier voltage transmitted, and as said before in INSTEON protocol, the carrier voltage is equal to 4.64 V.

Practically as the carrier voltage suffers serious distortion as the message passes through each node, the researchers term it infeasible to send direct message to the nodes, so it is a compulsory need to have many nodes between the sender and the receiver, such that the intermediate nodes as repeaters to energize the carrier and step up the carrier voltage each time the repeater receives and send it towards the receiver. In this way, there will be successful transmission.

Advantages of Powerline Communication
1. Additional wiring to setup the network and the devices is not required. This protocol takes the existing power transmission cable to transport data embedding the information into the power.
2. The devices involved are very simple. Installation and maintenance are very easy and require no trained personals. The method of installation is just plugin and use, which attracts users.
3. Fast internet connectivity can be provided at any nook and cranny of the house, by passing internet signals over powerline and without the requirement of separate Ethernet cables. This type of communication is actually faster and reliable than wireless internet
4. Powerline communication can involve encryption to secure the data passed over the powerline cable.

Limitations of Powerline Communication

1. As the current passes through the powerline cable conductor, magnetic field is produced. This could interfere with the carrier signal carrying the data that would lead to the distortion of the data.
2. The information could be subjected to various noises, traveling in the powerline conductor, because the cables used are high voltage carrying cables.
3. Although the technology is very advantageous in various aspects, it suffers standardization issues. Many technologies like ZigBee, Ethernet, and all the computer communication-related protocols have a proper standard, whereas powerline communication faces a serious issue of standardization.

References

[1] C. Gomez and J. Paradells, "Wireless home automation networks: A survey of architectures and technologies," IEEE Commun. Mag., vol. 48, pp. 92–101, Jun. 2010.
[2] M. A. Zamora-Izquierdo, J. Santa, and A. F. Gãşmez-Skarmeta, "An integral and networked home automation solution for indoor ambient intelligence," IEEE Pervasive Computing, vol. 9, pp. 66–77, Jan. 2010.
[3] Smarthome. (2015). Remote controllers. [Online]. Available: http://www.smarthome.com/controllers-apps/remote-controllers.html
[4] M. H. Mazlan, F. Mohamad, R. A. Rashid, M. A. Sarijari, and M. R. A. Rahim. "Realtime communication routing protocol for home automation via power line," paper presented at 7th Student Conf. on Research and Development, Johor Bahru, Malaysia, 2008.
[5] Insteon, "Insteon whitepaper: The details," Insteon, Irvine, CA, version 2.0, 2013.
[6] Yu-Ju Lin, H. A. Latchman, M. Lee, and S. Katar, "A power line communication network infrastructure for the smart home," IEEE Wireless Communications, vol. 9, pp. 104–111, Dec. 2002.
[7] L. Lampe and A. J. Han Vinck, "Cooperative multihop power line communications," presented at IEEE Int. Symp. on Power Line Commun. and App., Beijing, China, Mar. 2012.
[8] G. Bumiller, L. Lampe, and H. Hrasnica. "Power line communication networks for large-scale control and automation systems," IEEE Commun. Mag., vol. 48, pp. 106–113, Apr. 2010.

[9] C. J. Kim, and M. F. Chouikha, "Attenuation characteristics of high rate homenetworking PLC signals," IEEE Trans. on Power Delivery, vol. 17, no. 4, pp. 945–950, Oct. 2002.

[10] M. M. Rahman Mozumdar, A. Pugelli, A. Pinto, L. Lavagno, and A. L. Sangiovanni-Vincentelli, "A hierarchical wireless network architecture for building automation and control systems," paper presented at 7th Int. Conf. on Networking and Systems, Venice, Italy, May 2011.

[11] M. E. M. Campista, L. H. M. K. Costa, and O. C. M. B. Duarte, "Improving the data transmission throughput over the home electrical wiring," paper presented at IEEE Conf. on Local Computer Networks, Sydney, Australia, Nov. 2005.

[12] J. Heo, K. Lee, H. K. Kang, Dong-Sung Kim, and W. H. Kwon, "Adaptive channel state routing for home network systems using power line communications," IEEE Trans. on Consumer Electronics, vol. 53, pp. 1410–1418, Nov. 2007.

[13] H. Li and L. Fen, "An improved routing protocol for power-line sensor network based on DSR," paper presented at Second International Conf. on Future Computer and Communication, Wuhan, China, May 2010.

[14] O. Mirabella and A. Raucea, "Tree based routing in power line communication networks," paper presented at 36th Annual Conf. on IEEE Industrial Electronics Society, Glendale, The USA, Nov. 2010.

Index

About the Authors

Prof. Dawoud Shenouda Dawoud has a BSc (1965) and MSc (1969) from Cairo University in Communication Engineering. He completed his PhD in Russia in 1973 in the field of Computer hardware where he succeeded to own 3 Patents in the field of designing new types of memory which was the beginning of the FPGAs. In 1984, he was promoted to full Professor at the Egyptian Academy of Science and Technology, National Electronic Research Institute. During the period from 1973 to 1990, he supervised more than 5 PhDs and 15 MSc degrees all of them focused in the fields of computer and embedded system designs. During the period from 1990-1999 he established the Faculty of Engineering at the University of Botswana. During this period he supervised 3 PhDs and 7 MSc degrees. In the year 2000 he became Professor of Computer Engineering and Head of the Computer Engineering Department at the University of KwaZulu Natal, Durban, South Africa. For 10 years he was supervising research in the field of Security of Mobile Ad hoc Networks. He supervised 2 PhDs and many MSc degrees in this field and published more than 30 papers. At the same period, he was visiting the National University of Rwanda to run an MSc program in Communication. He supervised about 15 MSc students during these 4 years before moving to the National University of Rwanda in 2010 to become the Dean of the Faculty of Engineering.

In 2011 he moved to Uganda where he became the Dean of the Faculty of Engineering at the International University of East Africa (IUEA), where he currently remains. During this time, he also served as the Vice Chancellor of IUEA, for a period of 3 years.

Across his career, he has published over 200 Journal and Conference papers, as well as books in the fields of computer engineering, microcontroller system design, embedded system design and Security of Mobil Ad hoc Networks.

Peter Dawoud has been working in the area of biometrics and computer vision for the last six years, leading research and products within Microsoft in the space. After completing his Bachelor's degree in computer engineering, Peter spent time researching applied cryptographic and steganographic systems for his Master's Degrees. Upon completion of these degrees, Peter has been focused on the productization of biometric systems for consumer and enterprise scenarios and later focused on the principles and use of centralized cloud biometrics systems in the enterprise and commercial scenarios.